New Fields of Adventure

New Fields of Adventure

The Writings of Lyman G. Bennett, Civil War Soldier and Topographical Engineer, 1861–1865

Edited by M. Jane Johansson

Voices of the Civil War
Michael P. Gray, Series Editor

The University of Tennessee Press / Knoxville

The Voices of the Civil War series makes available a variety of primary source materials that illuminate issues on the battlefield, the home front, and the western front, as well as other aspects of this historic era. The series contextualizes the personal accounts within the framework of the latest scholarship and expands established knowledge by offering new perspectives, new materials, and new voices.

 Copyright © 2024 by The University of Tennessee Press / Knoxville. All Rights Reserved. Manufactured in the United States of America. First Edition.

Unless otherwise noted, the figures are Courtesy of the *Lyman G. Bennett Collection* (R0274), The State Historical Society of Missouri Research Center–Rolla.

Library of Congress Cataloging-in-Publication Data

Names: Bennett, L. G. (Lyman G.), author. | Johansson, M. Jane, 1963- editor.
Title: New fields of adventure : the writings of Lyman G. Bennett, Civil War soldier and topographical engineer, 1861–1865 / Lyman G. Bennett ; edited by M. Jane Johansson Other titles: Writings of Lyman G. Bennett, Civil War soldier and topographical engineer, 1861–1865
Description: First edition. | Knoxville : The University of Tennessee Press, [2024] | Series: Voices of the Civil War | Includes bibliographical references and index. | Summary: "Lyman Gibson Bennett (1832–1904) was a Federal soldier who saw extensive service in the Trans-Mississippi Theater. A writer of considerable energy, wit, and intelligence, Bennett's wartime diaries recount his diverse and wide-ranging military record, stretching geographically from the prairies of Illinois to the Rocky Mountains, while a postwar account details, among other things, his labors to recruit "Mountain Feds" in the Ozarks. This volume provides the perspective of an individual who was both a topographical engineer and a common soldier. As a member of the Thirty-Sixth Illinois Infantry, Bennett provided one of the most detailed contemporary accounts of the pivotal Battle of Pea Ridge, March 7–8, 1862. By December 1863, Bennett was promoted to first lieutenant in the newly formed Fourth Arkansas Cavalry (US) and wrote an invaluable first-person account of guerrilla fighting in the Ozark mountains. M. Jane Johansson's critical presentation of his writings will prove useful to scholars of the Ozarks, landscape studies, and the Civil War in the West"—Provided by publisher.
Identifiers: LCCN 2023058432 (print) | LCCN 2023058433 (ebook) | ISBN 9781621908616 (paperback) | ISBN 9781621908623 (pdf) | ISBN 9781621908630 (kindle edition)
Subjects: LCSH: Bennett, L. G. (Lyman G.)—Diaries. | United States. Army. Illinois Infantry Regiment, 36th (1861–1865)—Biography. | United States—History—Civil War, 1861–1865—Personal narratives. | West (U.S.)—History—Civil War, 1861–1865—Campaigns. | Soldiers—United States—Diaries. | Surveyors—United States—Diaries. | West (U.S.)—Description and travel. | United States. Army—Military life—History—19th century.
Classification: LCC E505.5 36th .B47 2024 (print) | LCC E505.5 36th (ebook) | DDC 973.7/81—dc23/eng/20240108
LC record available at https://lccn.loc.gov/2023058432
LC ebook record available at https://lccn.loc.gov/2023058433

To My Husband,
Richmond Brookshire Adams
Scholar, Pastor, and Fellow Adventurer

Contents

Foreword   xi
Michael P. Gray

Preface   xvii

Acknowledgments   xxi

Editorial Method   xxiii

1. Before the War   1
2. Off to War: August 21, 1861–September 23, 1861   7
3. Trip to Rolla, Missouri: September 24, 1861–September 29, 1861   21
4. Camp Life in Rolla: September 30, 1861–October 13, 1861   29
5. Engineering Work in Rolla: October 14, 1861–November 3, 1861   39
6. Surveying Work in Rolla: November 4, 1861–December 15, 1861   53
7. St. Louis: December 16, 1861–January 12, 1862   83
8. Work in St. Louis: January 13, 1862–January 30, 1862   109
9. Furlough: January 31, 1862–February 8, 1862   123
10. Return to St. Louis and Rolla: February 9, 1862–February 15, 1862   127
11. To Springfield: February 16, 1862–February 22, 1862   133
12. From Springfield to Pea Ridge: February 23, 1862–March 5, 1862   141
13. The Pea Ridge Campaign: March 6, 1862–April 4, 1862   153
14. Pea Ridge to the Mississippi River: April 5, 1862–August 17, 1862   171
15. Recruiting in Dixie, Part One: November 26, 1862–July 5, 1863   181
16. Recruiting in Dixie, Part Two: July 6, 1863–November 13, 1864   203
17. Mapping in Kansas and Missouri: January 1, 1865–February 14, 1865   235
18. Fort Leavenworth, Kansas, to Fort Kearney, Nebraska Territory: February 15, 1865–March 3, 1865   253
19. Fort Kearney to Denver: March 4, 1865–March 22, 1865   263
20. A Tour of Colorado Gold Mines: March 23, 1865–March 24, 1865   277

21. To Fort Laramie and Back to Denver: March 25, 1865–April 15, 1865	283
22. After the War	295
Notes	299
Bibliography	351
Index	359

Illustrations

Figures

Carte de Visite of an Unidentified Topographical Engineer	57
"Camp of 36 Illinois Regt Rolla Mo."	96
"Columbia's Banner!"	98
"My Childhood Home"	107
One of the "Spoiled Photographs" of Lyman G. Bennett, February 1862	130
"Fort Wyman, Mo."	135
"Sketch of the Battle Ground"	145
"Bobs Knob. McDonald Co. Mo."	166
Lieutenant Bennett in the Fourth Arkansas Cavalry, 1864	207
Bennett family, mid to late 1890s	297

Maps

Missouri and Arkansas, ca. 1862	22
The West, ca. 1864	234
Westport and Big Blue	237
Battle-Ground of Westport	238
Osage or Mine Creek	243
Battle-Ground of Charlot	245

Foreword

The value of the written records of Civil War military engineers are still being explored by modern day scholars. Although engineering dates to prehistoric times, the vocation evolved from building pyramids and aqueducts to assembling infrastructure of highly populated ancient cities. Present-day historians are just touching the surface with their analysis of data that was saved throughout posterity. From their timeworn design endeavors, engineers have passed oral histories, letters, journals, diaries, recollections, and reports, which in turn, have become a significant foundation in laying the application of new methodologies in the historical profession. For the Civil War historian, this record of constructing and destructing projects also led to the rise of Civil War subfields, namely environmental history, which has transformed the field and advanced the historiography.

However, it was not until 2001 and the vision of a Miami of Ohio scholar that the subject of the Civil War and "controlling" nature was truly broached, if not attacked, by the renowned environmental historian Jack Temple Kirby. Kirby sagely argued in a thought-provoking essay, "The American Civil War: An Environmental History," that two genres, environmental history and Civil War history "ignored" one another, when rather, they should "assimilate." By applying a modern hypothetical environmental impact report, Kirby addressed a variety of themes such as disease, death, forests, farms, and cities. He also attributed the dearth in scholarship to the placement of western and environmental history together, which reflected the reality that the majority of environmental histories were also western histories. Kirby would have been undoubtedly pleased to find that the study of nature and the Civil War has come a long way since his "call to arms," as enormous strides have been made since his passing in 2009, when he still served as the president of the Southern Historical Association.

Since then, Megan Kate Nelson's *Ruin Nation: Destruction and the American Civil War* (2012) takes readers on a journey through the literal groundbreaking of war in the South, beginning her story with an Andersonville Prison survivor, writing about his travels back home. In essence, the landscape was torn up by civil war. Nelson writes, "Some areas of the eastern theater were entirely cleared of trees by the end of the war," looking especially at the Virginia wilderness. Nelson investigates how contemporaries saw ruins caused by their fighting, culturally, which resulted from "savagery" prompted by the newfound technologies of a war machine, and was thrust into their lives—a novel and collective experience, "blown apart [with] the obliteration of cities, houses, trees, and men."

The same year Nelson's work appeared, so did Lisa Brady's *War Upon the Land: Military Strategy and the Transformation of Southern Landscapes During the Civil War* (2012). Brady chronicled the administrative strategies put in place for tearing up land as primary military objectives. She poignantly discerned that man attempted to control nature in building plans of advance or defense, focusing on engineers. The author follows the trials of General Ulysses S. Grant and his troops along the Mississippi during the 1863 Vicksburg campaign, which diverted waterways through canals, oftentimes failing to control mother nature. The work also passes into the 1864 Shenandoah Valley campaign, where General Phillip Sheridan wrought "fertile farms into wastelands," and the Georgia and Carolinas of 1864–1865, where General William Sherman supervised the "foraging, fire, and the science of engineering." According to Brady, Sherman's attempts in dominating the terrain in his "tactics of eradication" shaped "nineteenth-century ideas about nature and influenced strategic planning." The author surmised that nature can be used as a "historical agent" in demonstrating the "power to shape human decisions." The Civil War, to Brady, "was a transitional moment" and that "Americans increasingly assumed that, rather than improving nature, they would win a war against it and master it in the process."

Kathryn Shively Meier's *Nature's Civil War: Common Soldiers and the Environment in 1862 Virginia* (2013) explores the Shenandoah and Peninsula campaigns of 1862, and how the common soldier combated nature in a variety of manners, from foul water to infestation of mosquitos, which eventually bred disease in these moving armies. According to Meier, these soldiers utilized "self-care," both physically and emotionally, in attempting to adapt to their changing environs. A collaborative book was published two years later, *The Blue, the Gray, and the Green: Toward an Environmental History of the Civil War* (2015), edited by Brian Allen Drake. Drake demonstrates how environmental effects during the war, from food to fog, to drought and microorganisms, weather and illness, all affected "the Great Divide" of North versus South. Drake links the essays he assembled as a "methodological smorgasbord" that when served, will inspire "more substantial meals to come." Indeed, more meals were to come, including Erin Stewart Mauldin's *Unredeemed Land: An Environmental History of the Civil War and Emancipation in the Cotton South* (2018), which examines "King Cotton," capitalism, labor, and the processes of cultivation, including the diversion of water systems in attempts to harness the environment. Southerners overvalued the notion of cotton being king, as a single crop, in hopes it would lead to their fortune; but, in reality, it stagnated the land, and in turn, their economies and ecosystems.

A more rounded examination of sections beyond the South was produced by Judkin Browning and Tim Silver in *An Environmental History of the Civil War* (2020). They fastidiously argued that the conflict brought on "great transformation within nature." Their study especially yields revelations in the "new disease environment," which details how microbes traveled, poisoning many in their track, from animals to mankind. Like the previous scholarship, their introduction begins south of the Mason-Dixon line, during the muddled Virginia "Mud March" in early 1863. Rainstorms were especially emphasized in Kenneth's Noe's groundbreaking *The Howling Storm: Climate, Weather, and the American Civil War* (2020). Noe methodically establishes how weather defined much of the war. Unique weather patterns of the period show that the war was protracted by persistent rainstorms and flooding, and as a result mud did more than slow down armies tactically and strategically, it led to the demise of Southern crops, weakening yields—so much so that the Confederacy could not feed soldiers nor civilians. Noe even questions the likelihood of emancipation if it were not for the weather.

From press director Scot Danforth, down to series editors, the University of Tennessee Press has been very purposeful in publishing works that align to current trends in progressive scholarship. The Voices of the Civil War (VCW) series has already given readers accounts from soldier-engineers that made decisions in analyzing, interpreting, and narrating the landscape, from soldiers that ventured in the traditional theaters, and from soldiers that participated in unusual regions, hoping to enrich our audience with environmental, social, and cultural elucidations, as well as the more customary political, religious, and military approaches. Civil War engineers, cartographers, topographers, and surveyors left a panorama of information, and built upon or destroyed the landscape while they were at it; and, generations later, scholars would write about it.

The series introduced its readership to *Training and Campaigning with John Morris Wampler* (2000). Editor George G. Kundahl features Wampler's role in the Confederacy, and, at the time, provided the first look at Confederate engineering in more than forty years. Wampler served as a topographical engineer for the Confederacy's provisional army, ultimately rising to be the chief engineer for the South's Army of the Tennessee. Wampler also had the unique perspective of serving in both the Eastern and Western theaters of war, and during that time, he had construction projects in eight Confederate states and two border states. His adept cartography led him to oversee the building or repair of bridges, railroads, earthen works, and fortifications. Kundahl details the US Coast Survey as a type of professional preparatory school for both Union and Confederate engineers, he highlights the battle of Perryville, as well as the defense systems of Charleston's Battery Wagner, made famous by the storming of the African American 54th Massachusetts on Morris Island.

In addition to linking fashions in historiography with contemporary accounts, this series attempts to recruit scholars who are experts in their field, mining for subjects in previous work in order to amplify a voice from the past. Mark Smith's highly acclaimed *Engineering Security: The Corps of Engineers and Third System Defense Policy, 1815–1861* (2009) gave a macro-investigation of defense systems during peacetime. Eleven years later, Smith provided readers with *The Civil War Journal and Memoir of Gilbert Thompson, US Engineer Battalion.* Smith dives into the life of his subject, whose assignment centered on the Union engineer battalion, a highly selective unit that provided keen insight for the Federal high command. Editor Smith provides fresh information while inviting the reader along with Thompson, who engaged building roads, bridges, and surveying the topography, as well drawing up maps. The volume includes Thompson's sketches of social life and images of war, with his trained eye drafting scenes from his experiences, a common pattern developed by engineers in their accounts. Less common by far are the photographic images of landscape included in the volume.

The Civil War should be seen beyond the North-South regional context, and beyond the staple military operations occurring in the Eastern and trans-Mississippi Western theaters, too. Scholars have begun to venture past the Kansas plains by tackling the war's ramifications from the desert Southwest to the mountains in the Pacific Northwest, once dominated by frontier, western, and environmental historians. Editor James Robbins Jewell helped fill this void in the VCW series with *On Duty in the Pacific Northwest during the Civil War: Correspondence and Reminiscences of the First Oregon Cavalry Regiment* (2018). Editor Jewell masterfully takes readers on expeditions with the First Oregon Cavalry through untouched and untamed areas of the Pacific Northwest. Jewell argues his subjects, cavalry troopers and soldiers, were "embedded reporters" that became instrumental to the US government and its growth in preserving the Union. After the gold rush and growing population out west in the later 1840s and 1850s, the protection of miners and the new emigrants became critical in the nation's development. As civil war broke out, and soldiers were stationed in the far West, they encountered free African Americans, indigenous peoples, and Mormons. The First Oregon documented and pursued wildlife as if they were trackers. All told, the

First Oregon Volunteer Cavalry were trailblazers and mapmakers, surveying the countryside for good water, timber and fertile land for future settlement. A compass for them was as important as a weapon, while their military operations went hand in hand with agricultural exploration. They scouted, built barracks and established forts, were peacemakers and warriors, and protectors of the "iron horse," but always on the lookout for what one Oregon cavalryman termed "the hidden resources" of the far West, which were, as he wrote, highly "treasured."

The latest VCW volume, *New Fields of Adventure: The Writings of Lyman G. Bennett, Civil War Soldier and Topographical Engineer, 1861–1865*, interplays such themes and offers more "treasures" for our readership with "new fields" in Civil War historiography. Lyman Gibson Bennett led an extraordinary life, and his service illustrates the landscape, weather, climate, and encounters with native and migrant populations. Editor M. Jane Johansson found a gem in chronicling Bennett's prose, adeptly organizing his heretofore uncultivated writings and tracing her subject through a wide range of regions; from the steppes of Illinois and Missouri, to the Kansas grasslands and the hilltops of Arkansas, moving to the plateaus of Wyoming, forging further west, culminating at the Rockies of Colorado. Observant and curious, a voracious writer—though a better sketch artist then speller—Bennett detailed a geographically wide, if not diverse, service record of his explorations. Surveyor by profession, Bennett wed in December 1859, some sixteen months before the firing on Fort Sumter, and thereafter left his wife and daughter to enlist as a corporal in the Thirty-Sixth Illinois Infantry. Also, a devout "soldier" of the Temperance movement, he played an active role in the International Order of Good Templars, an organization founded in 1851, rare in the fact that it also recruited women. Coming off the economic woes of the later 1850s, and a yearning for patriotism, coupled with a need for steady pay, war motivated him to enlist so he could support his family and serve his country.

Bennett both wrote and sketched the societal order of army camps as well as the campaigns he participated in. As the corporal traveled, he perceptively logged the environment, people he encountered, sounds and scenery, weather conditions, and his adventures. According to Editor Johansson, "his writings can be described as a type of travelogue" and "Many of the diary entries . . . served as a type of map memoir that topographical engineers created as a written record to supplement their drawings." A romantic at heart, his prose was likened to poetry and during his exploits, he even found time to pick flowers for his beloved wife, carefully inserting them into his accounts before sending them home to Melissa. As Engineer Thompson highlighted the battle of Perryville, Engineer Bennett gives one of the best accounts of the early March 1862 Battle of Pea Ridge; crucial in opening up Arkansas for Federal occupation, and locking up border state Missouri for the Union. His narrative in the Ozarks is particularly compelling in recruiting "Mountain Feds" for Northern support. By the end of 1863, Bennett was transferred to the Fourth Arkansas Cavalry, stationed at Little Rock, then in spring of 1864, he was promoted from a lieutenant to that of a major. Seemingly always on the move, he then embarked in a nearly a four-hundred-mile trek from Fort Kearney to Denver, which took him on a journey just over two weeks.

All told, Editor Johansson follows Bennett from beginning to end, as he encountered famous and not so famous civilians such as the nieces of Nicholas Biddle (made prominent by the Second National Bank), soldiers like Grenville Dodge and Samuel R. Curtis, mountain men, and, later, Cheyenne, Arapaho, and Sioux warriors. Bennett describes the gold mines of Colorado and his ventures onto the Overland Trail, where he described "impressive natural formations." Editor Johansson painstakingly compiles his story, filling voids with her editorial pen, such as places like Sand Creek, the Powder River Expedition in Montana Territory, and a meeting with renowned mountain man, Jim Bridger. Lyman Gibson Bennett lived a long life after the war, into the early twentieth century, eventually retiring in Springfield, Missouri. But his high-paced army career ultimately caught up with him and he suffered a variety of maladies in late life, but his wartime service granted him a full pension. Weak in legs, with throbbing varicose veins, painful heart ailments, a sore bullet wound from the war, and finally, a debilitating fall from a horse, combined and he succumbed in 1904, at age 71. His cherished Melissa who he had enclosed flowers to in his war time diary, passed away just one week later. The obituary commented that their service in the community, and beyond, even in retirement in Springfield, "contributed largely to the growth and upbuilding of the city." An appropriate epitaph for a Civil War engineer and his treasured wife.

Michael P. Gray
East Stroudsburg University

Preface

Whenever Civil War soldiers enlisted or were drafted, they, of course, could not foresee the future. Some soldiers spent their enlistment period performing relatively mundane, yet important, duties such as protecting a supply line, guarding a prisoner of war camp, or fulfilling their station at a fort. The truth is that some soldiers never experienced combat or engaged in active campaigning. At the other end of the spectrum were those who saw much fighting, such as in the extended and bloody Overland Campaign. If a soldier survived these encounters, he had many stories to share, if he felt inclined to do so. Many unlucky men died after only a few days or weeks of camp duty. Although each soldier had a unique record, Lyman G. Bennett's was more diverse and wide-ranging than most of his comrades.

Bennett's wartime narrative stretched from the Illinois prairies to the Ozarks of Arkansas and Missouri, the Rocky Mountains of Colorado, Fort Laramie in Wyoming, and the eastern Kansas grasslands. His observant eyes and curiosity took in these varied landscapes and captured those experiences in his rich diaries and postwar account. Not only was his perspective geographically broad, but his duties proved to be diverse as well.

Married in December 1859 and the father of a daughter, Bennett attempted to enlist in the Federal army soon after the attack on Fort Sumter in April 1861. Unable to join one of the first Illinois regiments, he ended up in the Thirty-Sixth Illinois Infantry and was mustered in as a corporal. His writings chronicled early, undisciplined camp life punctuated by controversies between officers as well as periodic camp violence. Corporal Bennett soon yearned to leave the Thirty-Sixth Illinois and got his wish while his regiment was in Rolla, Missouri, in the fall of 1861. A surveyor by profession, Bennett was detached to help design Fort Wyman and survey the land around the town's immediate area. He also "saw the elephant" as part of the Thirty-Sixth Illinois at the battle of Pea Ridge between March 6 and March 8, 1862. His account of the battle is one of the most extended contemporary accounts written by an enlisted man. Following Pea Ridge, he settled into his duties as a topographical engineer on Major General Samuel R. Curtis's staff during an expedition to the Mississippi River in the summer of 1862. Tedious office work drawing maps followed in St. Louis.

Bored in St. Louis, Bennett reentered field service in April 1863 with a promotion to first lieutenant and adjutant in the newly forming Fourth

Arkansas Cavalry. In his postwar account, "Recruiting in Dixie," Bennett described the high adventure of slipping into the Ozark Mountains of north Arkansas with a small group of men as they attempted to recruit Unionists, more colorfully known then as "Mountain Feds." Following this excitement in the spring and summer of 1863, Bennett campaigned with the Fourth Arkansas Cavalry around Little Rock. For various reasons explained later in this book, he resigned his commission on August 22, 1864, after reaching the rank of major. Civilian employment by the army followed with mapping duties and an inspection tour of forts and other defensive sites along the Overland Trail from Fort Kearney, Nebraska, to Denver, Colorado, and on to Wyoming's Fort Laramie. Bennett's employment ended as chief engineer on Colonel Nelson Cole's column of the Powder River Indian Expedition between July and September 1865. Bennett's account of this expedition has been previously edited in David E. Wagner's *Powder River Odyssey: Nelson Cole's Western Campaign of 1865: The Journals of Lyman G. Bennett and Other Eyewitness Accounts* (2009) and is not a part of the present book. Intermixed with the above activities were some unusual occurrences such as his three-week hospital stay in St. Louis, a bizarre arrest for failure to carry a pass, the exploration of a Missouri cave, and a tour of Colorado gold mines.

Writing is often a reflection of the writer's personality and interests. Like many diarists, Bennett failed to overtly describe his motivation for keeping a personal record, although he had kept a diary in 1857. He sent the diaries he kept while he was a soldier to his wife, Melissa. A simple desire to chronicle an important time in his life for Melissa and himself may have been a primary motivation. In his 1865 diary, he addressed "those that are interested in the personal history of the humble writer of these pages. For such and for my own personal gratification[,] I shall endeavor to keep a faithful record of the transactions and observations of each day, and trust that in so doing[,] I will not leave behind me a record of shame[,] misfortune, and crime."[1] There are certain topics, as might be expected, on which Bennett did not focus. Extended discussions of politics or events outside of his purview rarely crept into his narratives. He also gave no explanation for why he took up civilian employment with the army. Perhaps limited employment opportunities at home, coupled with the enticement of steady military pay, was the reason. Also, regrettably, it is nearly impossible to get a sense of his wife's personality from his diary. Although Lyman obviously loved and respected her, much about Melissa and her life while Lyman was absent remained unspecified.

Conversely, extended and often perceptive descriptions of the environment, people he encountered, sounds, weather conditions, and a sense of adventure pervade Bennett's writings. In many ways, his writings can be described as a type of travelogue. His diaries, though, revealed him as a relatively conventional man of his time in the sense that his prejudices were typical of most white Americans. His most scathing comments, for example, were reserved for Native Americans. Derogatory phrases such as "dirty squaws," "nasty pappooses," "lazy indians," and "cowardly & dirty *Indians*" appeared in his 1857 diary as he worked in Wisconsin and Minnesota.[2] On his 1865 inspection tour from Nebraska to Colorado, Bennett traveled in the wake of native attacks, and he frequently expressed his understandable fears that they would assault his party. Bennett's diaries sometimes reflected all too common negative attitudes toward German Americans. He rarely mentioned African Americans, with his few comments running the gamut from a crude sexual allusion to support of a man who killed a white soldier in self-defense.

The best diarists have a sense of curiosity, a trait revealed frequently through their pages. At times, Bennett's diaries become almost journalistic as he sought people to question about matters of interest to him. Lyman was the type of person who liked to write, and he was a remarkably faithful diarist. Out of the 332 days in the diaries published here, Lyman wrote on 313 days, thus missing only nineteen. Beyond the diaries, as mentioned in the editorial section, Bennett also wrote a handful of newspaper articles as well as a postwar account. An interesting question is whether he ever kept any other diaries. It is likely that he did, as his postwar "Recruiting in Dixie" piece has a richness that probably cannot be ascribed only to Bennett's memory.

Many documentary editors reach a point where they feel like they are acquainted with the writer whose works they are editing. This editor admired Bennett's curious mind, his ability to pen vivid word pictures, his sense of humor that pops through at unexpected moments such as his hilarious description of St. Louis ladies walking through mud, his wry comments on the lowly chigger, and his rather cutting words about some of his fellow hospital patients. Although Bennett was certainly struck by the pathos of war, he thrilled as he rode horseback through a Missouri snowstorm and acted like a boy when he caught his first glimpse of the Rocky Mountains. There was a man who lived life to its fullest.

Acknowledgments

In the search for a new project after the publication of my last book, *Albert C. Ellithorpe, the First Indian Home Guards, and the Civil War on the Trans-Mississippi Frontier* (2016), I happened across Lyman G. Bennett's diaries and a postwar account available in a digitized format on the exceptional website *Community & Conflict: The Impact of the Civil War in the Ozarks*. Like Ellithorpe, Bennett resided in Illinois as the Civil War started, but the two men were dissimilar in personality and experiences. More conventional than Ellithorpe, Lyman served in the trans-Mississippi, but experienced a wider range of activities during the war.

Most of Bennett's writings are housed at the State Historical Society of Missouri's Rolla Research Center. During a visit of two days, Senior Archivist Kathleen Seale, assisted by Carole Goggin, was helpful and accommodating in allowing me to examine Lyman's original diaries. She also promptly answered my numerous email questions and granted me permission to publish their Bennett holdings as well as a photograph of him and several of his drawings. The previous senior archivist, John F. Bradbury, Jr. (now deceased), helped secure Bennett's papers for the Rolla Research Center and offered assistance on the project. This book would not have been possible without the help of the staff at the center.

Although primarily a teaching institution, my employer, Rogers State University, once again generously allowed me a one-semester sabbatical during the spring 2021 semester. Thank you to Dr. Keith Martin, the dean (now retired) of the School of Arts and Sciences, and Dr. Kenneth Hicks, chair of the History and Political Science Department, for encouraging me to apply for the sabbatical. I'm also grateful for the assistance of Dr. Richard Beck, vice president of academic affairs, Dr. Larry Rice, president of Rogers State University, and the University of Oklahoma Board of Regents. Kelly Ewing, coordinator of interlibrary loan and circulation, once more skillfully tracked down needed articles and books for the project.

The Huntington Library, Art Museum, and Botanical Gardens owns a photocopy of Bennett's handwritten postwar memoir, yet the institution was closed to researchers during the height of the COVID-19 pandemic, which meant that I was unable to obtain a copy. All was not lost, however. Reader services coordinator, Karina Sanchez, offered me a virtual reading room appointment, and through a Microsoft Teams meeting, Morex Arai, a reader services assistant, showed me every one of the 599 pages in the document. Being able to see and have images of some of the pages was extremely helpful.

Many thanks to Scot Danforth, director of the University of Tennessee Press, for seeing the value of Bennett's writings and for making the publication process such an easy one. Thanks as well to staff members and contributors Jon Boggs, Stephanie Thompson, Katie Little, and Alex Mendoza for making this a better book.

My husband, Richmond Brookshire Adams, provided constant support and love. During his three readings of the manuscript, he deployed his copyediting skills and deep knowledge of the Civil War era with kindness and encouragement. It was an easy choice to dedicate a book about a nineteenth-century gentleman to one of the twenty-first.

Editorial Method

During the course of this work, I have transcribed Lyman G. Bennett's writings as faithfully as possible. Bennett misspelled many words, but most have been left unchanged in order to leave the flavor of his work unaltered. Bennett rarely denoted the possessive case correctly, and those errors were silently corrected in the book. Words that Bennett underlined have been changed to italics to aid in readability. If the meaning of a word was deemed unclear, then the correct spelling of the word has been inserted in brackets. Occasionally, a word is followed by [?] to denote that the editor is uncertain that a word has been transcribed accurately. Additionally, Lyman Bennett used superscript letters and numbers occasionally. These have been changed to regular type to aid in readability. Because Bennett's experiences were wide-ranging and diverse, I chose to write full introductions for most chapters rather than consigning complex material to the footnotes. As much as possible, I have also included biographical information about Bennett to aid in a fuller understanding of this adventurous man. People that Bennett referred to have been identified to the greatest extent possible, but, regrettably, some of the individuals that he referenced could not be located in census or military records.

Bennett's body of work listed below, unless otherwise noted, is part of the Lyman G. Bennett Papers (R0274) at the State Historical Society of Missouri, Rolla Research Center. The full text of all the sources below is included in this book unless otherwise stated. Bennett's 1857 diary was used for reference purposes and background material.

1. Diary (August 19–December 20, 1861). This is the only source where only a typescript transcription is available. The typescript was produced by Lyman L. Bennett, a grandson, and may be viewed on the *Community & Conflict: The Impact of the Civil War in the Ozarks* website (https://ozarks civilwar.org).

2. Diary (December 21, 1861–April 4, 1862). A transcription was made from the original diary.

3. "Route of the Army of the Southwest from March 1st to July 10, 1862." This set of maps, drawn by Bennett for Major General Samuel R. Curtis, is in the collections of the Kansas State Historical Society. The text for the majority of Lyman's map memoir is in this book. No effort was made to reproduce Bennett's map set, although Robert G. Schultz used some of them

as illustrations in *The March to the River: From the Battle of Pea Ridge to Helena, Spring 1862* (2014).

4. "Recruiting in Dixie." This is the only postwar account by Bennett included in the current book. This account of Bennett's recruiting duty in Arkansas during the summer of 1863 is one part of a much larger work titled "Personal Reminiscences," which included sections about his family history, reflections on the Indian wars after the Civil War, and participation in the 1865 Powder River Expedition. There are ten to fifteen drawings by Bennett in this source. The Huntington Library has a photocopy of this work, handwritten by Bennett, but it was unable to legally grant permission to reproduce any of the artwork in the current volume. The location of the original manuscript is unknown. My transcription was prepared from a typescript done by Roger R. Bennett, a great-grandson of Lyman's, with the Huntington manuscript used to make corrections. The account itself, judging from internal evidence, was written around 1875. "Recruiting in Dixie" has been published twice previously in historical journals. Transcribed by Helen C. Davenport, the memoirs were published across several issues in volume four of the *Christian County [Missouri] Historian*. In 1999–2000, historian James J. Johnston republished the Davenport transcription with a brief introduction in the *White River Valley Historical Quarterly [Missouri]*. However, neither the Davenport nor the Johnston version attempted to identify all people mentioned by Bennett or contextualize his account in its entirety. For those reasons, and because the previous transcriptions were published in journals with limited circulation, I decided to publish "Recruiting in Dixie" again here.

5. Lyman Bennett's letter to Owen Lovejoy (February 17, 1863) and Lovejoy's subsequent letter to Abraham Lincoln (February 21, 1863). These transcriptions were made from digitized copies of the originals from the Abraham Lincoln Papers at the Library of Congress.

6. The *Chicago Tribune* published at least three letters Bennett wrote. One of those letters was published on August 28, 1862, and has been included in this book. The other two letters, published on March 18 and March 23, 1862, have not been reprinted in this book as they closely follow the content of Bennett's diary. However, interested readers can easily locate these articles on Fold3.com or other newspaper databases.

7. Diary (January 1–October 4, 1865). My transcription was made from the original diary. This diary has been digitized and is available for viewing on the *Community & Conflict* website.

New Fields of Adventure

1 | Before the War

Charles M. and Louisa Bennett welcomed a child into the world on August 1, 1832.[1] Lyman Gibson Bennett was the first child in a family that grew to eleven, consisting of six boys and five girls, with eight surviving at least until the 1870s.[2] Lyman's father was descended from a Rhode Island family that migrated into New York's Wyoming Valley in 1775, a place that became the scene of intense and brutal warfare during the Revolutionary War between white settlers and Native Americans allied with the British.[3] At least one family member, James Bennett, was scalped during the fighting, but survived the ordeal until finally succumbing sometime after the conflict.[4] Lyman Bennett's negative attitudes toward Native Americans possibly took root while listening to stories of his family's struggles on the New York frontier.

Lyman believed that his mother, Louisa Canfield, was born in Connecticut and descended from a Loyalist, Oliver Canfield, who died in poverty after his "lands were confiscated."[5] Lyman Bennett neatly summed up the history of his ancestors by writing that "like the Smiths and Jones', the name [Bennett] occurs more often in the census tables than in the Halls of Congress . . . [and all] followed the same profession of honesty and poverty."[6]

Judging by scattered comments in his *Reminiscences* and his diaries, Bennett's parents provided a stable, loving atmosphere for their children. Like most Americans of the time, Bennett grew up on a farm and experienced limited educational opportunities in country schools where the teachers believed strongly in using "the rod" in punishment.[7] These strict schoolmasters, however, taught him effective communication skills and gave him a solid mathematical foundation.

At some point Charles Bennett got the urge to move West, and in 1849 he uprooted his large family and migrated to Kendall County in the northeast part of Illinois.[8] There, he purchased a 250-acre farm "on both sides of Fox River," and the family settled into their new home near the village of Oswego.[9]

Little is known about Lyman's life between 1849 and 1856, but it is safe to assume that he helped his parents and siblings on the farm that a census enumerator assessed at $5,000.[10] According to the 1850 census, nineteen-year-old Lyman attended school along with his siblings, down to three-year-old Charles. Only one-year-old Franklin was excluded from school.[11]

Between ages nineteen and twenty-four, Lyman plunged into a variety of activities that came to characterize his adult life. If a county history is to be believed, voters elected him at age nineteen as one of five county circuit clerks.[12] In 1854, he became one of the first secretaries of the Kendall County Agricultural Society.[13] Sometime during these years, Lyman also became a fervent supporter of the burgeoning temperance movement, and turning to the classroom, he taught school four times at such country schools as the "Stebben's or Gorton" and the "Wormley, District No. 3."[14] Although silent on the subject in his writings, Bennett also learned the craft of land surveying, a skill that required an understanding of "plane trigonometry, logarithms, and calculating square roots" and that would serve him well throughout his adult life.[15]

An examination of his sole prewar diary, one kept in 1857, offers a glimpse into the life of an active and intelligent young man in search of steady employment. When coupled with his occasional comments about courting, one can conclude that Bennett believed that steady employment was necessary before marrying and starting a family.[16] To this end, Lyman left home on January 5, 1857, for the Minnesota Territory. This was not the first time he had traveled to these lands, as his diary mentioned that he had previously visited Winona, a Mississippi River town in the territory.[17] After a brief visit to relatives in Madison, Wisconsin, Bennett traveled again to Winona and used it as a starting point for other travels.[18] Before going to Winona, he had crafted maps, apparently of the area, and signed up subscribers who agreed to purchase them.[19] By delivering these to his subscribers and peddling the maps on the streets, he netted $150 by mid-February and lesser amounts later in the month.[20] Bennett's customers provided mixed feedback with some buyers expressing "themselves satisfied with the mechanical execution of the maps, while others are dissatisfied."[21] With the economy beginning a significant decline, Bennett observed that "everybody complains of hard times[.] An occasional sale of a map encourages me to go ahead."[22] Caused by a rebalance of trade at the end of the Crimean War, Americans labeled this financial crash as the Panic of 1857.[23] American farmers had grown accustomed to selling their wheat in Great Britain and France during the war, but the conflict's end opened up the market once again to Russian farmers.[24] Gold, which theoretically backed the bank notes of the time, dropped in supply, and economic dislocations rippled across the states.[25] Recovery began in 1858 and continued until the end of the year.[26]

Bennett's temperance activities achieved greater success than his efforts to sell maps. Bennett participated actively in the International Order of Good Templars (IOGT), an organization created in New York State in 1851.[27]

Founded as a fraternal organization in a time when such groups experienced a surge in popularity, the IOGT developed sophisticated rituals, regalia, elaborate laws, and secret passwords.[28] Unlike other temperance societies of the time, the Good Templars encouraged women to join, which helped push membership to 78,185 by 1860.[29] In many ways, the "Templars saw themselves . . . [as] a missionary organization to redeem that world."[30] Bennett served as secretary of the Minnesota organization, an office that kept him busily creating reports and answering questions.[31] He helped establish IOGT lodges in Winona, Owatonna, and Wabasha.[32] Sprinkled throughout his diary are accounts of alcohol-fueled fights that broke out on election days as well as mentions of "liquor dens" and grocery stores that peddled whiskey.[33] On one of these occasions, Bennett stepped in and pummeled the face of a drunken Mr. Kennedy, who hurled "denunciations of the temperance men" distributing ballots that opposed alcohol during Winona's city elections.[34] Lyman commented wryly that once Kennedy sobered up, he would be "chagrinned" that such a small man had beaten him.[35] In all, Bennett's temperance work mostly consisted of calmer activities such as giving speeches and lectures, presiding occasionally over meetings, and helping to organize a "state . . . Association based upon the old Washingtonian principle" in Minnesota.[36]

Unable to find a "permanent business" along with a friend named McDowell, the two men decided to trek further into Minnesota "& use our pre-emption right in securing farms."[37] Mostly walking the seventy-three miles to Ashland, Bennett and McDowell each selected a quarter section of land eight miles southwest of the town.[38] While building crude houses and working on some improvements, the two friends boarded at an Ashland hotel whose proprietor was an "old *deaf* & devilish" person.[39] Bennett and McDowell did not limit their land acquisition to Ashland; they also claimed land near Faribault.[40] Vandals, though, tore apart Lyman's crude "house" on the Ashland claim while he worked on a railroad surveying crew.[41]

Land surveying proved to be Bennett's steadiest form of employment in 1857. From late May until early August, he worked as part of the "engineer corps on the Transit Railroad survey."[42] Starting at Minnesota City, the initial survey went to Stockton City, eight miles from Winona.[43] His diary described several survey expeditions with some routes deemed as impractical for railroad use, such as one that had "sloughs" full of water.[44] Working as a "draughtsman" and, at times, using a compass, Bennett labored as part of a small team.[45] Battling mosquitoes and the elements proved challenging, but eventually the survey work ground to a halt due to an Indian threat.[46] When the group's leader asked soldiers at Fort Ridgley for more percussion caps, the soldiers advised that the surveying work stop because Native Americans from the Redwood and Yellow Medicine agencies "were exasperated against the whites."[47] Heeding the warning, they stopped the work.[48] Lyman then found a railroad survey job along the Straight and Bear Valleys near Winona, but he quit because he both disliked his boss and believed the work was too hard for the pay.[49] He used some of his earnings to purchase a sixty-five-dollar surveying compass and set out to contract himself.[50] Making his way to Ashland to check his land claim, Bennett staked out lots there in spite of threats from illegal squatters and then worked on various other surveying jobs in the area for several weeks.[51]

The following month, Lyman wrote that "the financial crash is felt with rigid severity here. In short it is utterly impossible to get debts of the smallest amount cashed." In those hard times, the Transit Railroad Company owed him thirty dollars, and between it and his other surveying jobs, he calculated that customers owed him nearly $300.[52] Disappointed by his prospects in Minnesota, Bennett returned to Oswego in late October and took up teaching at the District No. 4 school in early December.[53] Back in the classroom, he vowed not to use the rod, but instead to act as a parent toward his students.[54] Miss Susan Gorton troubled him with her whispering during class, and his students whose heads "are so full of Santa Claus" wanted school to be cancelled on Christmas Day—something the tough-minded teacher refused to do.[55]

Being home again also allowed Bennett to pursue romantic opportunities, not that these had been absent during his forays into Minnesota and Wisconsin. Near Ashland, Bennett penned a letter "to the *black eyed school mistress*, whose absence has not diminished the warm attachment of *one* admiring friend."[56] In Winona, he confessed in his diary that he and a Miss Perry "came near *sparking* each other to sleep" before friends returned from a prayer meeting.[57] After returning to Oswego, Lyman dressed up to call on a Miss Sanderson, possibly the black-eyed school mistress, but she was away from her house and apparently her interest had faded, if one is to judge from his behavior at the "Congregational Sociable" in early November.[58] Feeling "bashful & diffident in the extreme" after being "so little in the society of ladies," Lyman was not quite prepared "when the *kissing* commenced."[59] "At first my great big feet were in every bodies way, & I was constantly getting tangled among the silks & hoops of the *fair ones*. . . . But before the evening was over, my big boots were out of mind, & I could kiss with as much *gusto* as the rest, as was evident from a *charge* I made among the ladies, *kissing right & left* without 'any respect to persons.' Doubtless I would not have been so bold, had not Miss Sanderson's presence stimulated my bravery."[60]

At some point in 1859, Lyman stopped courting Miss Sanderson and turned his attention to Miss Melissa Emma Lyon (or Lyons), a woman who was six years his junior. By 1859 he had become the Kendall County surveyor, and his financial position had stabilized.[61] Unfortunately the historical record does not reveal how Lyman and Melissa met or how they courted, but the couple married on December 28, 1859, in Kendall County, Illinois.[62] Little is known about Melissa, although like Lyman, she was a native of New York.[63]

Just a few weeks after their marriage, Melissa's father, a Scotsman, died in Cherokee County, Kansas. In late 1859, he had gone to Cherokee County to seek employment as a teacher. He successfully landed a position to teach the family of Mr. James Rawles, but according to a letter written by Rawles, Mr. Lyon had a "fit of ague every night and a severe cough" and died on January 21, 1860.[64] Rawles assured Melissa and her brother, Edgar, that Mr. Lyon was "followed to the grave[site] by the whole neighborhood . . . [and] buried in plain view. . . . [of] the vast Prairies of Kansas above the thick forest which overhang the river and spread over a portion of its valley. It is one of the most beautiful spots of creation . . ."[65] After Mr. Lyons's death, one Mary Lyons, age sixty-five, and likely Melissa's mother, lived with the young couple according to the 1860 census. One of Melissa's brothers, Thomas—a carpenter by profession—also lived with them during 1860.[66] Her two other brothers (Edgar and George) worked as clerks.[67] The young couple welcomed their first child, Minnie, on October 24, 1860.[68] In this immediate prewar period, Lyman continued to provide for his family through his work as a surveyor, and with an eye to the future he claimed land near Dodge, Minnesota.[69]

2 | Off to War
August 21, 1861–September 23, 1861

Our glorious flag is trailing in the dust and trampled under the feet of traitors, Who, that has a drop of our fathers' blood surging in his veins can remain a disinterested spectator until the angry tide of war overwhelms us all. I can not! I will not!

—Lyman G. Bennett, introduction to his diary

The outbreak of war at Fort Sumter on April 12, 1861, sent a shock wave across the nation. In its wake a surge of patriotism, dubbed by the French as *rage militaire*, rattled every community.[1] The scene that played out in Oswego, Illinois, differed little from what occurred in other towns in the North and South. On April 15, 1861, President Abraham Lincoln called for seventy-five thousand militiamen to serve for ninety days.[2] That evening, Lyman Bennett joined an Oswego crowd that had gathered at the Kendall County Courthouse.[3] Various local dignitaries gave speeches, and then Lyman asked to speak to the crowd.[4] According to a postwar county history, he electrified the audience by asking "how many would then and there volunteer for their country. He held in his hand a paper with one name on it—his own. Who would go, if need be? The spark of patriotism caught like fire in dry tinder, and in a few minutes eighty names were enrolled."[5]

Like many Union volunteers, Bennett explained his enlistment by alluding to the Revolutionary generation and his desire to end the threat of secession to the nation. Although the controversy over slavery had torn apart the United States, Lyman was silent on that topic, as were many other Union soldiers at that stage of the war.[6]

It is possible that economic factors also motivated Bennett to enlist. According to historian William Marvel, secession triggered another round of financial problems as Southerners cut orders to Northern factories and trade became increasingly disrupted. Cash grew scarce, many banks stopped lending money, and unemployment rose. In Illinois, the bitter cold winter of 1860–1861 led to a late planting season, and then cutworms infested the wheat crop. Sixty-eight percent of Illinois's ninety-day volunteers were below the $815 median household wealth, hinting at the role of economic factors in enlistment decisions.[7] According to the 1860 census, Lyman's real and personal estate totaled $800, which placed him only slightly below the median.[8] With a wife and a baby to support, it is not unreasonable to suppose that Bennett had at least some economic incentive, along with his evident patriotism, to enlist.

The surge of Northern volunteers far outstripped Lincoln's initial call of seventy-five thousand men. The president asked for forty-two thousand three-year volunteers on May 3, 1861, and enlistees in Illinois exceeded the State's quota by 70 percent.[9] Like many other men in Illinois, Bennett had not found a place in the six regiments Illinois organized under Lincoln's initial call.[10] Not long after the Federal defeat at the First Battle of Bull Run, the "Fox River Regiment" began organizing in Geneva, Illinois, on July 29, 1861.[11] Eventually designated as the Thirty-Sixth Illinois Infantry, officials organized fifteen companies with twelve (ten infantry and two cavalry) accepted into the Thirty-Sixth.[12] Lieutenant Colonel Nicholas Greusel of the Seventh Illinois Infantry was promoted to Colonel and assigned to command the new regiment on August 14.[13] Numbering 965 soldiers, the new volunteers began their training at a location near Aurora named Camp Hammond in honor of the superintendent of the Chicago, Burlington and Quincy Railroad who took a "warm interest" in the regiment.[14] Lyman was a member of Company E and listed as the third corporal out of eight in the company.[15]

Bennett described a "gay and happy life"[16] within the Thirty-Sixth Illinois Infantry during his first month's entries as he began to learn the duties of a soldier. He also managed to pay regular visits to his family since strict discipline was not yet imposed. Camp life was not entirely calm, however, as a troubling controversy involving First Lieutenant William Walker escalated into a saga that played out over several months.

My dear wife:

To thee I dedicate this diary. I go from you to the field of battle, and however bitter the strife, for your sake I will be strong and brave. The memory of you will give me hardi-hood to endure privation, courage in danger, and patience and fortitude in the hour of trial. May God be with Thee and me, protect us both, and grant a happy meeting both in this world and in the world to come.

Your affectionate,
L. G. Bennett

The history of nations is but a history of men and though full of instruction and of warning, how few profit from the perusal. Our nation like others was cradled in war and carnage, but from the strife we issued free and independent. Within the lifetime of many who fought to gain that independence three wars with foreign nations have been successfully waged, and now another far more bitter is raging between members of our own sisterhood of states. Our glorious flag is trailing in the dust and trampled under the feet of traitors, Who, that has a drop of our fathers' blood surging in his veins can remain a disinterested spectator until the angry tide of war overwhelms us all. I can not! I will not! I have a wife, a lovely wife, the daughter of a race of heroes, who tells me "Go and may God be with thee." God bless her! Heaven be with my dear ones. I will go to return victorious or a corpse.

The disaster at Bull Run[17] opened the eyes of the nation to the magnitude of the danger which threatened it, and again all the patriotic fires of Columbia's free born sons burst out afresh and companies and regiments were raised with astonishing celerity and hurled upon the foe. Fox River valley was all on fire and stalwart sons of toil were eager to grasp their firelocks and march off to the fray. At a meeting held at Geneva in July preliminary steps were taken to organize the Fox River Regiment. Major Greusel[18] of Aurora was designated as the leader. With this manifestation of confidence in him he went to St. Louis and procured the acceptance of the regiment from Gen. Fremont[19] together with means for an outfit, and an order for its immediate assemblage at Aurora. A site for a camp was selected east of Montgomery but the owner of the land, with more selfishness than patriotism would not allow his ground to be used for a camp without an exorbitant consideration. Another site was selected on the west side of the river near a large spring with the exception of being on the naked prairie fully as good a location as the first. On Saturday the 19th of August 1861 the Young America Guards[20] went into camp and were the first company on the ground. They were a fine body of men under the command of Capt. Baldwin, athletic and strong and looked as if used to toil. Their tents were all arranged and when the other companies arrived they were in comparatively comfortable quarters. It was my good fortune to be attached to the Bristol Company.[21] This company was composed mostly of Farmers' sons, and all embued with sterling patriotism. On Saturday our officers were elected which resulted as follows: Charles D. Fish,[22] Captain, without a dissenting vote; Albert M. Hobbs,[23] First Lieutenant; Wm. H. Clark,[24] 2nd Lieutenant. The good citizens of Bristol and Little Rock with commendable zeal turned out in mass, in greater numbers even than to Republican meetings and escorted the boys to camp on Monday the 21st of August, 1861.

Monday Aug. 21 1861. [Tuesday, August 20][25]

Awaited the Bristol co. at quarters until 12 o'clock and then with the whole crowd proceeded to Camp Hammond, a half mile north of Montgomery. This has been named in honor of Mr. Hammond the Superintendent of the C B & Q Railroad under whom Col Carnisal conducted a train until the breaking out of the war. The Young America Company marched out to receive us and the whole reception was conducted in fine style. Ours was the 2nd company on the ground. A picnic dinner got up by the good Bristol dames was served up to the boys, after which the tents were pitched, our blankets and rations secured and a supper of bread, coffee, pork, &c prepared by the boys, who showed themselves anything but adepts in the culinary department. One poor fellow's meal caught afire and burned up in his awkward efforts to cook his supper, after which preparations were made for the first night in camp.

Tuesday [Thursday] Aug. 22nd 1861.

The first night in camp was one to be long remembered. Few had ever experienced its luxuries & conveniences. A few retired for the purpose of rest but their effort was a vain one as the greater part under about as much restraint as the forest deer gave way to joyous hilarity, some even running, leaping, shouting, and laughing out at every prank which could be invented to make night hideous and disturb the more serious ones, who were vainly trying to coax a little sleep from this new Pandora's box of fun. From one tent was heard the slow and measured notes of psalm singing, and swearing, laughing, and conversation in a perfect babble of confusion. Day dawned again after what seemed to me an unusually long night and with it, the work of preparation for living, and for the manufacturing of soldiers. Three companies were today added to our numbers and soon after noon a U. S. Military officer came upon the ground and mustered seven companies into the service of Uncle Sam. It was a fine sight to see the boys with hands high uplifted to Heaven solemnly swear to protect and defend our loved country from all foes. The Newark company[26] came in on the 4 p.m. train and immediately formed at the ammo depot and officer Webb swore them in and went to Chicago on the same train the company came on. In the evening I asked for a furlough for the remainder of the week but no one appearing to know anything about it, I went on my own responsibility and found Melissa and the little one in fine health and spirits.

Wednesday [Friday] Aug. 23rd 1861.

Drew cobs for firewood and prepared to remove the old depot stable on my own premises for the purpose of keeping a cow during the winter.

Thursday [Saturday] Aug. 24th 1861
Worked hard at stable and pig pen, father assisted.

Friday [Sunday] Aug. 25th 1861.
Cut wood and done chores.

Saturday [Monday] Aug. 26th. 1861.
Went to Aurora, purchased groceries and set for my picture for Mellie and she and Minnie set for a picture for me. Minnie could not set still, but a tollerable picture was taken of the little one but Melissa did not look well, I may try again.

Sunday [Tuesday] Aug. 27th 1861
About 4 o'clock started for camp, found the boys enjoying themselves hugely. All the companies were on the ground, more than a hundred tents were pitched, and though a large number were off on parell [*sic*] a large number were on the ground. Services were held at 9, at 3, and in the evening. The Rev. Mr. Fowler from Newark preached the evening discourse, which was good. I shall try and get him appointed Chaplain.

Monday [Wednesday] Aug. 28th.
Learned that I was appointed to Corporal of our company. And in the morning was dispatched with a squad to dig a privy cellar. As boss I did but little of the work. Officers did relieve me of this job after which nothing of moment occurred.

Tuesday [Thursday] Aug. 29th. 1861
Was detailed as Sergeant of the guard and could not go home as I had designed. Guards are appointed and arranged as follows: The evening previous on dress parade the Col. gives his orders for the succeeding day among which is the number to be detailed for guard. The orderlies of each company selects the men, generally in rotation and at 9 o'clock the men detailed from the several companies assemble at the headquarters and are arranged in ranks and divided into three reliefs. These reliefs are numbered 1, 2, 3, &c. The stations for guards are also numbered 1, 2, 3, &c. At 9 a.m. the 1st relief marches in order to the several stations and leaves a guard at each one and the guard who is relieved takes his place in the rear and when all are relieved the old guards are marched to headquarters and dismissed. At 11 the 2nd relief is called and in the same order marched to the several posts and the guards there are relieved. It is the sergeant's duty to oversee and take charge

of the guards & hence is obliged to go the rounds every two hours. The day guards only watch the property of the army, and fences & also preserve order. The night guards are instructed with the countersign and allow no one to pass without it. If a person advances to a guard without it and has a good reason for passing, the sergeant of the guard is called and he can give permission for him to pass. Of course I got but little sleep. After posting the first night relief and nicely layed down for sleep the sergeant was called to Station no. 12. I went in a hurry and found a person who wanted to go to Aurora for a sick comrade. Of course I let him pass. Several alarms were given which kept me on the run all night. I got but little rest. One alarm at station number 4 I found a cavalry horse down and unable to get up: cutting his halter I led him around awhile and then ride him again. Guard duty for a sergeant is easy only it deprives one of sleep. I suppose our duties are not as great or the guards not as strict as when in actual service.

Wednesday 28 Aug. 1861.[27]

A gay and happy life we are leading at Camp Hammond. No dullness or lack of excitement ever creeps into our camp. I have drilled but twice today for guard duty does not fit a person for very active work the succeeding day. Our camp is thronged by thousands, mothers with anxieties about their sons now on their first advent in the world. Sisters with sweet meats and sweet words for loved brothers, and sweethearts, with words of endearment for their cherished idols, all throng our camp and help to break the monotony and drive dull care away.

Thursday Aug. 29th.

Another bright and sultry day has come and gone and the same routine of drill and dress parade is all that has disturbed the monotony of camp. Not all: for among 800 or 1000 men monotony cannot come. I was amused to see a pompous commissioned officer attempt to pass the guard and when ordered to stop paid no attention to the guard who forthwith commenced a vigorous pursuit. Commissioned officer started on the run with guard close at his heels and a bayonet in close proximity to his coat tails. When officers disobey orders privates will imitate their example.

Friday Aug. 30 1861

The heat of the sun and the dust of the parade keeps many within their tents when not on duty. An abundant supply of reading material furnished by the Bristol people serves to employ our leisure time very pleasantly. Went home in the evening and passed an agreeable evening with wife and baby.

Saturday Aug. 31.

Small crowds of people have not before been present at one time as today. The dust is perfectly suffocating. A copious shower would be hailed with gladness by us all. Home again. There is no place like home. That laughing sunburn [sunbeam?]—(Minnie) has a sort of magnetic influence to draw my thoughts and person home.

Sunday Sept. 1st. 1861.

It was my design to go with wife and baby to camp in time to attend the religious service of the day but dark threatening clouds in the West, the sure procusor of a storm frustrated my design and I accordingly proceeded to camp on foot and alone & arrived just in time to escape a heavy storm of rain accompanied with wind. Our cloth tents flapped and swayed before the blast and I expected every moment they would be swept away but for two hours they stood firm and securely protected us from the raging storm. The guards maintained their posts and though drenched to the skin kept up their respective rounds. At sunset the storm clouds were hurrying away in the east and patches of clear sky here and there betokened that the storm was over. But far in the west one black cloud arose in heavy masses. The faint gleams of lightning illuminating the deep recesses of cloud was portentious [portentous] of what was to come. The almost unearthly quiet which pervaded the camp was significant, full of articulate meaning. The cloud drew nearer and at midnight the rushing wind shrieked among the tents and with resistless fury went howling by. Here and there a tent would careen and finally fall upon the suddenly aroused inmates who would scamper through the rain to other quarters. I expected our tent would fall every moment and was ready with my blanket to flee from the rains. The cloth surged and flopped in the wind but stood fast. The wind soon was over but the rain continued until daylight in an unceasing patter the music of which lulled us to sleep.

Monday Sept. 2nd. 1861.

A hasty stroll through the camp just after daylight discovered eight tents in ruins and 4 or 5 badly shattered. The Col.'s tent suffered more than any other for the rain had beat it into the earth until there was but little difference in the color of each. Men were detailed from each company to wash it. The earth was completely filled with water, so much so that there was no drill or even dress parade. The day was passed in lounging about camp. I wrote two letters and read the Indian History of the Revolution. The Bristol library is a fine institution.

Tuesday Sept. 3rd 1861.

Boxing and wrestling has been the amusement of the day during which the 1st. Lieut. of the Young America was injured in his shoulder and taken to the hospital. Very many have a slight disentery owing to change of life and food. There are but three however in the hospital & they not seriously ill. Went home in the evening. Home! Sweet home! how I shall miss thee.

Wednesday Sept. 4th.

Transacted a little business at Oswego in the morning and at noon returned to camp. We had a battallion drill in the afternoon. It went off finely so our officers and spectators said. This was the 2nd since coming to camp.

Thursday Sept 5th.

Last night the Rev. Mr. Hocker gave us a first rate sermon. It was the best we have had since we have been in camp. Our parade ground is now in excellent condition—no more dust and dirt. It is a pleasure for visitors now to come here, we look clean and respectable. Before noon as an invited guest of the Young America boys I marched with them to the old camp meeting ground in the grove to partake with them of the goodies prepared by the citizens of Montgomery and neighborhood. I expected to find Melissa and my little "angel" there and was much disappointed to find they had not come. Their absence made me sad and though I ate as heartily as any I could not enjoy myself with the others. Was visited by father and mother in the afternoon but even that pleasure was shortened by battallion drill. The guard is now doubled and increased watchfulness enforced. No soldier is allowed to pass the line outside of camp and visitors only admitted at the gate, and today an order has been issued against picnics outside of the camp. The rules are quite military indeed. The boys are quite friendly—I hope I shall make friends of all deserving boys. We shall all perhaps see the time when we shall want friends. Some of my brother officers are a wild set and long after the more sober ones had curled down for sleep the coarse jest and heavy laugh resounded from our tent. I should not wonder if it would be carried too far some time and the whole batch put under guard.

Friday Sept. 6th 1861.

Has been the most dull day since coming to camp. A military funeral for an officer in Wyman's Regiment[28] has taken most of our officers away so that we have had company drill but once. I am anxious to see wife and little one. I will not acknowledge I am homesick.

Saturday Sept. 7th 1861.

Was out on drill by sunrise. Soon after breakfast it commenced to rain and supposing it would prevent drilling I wrapped my blanket around & about me and went home, found all well and happy & in the afternoon took mother and Melissa to Oswego & went myself collecting but had no luck.

Sunday Sept. 8th.

Went to church in the morning and passed the remainder of the day with the "Loved and Best." One's own fireside is the place to look for love. I fear I cannot come home again before we leave.

Monday Sept 9th. 1861.

Lounged around Oswego until noon expecting the payment of some money but being disappointed I kissed my loved ones good by and proceeded to camp. About half the regiment were away on furlough and the battallion drill was a meager affair. In the evening I went with a large number of comrades to Montgomery to a temperance meeting. After a few speeches it was announced that all soldiers who wished would be initiated into the Division of Sons of Temperance.[29] 76 accepted the invitation and were duly initiated without the payment of a cent. A good meeting was held after the ceremony. And measures were taken to organize a temperance society in camp & for that purpose two delegates were appointed from each company to meet the next evening at Capt. Price's headquarters.[30] It was near mid night when we returned to camp. If a thorough working temperance society can be organized and the proper influence exerted it will prove the salvation of the regiment.

Tuesday Sept. 10th.

Yesterday's bright sunshine was succeeded by a dark and lowering sky. A short quick step and a hasty breakfast was succeeded by a heavy rain storm driving the boys to their tents. Being thus closely thrown together means for amusement was soon brought into requisition. A lively and happy company were in the officers' tent. While I am writing a glance around the tent discovered one reading a bible which perhaps a loved wife, sister, parent, or friend has given him. A few are writing. One group is playing the enchanting game of euchre—a few are fast asleep upon the straw. Thus passed away the fore part of the day while all outdoors is drenched with rain and the unceasing patter upon the tent woos one on to sleep and to pleasant dreams. The rain ceased after noon but the ground was too wet for drill. Orders now are to allow no soldier to pass out—an indication that we will soon be on the wing, for one I am as near ready now as I shall be. O how I wish my dear ones could go along.

Wednesday Sept. 11th. 1861.
The morning was cloudy but the thin vapory clouds scudding across the sky revealed many broken patches through which the sun occasionally smiled and spred [spread] a genial influence around, warming the air until a coat was a burden more than a luxury. We had but one drill, aside from dress parade. Dress Parade is at 6 o'clock each day at which time the whole Regiment is drawn up in line and at times put through with the facings &c. The music then marches up and down the lines in full blast after which the Orderlies report the condition of each company. The Adjutant then gives the orders for the next day and the Parade is then dismissed and the companies marched back to their quarters.

Thursday Sept. 12, 1861.
Today a large number of Little Rock people were over to see their sons, brothers, & husbands for the last time. Their joyous hilarity when they first arrived contrasted strangely with the tears and half suppressed sobs at parting. I could not endure it and wandered to the upper part of the camp and at times it was more than I could do to control my own feelings. On my return I found one young and beautiful wife who could not tear herself away. She clung to her husband and no effort of theirs could hide the struggle going on within. A few words of devotion and love, a silent look, and then the unbidden tears would flow. At two o'clock [we] were drawn up and examined by the surgeons, or rather passed between two doctors & if any by their walk betrayed any signs of lameness they were marched aside and examined. One was thrown out of our company, others should have been but were not. 101 were thus examined and 4 were in the hospital sick. Col. Webb of the U.S.A.[31] was present and after examination mustered us all in by again administering the oath. One in the Newark Co. would not swear and was kicked off the grounds. Two dutchmen in our company also refused to be sworn in and Col. Joslyn[32] jerked them higher than a kite & gave them a farewell blessing with his boot. As soon as we were dismissed nearly [all] our company rushed into the road and overhauling the dutchmen each one gave them a kick. I was among the first to reach them and of course gave them the benefit of my boot, but when I saw a shouting crowd of 100 men rushing up and each one pitching in, and the blood flowing from the poor fellows I really felt sorry for them and regretted that I had a hand in it. The dutchmen were pretty severely handled and one could scarcely go. To close

the scene a horse whip was procured and both severely lashed. Their piteous cries unmanned me but awoke no sympathising response in the infuriated crowd around them. I hope we will have no more such scenes in the whole campaign.

Friday Sept. 13th. 1861.

Was visited by Melissa and Minnie and enjoyed a happy hour with them. I wish I could be with them or they with me. I never knew until now how much I loved them. Mother came up in the afternoon but battallion drill cut our visit short. There were many tears [falling] today. A severe rain prevented dress parade. I trust we will leave this place soon. I am tired of Camp Hammond.

Saturday Sept. 14th. 1861.

I have been detailed as Corporal of the guard and can not go home as I wished. The ground being wet and muddy has prevented everything but dress parade. We really had a severe rain storm. I think we have an unusual amount of rain for the season. Wm. Walker, 1st. Lieutenant of the Oswego company is in trouble. The surgeon of the regiment pronounced him unfit for the service and the Col. has appointed one Murrell[33] in his place. Walker, not being satisfied, went to Chicago and was examined by Dr.——— who sent him back to his post. Upon coming to resume his duties the Major ordered him off the grounds or he would arrest him. Walker is in a bad dilemma. I am thinking I should test the validity of an arrest under the circumstances.[34] This afternoon a private in the Lisbon Co. came reeling on the grounds with a stomach and bottle full of whiskey. Lieut. Van Pelt emptied the bottle and then broke it, which Whitham resenting began to speak in no very respectful way of his superior. Van Pelt then put him under guard, which cooled him considerably. He was just drunk enough to feel well and whenever the drums would beat he would dance until exhausted, then fall.[35] One of the guards by some means was discovered with a bottle of whiskey and was beginning to feel the effects of it when the contraband article was siezed and confiscated. O! when will men learn wisdom! Why do not our officers stop entirely the use of the accursed stuff? Each day but more strongly confirms me in my temperance principals.

Sunday Sept. 15 1861.

Taking advantage of my position as Corporal of the guard I came home in the morning to pass one more day in the circle of my own family. Neither Melissa or Minnie are well. I can not be too thankful that I have been always free from the tooth ache. One rotten shell is the parent of a dozen ills to Meliss[a][.]

Monday Sept. 16th.
Early the clouds began to shed their tears and everything indicated a day of storms. Some one has described a rainy day as follows:

> There is a gloom on the sky & it's shadows
> Lie chill on the morning's pure breast.
> The sunshine is hid from the meadow
> And nature with tears is opprest.

Notwithstanding the drizzling storm I went with Melissa to Aurora for the purpose of permitting a dentist tendering his compliments to Melissa's mouth. The culprit was found & duly extracted and I hope a stop to her misery. Returning I stopped at camp & found all military business at a stand still in consequence of rain.

Tuesday Sept. 17th 1861.
Pleasant weather again and & drill resumed. Went home after battallion drill and found Melissa nearly sick. Nothing definite in regard to our uniforms or time of departure. I fear some of our men are suffering from that terrible disease, home sickness. Dreams of the cheerful hearth—the comforts of home. The cheering smiles of wife, children, and friends is fast irradicating the stoical indifference of the soldier. I hope however that the love of home and the ties of friendship are not so deep as the feelings of patriotism which should stir the whole country.

Wednesday Sept. 18th.
Returned early to camp but not in time for roll call. Was not however reprimanded. Saw many friends but nothing more than usual occurred. At 9 in the evening I was agreeably surprised at the appearance of C.P. Drake.[36] We lived a few years of our lives over again. Drake came direct from Connecticut and I was more than glad to see him. Something definite has at last been received in relation to uniforms & our departure. We will go as soon as Monday, certain.

Thursday Sept. 19th 1861.
Captain Camp[37] not satisfied with the difficulties already in the Oswego company has assumed the prerogative to remove all the non commissioned

officers including their orderly, Gust Voss.[38] I wrote a letter to Col. Grusal & gave a full statement of his case & the Col. refuses to ratify Camp's appointments. Gust is very thankful to me & I am pleased if I have been of any service to him[.] A portion of our outfit came today—to wit—one pair drawers, one pair stockings, one shirt, & a cap. The boys look and feel 20% better and we all begin to think the time not far distant when we will be off. Melissa came to camp after noon & I went to Aurora with her & after dress parade went home. And passed a pleasant evening with Drake and the loved ones. A rain toward night completely plastered the ground making it disagreeable indeed, hundreds of ladies sadly soiling their white linen. Eleven car loads of people came from Elgin & Woodstock to visit friends. A happy time was passed during their short stay.

Friday Sept. 20th. 1861.

Another lot of equipment was rec'd today but not the so much needed & long expected coats & pants. It was oppressively warm in the morning and I left my coat at home, but the afternoon was cold and the Northwest wind whistled shrilly through the aisles and among the tents. I could not go out of the tent for the cold[.]

Saturday Sept. 21st.

A cold and chilly morning but as the day advanced it grew warmer. Some of our boys were transferred to other companies. I believe they were not the best ones we had and I cannot say I regret their loss. A fresh lot of supplies arrived, but only canteens were served to the soldiers. Each hour brought its rumors in regard to our departure, among which was one that we were destined for Washington. Mother & Father were in camp this afternoon & in the evening I went home again. The most rigorous camp duty cannot keep my thoughts from home, and every chance while I can will be improved to get home. The thought of leaving it is a sad one.

Sunday Sept 22nd 1861.

O! must I break away from the cherished idols of my heart! As I hold my little one in my arms its innocent and smiling face turned to mine, I see my own features glancing up from the watery depths of her eyes. As I think that soon I will be engaged in the turmoil and bustle of war and that I may go down beneath the swelling crests of its bloody waves forever, and no more mingle with the ones my heart holds dear I cannot restrain the tears that will flow unbidden from my heart. God bless all my dear ones, and may I be spared for their sake. Returned after dinner to camp, my dear wife going with me & remaining during the religious services of the afternoon. In the evening I attended a prayer meeting at the officers' quarters of the Newark company. How much need there is of God's mercy and love in this regiment. Sin abounds every where. The great vice of the camp is card playing.

Monday Sept. 23rd.

It has been a busy day of preparation. Our uniforms and outfit except arms were fully supplied to us and at dress parade we presented a fine sight with our blue uniforms. Tomorrow we leave Camp Hammond for Missouri. In bidding adieu to camp we bid farewell to our friends for a time if not forever. Mellie, mother of my little one came to see me. I think I can meet death upon the gory field of battle and my heart not fail me as when I bid my wife & little one good by. It is the hardest part of the campaign. I went to the gate and again and again kissed my loved ones and not all the manhood I possessed could suppress the choking sobs that came welling up from my heart. It was sundering the holy ties of husband and wife, of parent & child. Two years ago we set out together to win the prize of life, and to battle side by side. But now I am following ambition's call and seeking for glory and my country's greatest good. Should I fall in the bloody fray I know I shall have no kind friend, no loved companion near to soothe the hour of death. I know no wife or child will be there to shed a tear over my cold grave but still a sense of duty enables me to dare it all. Melissa, farewell but not forever. Hope whispers to me, if we do not meet here we will meet in the skies.

3 | Trip to Rolla, Missouri

September 24, 1861—September 29, 1861

As we passed Oswego I saw Melissa and Minnie in the yard gazing at the passing train, but not one exultant shout escaped the grief sorrowing family. I gazed at my loved ones and my home until borne from sight.

—Lyman G. Bennett, September 24, 1861

After the First Battle of Bull Run, much of the nation's attention turned to the trans-Mississippi, the area west of the Mississippi River, as the struggle began for control of a divided Missouri. Following the election of pro-South sympathizer Governor Claiborne F. Jackson, militiamen supporting his cause gathered at Camp Jackson in St. Louis.[1] A Regular Army officer, Captain Nathaniel Lyon, and Colonel Franz Sigel, led a force that captured these men on May 10, 1861.[2] As the captives were led through the city streets from Camp Jackson to the Arsenal, a riot began, and Lyon's men fired into the crowd, killing twenty-eight civilians.[3]

This incident sparked a political and military controversy, but also led to a compromise between General William S. Harney, the commander of the Department of the West, and former governor Sterling Price, the commander of the Missouri State Guard and a supporter of the new Confederacy. Under terms of the compromise, Federal troops would not be dispatched to the interior of the state. Missouri, in effect, would be allowed its neutrality.[4] These efforts broke down when President Abraham Lincoln allowed Francis P. Blair, Jr., an unconditional Unionist and important Missouri Republican, to remove Harney as department commander and replace him with Lyon.[5] The new department commander met with Governor Jackson and other officials at the Planter House Hotel on June 11, 1861.[6] The meeting did not go well for those advocating political moderation and compromise. Lyon refused Jackson's idea to disband both the Missouri State Guard and the pro-Union Home Guards and stubbornly insisted on "the supremacy of Federal authority in the state."[7] In Lyon's words, "'This means war.'"[8]

Jackson and Price began recruiting soldiers near the Missouri River town of Boonville in the central part of the state where pro-Southern

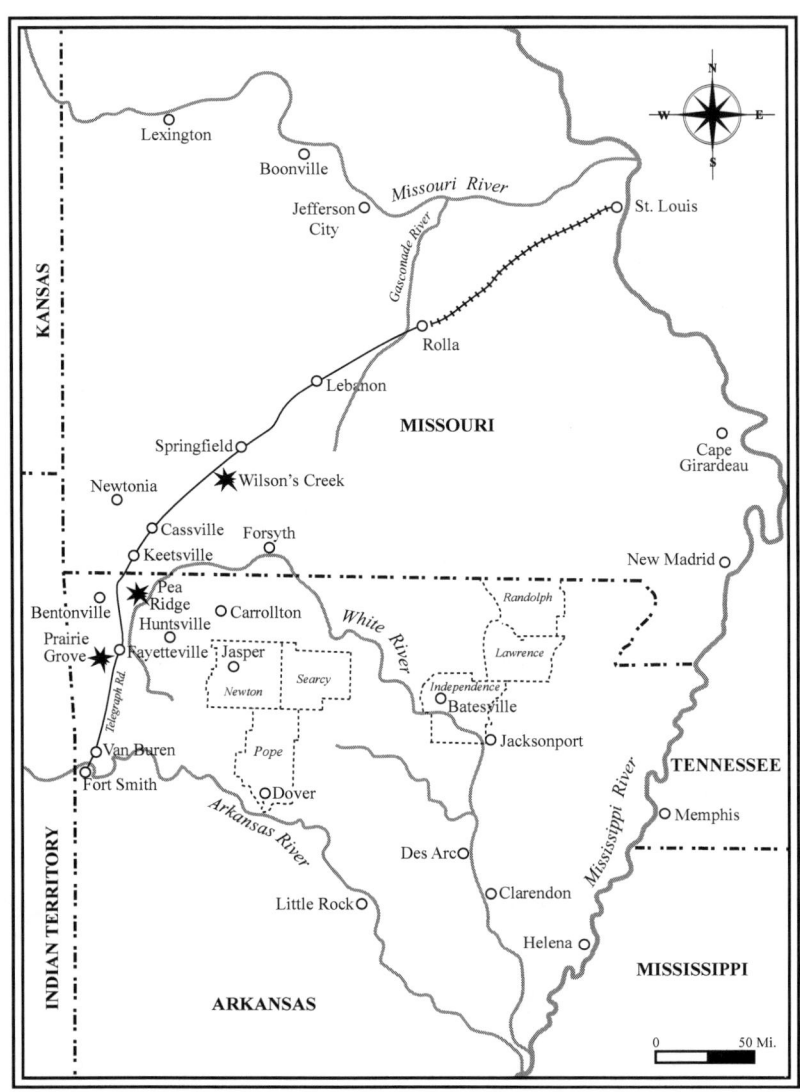

Missouri and Arkansas, ca. 1862.

sentiment ran deepest.⁹ Lyon quickly launched a two-prong strike that dispersed the guardsmen from Boonville and then ordered Brigadier General Thomas Sweeney to Springfield in the southwestern part of the state.¹⁰ Weeks later, on August 10, 1861, Lyon and Sigel (both now promoted to brigadier general) led their army in a surprise attack on Confederate and Missouri State Guard encampments along Wilson's Creek southwest of Springfield.¹¹ Lyon was shot through the heart and killed in the attack, and Missouri State Guardsmen and Confederate forces repulsed Federal troops in bloody fighting. This Confederate victory led to an advance by former Governor Sterling Price as he led seven thousand Missouri State Guardsmen north to besiege the Federal garrison at Lexington situated on the Missouri River.¹² Price's men forced the capitulation of Colonel James A. Mulligan's approximately three thousand soldiers after a three-day siege on September 20, 1861.¹³ Meanwhile, Federal authorities dispatched more units, including the Thirty-Sixth Illinois, to deal with the emergency in Missouri.

This set of Bennett's diary entries documents a six-day time span and describes the journey of the Thirty-Sixth Illinois Infantry from Camp Hammond to Rolla, Missouri. Bennett wrote a detailed description of the regiment's travel by the Chicago, Burlington and Quincy Railroad to the Mississippi River town of Quincy where the men boarded the steamer *Warsaw* that transported them to St. Louis, the center of the state's railroad system.¹⁴ Interestingly, no railroad lines connected "Missouri to the Confederacy."¹⁵ From St. Louis, the regiment traveled by way of the Southwest Branch of the Pacific Railroad to Rolla.¹⁶ This journey of approximately 535 miles gives insight into the substantial transportation network that had developed in the region by 1861.

It is in this part of Bennett's diary that the reader can first see the characteristic quality of his writings as a travel narrative. A curious and observant man, Bennett left vivid pen pictures of the countryside, civilians, and the actions of people around him. With William M. Haigh, he co-authored *History of the Thirty-Sixth Regiment Illinois Volunteers, during the War of the Rebellion* (1876). Bennett's diary served as the basis of the first fourteen chapters of that book, and yet the two accounts differ significantly. In his diary, Lyman recounted scenes of drunkenness, theft, fistfights, and other bad behavior by men of the Thirty-Sixth, but these rarely appeared in the more sanitized regimental history. This was not uncommon, as many regimental historians sought to present their regiment in the best light possible.¹⁷

Tuesday Sept 24th. 1861

At length the expected and much wished for day of our departure dawned and long before day nearly all in camp were astir. A few who were out late in the not very honorable or manly business of chicken stealing, or celebrating the event with soaking themselves with whiskey were in no mood to get up early.

At 9 o'clock the tents were struck and rolled up for carrying and boxes, bales, and other camp equipage were trundled into the cars. Lieut. Walker of the Oswego Co. came into camp with his commission from the governor and I thought he was now all right but Col. Greusal ordered him off the ground and he left with anything but exuberant spirits.[18] The people flowed in solid masses to camp in squads & hundreds and many a heart was saddened at the (perhaps) last farewell. Caroline & Louisa[19] were there and clung to me with all a fond sister's love. At noon we filled our haversacks with provisions for two days & then for four hours loitered around camp, complaining loudly that we did not move on. At 4 we formed in line and bidding adieu to Camp Hammond marched to Aurora, followed by at least two regiments of civillians. At sundown we embarked and amid the almost deafening shouts of thousands we swiftly glided away. As we passed Oswego I saw Melissa and Minnie in the yard gazing at the passing train, but not one exultant shout escaped the grief sorrowing family. I gazed at my loved ones and my home until borne from sight. I shall remember the scene and the position of all in the deafening roar of battle and until life shall cease to be. I did not notice a sad face in the whole regiment. All went away with joy & gladness, if the flashing eye, the loud laugh and still louder huzzah was an index of the feelings within. As for me I felt in good spirits and though the image of wife and child & home was ever before me, not a sorrowing tear escaped. That night I shall never forget, it seems as if the whole population were out and lined the road to bid us God speed. Bonfires blazed, guns were fired, and loud shouts made the welkin ring. I think we were expected for women & men lined the track at every station till long after midnight. At Arlington near Mendota we stopped for water and crowds of ladies came along to shake us by the hand and welcome the defenders of their country and in some instances when a smooth, good looking face was presented they did not hesitate to kiss the boys.

Wednesday Sept. 25th 1861.

We arrived at Galesburg at daylight and remained there near two hours watering horses &c. Large crowds came around to greet us. All along large crowds stood out in the rain waving handkerchiefs & cheering us on. Many of our boys had liquor in their canteens and though I saw none dead drunk, yet enough was taken to keep the steam up at fever heat. At one place when we stopped for water more than a hundred of the boys rushed into an orchard and nearly stripped it. They were carrying their depredations into a melon patch when Capt. Baldwin[20] went out and ordered them out. He kicked one impudent fellow, but none lifted a hand against him. I do regret we have so many pilferers. They act as if all they could lay their hands on was theirs by right. We got to Quincy at 3 p.m. and the horses & Ref. [regimental] stores were immediately transferred to the steamer Warsaw which was completed at so late an hour that the soldiers did not go on board but were quartered in a large warehouse, except such as chose to stay in the cars. A thousand or more of Mulligan's men[21] were at Quincy and each had their stories of suffering and hardship to tell. Many were known to our boys, and our fifer had a brother among them. A few joined our Regiment & will proceed to St. Louis with us. They are a hard looking set of irishmen, have fought well and deserve much of their country which has given them nothing.

Thursday Sept. 26th.

At 8 o'clock we left our lodgings (the cars) and went on board the Warsaw and amid loud cheers between the boat & shore we started for St. Louis. The strains of martial music from the band and the loud beating of drums which echoed along the woody shores filled the whole crowd with joyous enthusiasm, every object on shore was duly inspected & commented upon and whenever a darkey, or crowd of cheering inhabitants was passed a thousand shouts went up from the dense crowd which lined every part of our craft. Our rations consisted of hard sea biscuit and skippery [slippery?] cheese which were supplied in abundance. Nothing worthy of note occurred during our passage which was made without a single stop, except once when stranded on a sand bar for 20 minutes. The boys were kept on the run from bow to stern until she finally floated.

Friday Sept 27th. 1861.

Last night our company was marched below and occupied a part of the cabin for a bedroom. 100 men lay down for the night in an incredible small space & so closely were we crowded that not a single square inch remained unoccupied, but at daylight we were marched upon deck and found ourselves safely moored before the city of St. Louis, its crowded wharf and thousand steam boats was a fine sight indeed, and for hours the shouts from shore & from the boat was almost constant. After masticating a breakfast of hard sea biscuit I wrote a short letter to Melissa. About 9 the steamer fired up and dropped down the river 3 or 4 miles, nearly to the arsenal,[22] where we were landed and marched through many streets where nearly all the inhabitants were dutch who lined the streets and bade us welcome. Arrived within the stone walls of the arsenal, our arms were distributed, after quite a delay caused by the quality of the arms not being as good as we expected. The two flank companies received minnie [Minie] and Enfield rifles and the rest received only rifled muskets.[23] The Col. was very indignant that we should have no better, but no expostulation could procure different. Two companies would not take any because they were so rusty. The boys were all much disappointed except Co.s A & B. At a late hour we marched to the boat and again steamed up to the city. Here a few companies were marched to a wharf boat, and procuring plenty of hay to sleep upon we passed a pleasant night with the exception of a few alarms. About 2 one boy of the Lisbon Co. got the nightmares and gave three or four most unearthly screams which startled and aroused us all and until the cause was found out a considerable excitement ensued. The steamer Ben Taylor with a part of the Missouri 3rd[24] on board lay next to us and during the night one of the soldiers being pretty drunk fell overboard and was drowned. So much for the use of liquor. Would to God it would have the effect of making temperance men of all who are disposed to tipple. Our bill of fare has been various, mostly bread with no meat or butter to eat with it. Bread alone is a poor meal.

Saturday Sept. 28th. 1861.
We were early out of our bunks on the wharf boat and shoved on board the steamer which was closely guarded & none of the boys allowed to leave except those detailed to unload which occupied the whole fore noon and then our luggage was again loaded upon some cars which were shoved down to the wharf. Let me describe the arsenal, or as much as I saw of it. The grounds include I should think 15 or 20 acres and quite a number of forest trees are standing all through the grounds. A large number of brick and stone buildings are scattered around, some for machinery, for making and repairing guns. Some for officers and soldiers' quarters and some for store rooms. I was not allowed to enter the buildings but outside I saw 40 or 50 siege guns laying around, some of them 12 & 15 feet long, also a few batteries of lighter field artillery among which I saw 6 iron rifled 10 pounders. Many of the buildings had a garden attached with large fruit trees, grape vines, and flowers. Altogether the arsenal is a pleasant place and I should ask for no more pleasant quarters if placed in garrison. About 5 p.m. we were marched through the city to the Pacific R.R. and embarked for Rolla. The boys were all in fine spirits as they knew many who were stationed there in Wyman's

Regiment.[25] As we passed out of the city we were greeted by ten thousand cheers from those who thronged the depot and along the Railroad. Twenty miles were accomplished before darkness set in & I perched upon top of the cars to see what was to be seen. We passed through two tunnels and as the cars went thundering through the noise was almost deafening. In the cars the men were completely packed. The seats were rough boards placed across box freight cars, and every seat was completely filled. To lay down and sleep was impossible. Such a ride upon a rail I never before had the misfortune to be treated to. About 25 or 30 miles from St. Louis the country and road was so rough that the train was divided in two and even then it was with the utmost difficulty that the iron horse could pull the train up some of the steep upgrades. At times we would come to a full stop and a long time would elapse before the steep grades could be surmounted. The country out from St. Louis is all more or less rough but as we approach Rolla it becomes hilly but not mountainous. And the grade is mostly up, occasionally we go down hill and then the cars would leap along at a break neck speed. The portion of Missouri that I have seen is poor, scarcely worth the blood that has been spilled to defend it.

Sunday Sept 29th 1861.
At 4 o'clock in the morning I made my way out of the crowded car and resumed my old perch on top the train and as soon as daylight would permit obtained a fine view of the rough country we were passing through which appeared an almost unbroken wilderness. No signs of human habitations would appear for miles, and whenever a house was passed it was nothing more than a log cabin with mud chimneys built outside with generally a seedy butternut colored man and a homely and no less seedy woman and a crowd of dirty half clothed urchins protruding their uncombed heads through the cracks and chinks in the walls, or standing in front gaping at us. Very few cheered us showing clearly their secession proclivities. The towns we passed (for by that name I suppose they are called) are generally a collection of a dozen or more log cabins thrown promiscuously in every sort of disorder imaginable and of a character that would disgrace an Illinois hog pen. As we set out from St. Louis a suspicious looking four gallon contraband was smuggled into the officers' car, and the loud shouting and notes of bachinalian revelry showed clearly the part which the contraband was playing. Another proof of his presence was a large patch on the side of Lieut. Col. Joslyns's head, placed there to cover sundry bruises received in a fall from the top of the cars. Some thought it a pity that the fall did not terminate the career of one setting so pernicious an example to a regiment of men. Great God! Can it be that such

leaders and no others can be found to lead us to battle and victory. A few patch[e]s of prairie not more than a mile or so in extent was passed which looked like an oasis in a desert of wilderness. We reached Rolla about noon and was much disappointed to find Wyman's Regiment a day's march on their way west to cut off the retreat of any flying bands of secessionists from Lexington and the North. But soon word came that an order for his return had been received, and all who wished to see their long absent friends set up a shout of joy. We were marched to the old campground of the 13th and pitched our tents and were at home again. The 7th Mo. were in camp on our arrival but they shipped their camp equippage on the same train we came on and at midnight were on their way to St. Louis & Jefferson City. Their loud cheering on their departure aroused me & I supposed the 13th had returned and was for going out to meet them, but learning the true cause of the noise I returned to my bunk.

4 | Camp Life in Rolla

September 30, 1861–October 13, 1861

Have drilled in the manual of arms until my fingers ache. I feel like leaving as rapidly as possible and notwithstanding aches and fatigue I will not shrink from my task.

—Lyman G. Bennett, October 3, 1861

Corporal Lyman Bennett derisively labeled Rolla, Missouri, as "an insignificant one horse town," a wildly inaccurate observation, but one shared by many Federal soldiers.[1] Admittedly in 1861, it was not a pretty place. Its buildings, except for the new courthouse and jail, were unimpressive, and the streets were either dusty or muddy, depending on the weather.[2] A major supply center and the end point of the Southwest Branch of the Pacific Railroad, Rolla was the proverbial "jumping off" point to Springfield, 122 miles to the southwest.[3] In the Upper trans-Mississippi, armies could not survive by foraging only.[4] Instead, the army created supply depots, such as the one at Rolla, along the important four hundred-mile St. Louis to Fort Smith, Arkansas, transportation corridor.[5] Control of this corridor was vitally important for the Union cause since Confederate supremacy over it would threaten St. Louis, the base for Federal operations on the Mississippi River.[6] As department commander Major General Henry W. Halleck correctly perceived, the Missouri State Guard and Confederate troops there pulled Union soldiers away from the campaign to control the Mississippi River. Until this enemy threat was neutralized, Federal commanders would have to put that important campaign on hold.[7]

Like Bennett, most soldiers focused on events immediately around them rather than the conflict's grand strategy. In these entries, Bennett documented regular drills and the spread of camp rumors and diseases. Lyman also reflected on his enlistment in his October 5 diary entry: "Many at home thought me mad, or crazy to leave a comfortable home, a kind and loving wife, a lovely child to link me to my fireside—friends on whom I could depend, and all the care and responsibility of a husband and parent; to leave all these for the hardships of a soldier's life[.]" Stalwartly, he contended that "my country has claims upon me strong as that of home and family. I had rather feed upon roots, pillow my head night after night upon the cold ground, rather than keep aloof from the struggle that now menaces our country, and her liberties. If I should live to return I shall walk upon the soil that gave me birth and exult in the thought that I am not unworthy of it. I shall mingle with my friends & feel that I do not dishonor them. I shall look upon my country and her own sacred sights and rejoice that I did not basely desert them." Yet, he had confessed three days earlier a longing to be in his home with Melissa. Writing and receiving letters demonstrated the importance of those linkages between home and camp; like countless thousands of other soldiers, Bennett rejoiced upon receiving a letter and bemoaned the absence of one.

Monday Sept. 30th 1861.
Dress parade in the morning and general orders for drill given out. I conclude that our play spell is now played out, and that now we must come down to hard work. Six or eight hours drill is well calculated to take the starch out of some of our lazy struts. I hope every man not on duty or in the hospital will be forced into the ranks and right up to the rack, fodder or no fodder. We have had enough skulking and shirking from duty. I have been up to the fort that is being built about three quarters of a mile to the South of the town on the highest land in the vicinity. The trees & brush has been cleared from the contiguous land so that no enemy can find a hiding place within shooting distance. The fort is a small one but when completed [will] be strong. It is four 32 pounders, mounted, and from their position will make terrible havoc among an enemy. Rolla, the County seat of Phelps Co. is an insignificant one horse town with but one good looking house in it. That is the Court House, which is now used for a hospital. Near 200 sick are in it and the contiguous buildings. The sick are mostly Missourians showing conclusively that the natives who are supposed to be acclimated to the climate, from some cause are more subject to sickness than those from Illinois. That cause is whiskey, the bane of every camp and the fruitful source of loss of life than the sword and all other diseases combined. It is astonishing that with all it's dire effects before us, so many will guzzle and soak and among them very many who at home occupied respectable positions in society and pretended to be temperance men. I can come to no other conclusion than that an army can not be equaled in demoralizing a man, and that no greater calamity can happen to a nation than war. The rest of the town is composed

of log and board shanties, every other one being a whiskey shop. Before the place was occupied by troops with trees and undergrowth, but now several hundred acres are cleared for parade grounds &c. What particular natural inducements for a town site is here presented, I have failed to see. This is the most advanced post now in the federal hands and of course every day brings its rumors of secession encampments, of rebel advances, &c. But those who have been longest here give but little credit to these rumors, coming from the sources they do. The cavalry regiment who are camped next to us have 8 secession prisoners. What are their particular crimes I do not know. The day has been passed in cleaning up and putting the camp in order. About 10 Willet[8] & myself were passed out and went on the Springfield road two miles to meet our friends in the returning 13th. We first met the Iowa 4th,[9] and then came the 13th, and we were not long in finding our friends. I found Walker[10] and marched by his side into town. We were mutually glad to see each other and had a glorious visit. I found all the Oswego boys in fine spirits except John Martin,[11] he was in the hospital & I went to see him and cheered him up as well as I knew how. I know not how long the 13th will remain here. The probability is that it will be but a day or so, or until cars can be had to carry them east.[12] I wrote a letter home, and it seemed as if every man in camp was writing one or more letters. The reins are very much tightened upon us. It is by no means an easy matter to get outside the lines.

Tuesday Oct. 1st, 1861

Went early to drill in the manuel of arms but the dust was blown in such masses into our faces that after half an hour we were dismissed. The high wind and dust prevented any further drill except dress parade in the evening. I was passed outside the lines and I went into the camp of the 13th, found Walker so intently engaged with a game of poker and gambling. It seems as if nearly every one in that regiment gamble and very many get drunk. I saw two drunken fellows tied hand and foot and put in the guard house. John Martin I found improving and much cheered up to see old friends. Col. Greusal went with a file of men to hunt out and spill liquor. At one log hut a little was found and spilled and a few beer barrels broken. But the other shops got wind of what was up and secreted theirs in the bushes or some other place safe from the scrutinizing eye of the Col. Rations this morning were short and many of the boys complained bitterly. I find that complaining amounts to but little in the army. I shall be satisfied myself.

Wednesday Oct. 2nd.

We had a heavy rain during the night but our paraid being sloping it was soon dry after the rain ceased. Col Wyman came over with his regiment (the

13th) & had dress paraid on our grounds. It was a sight worth seeing. The boys are drilled with the precision of clock work. I was quite ashamed of my awkward movements. Col. Greusal went to Chicago today; the object of his visit I know not. In the afternoon I went with George Walker to a pool of clear water a half mile from camp and we had a good bath. I shall try and keep clean. During our ramble we had a good visit reminding me of home and its thousand endearments. My own Melissa, what would I give to be again with you. I feel that many weary months as well as miles separate us. "Home! Home! why did I leave thee, Dearest and happiest home." We found quite a quantity of nuts and returned in time for supper and dress parade, after which I retired to my straw and soon the hum of the camp gave way to dreams of those my heart holds dear.

Thursday Oct. 3rd. 1861
Have drilled in the manual of arms until my fingers ache. I feel like leaving as rapidly as possible and notwithstanding aches and fatigue I will not shrink from my task. I have grubbed among the roots and stone in the rear of our encampment two hours, which has somewhat fatigued me. Five or six acres are being cleared and grubbed in this manner. The work connected with this is worth more than all the land barren of every thing except grubs and stone. While not employed my thoughts have been of home. Hard has been the necessity for my leaving it. Cooped up as we are within the limits of the camp I can hear nothing of the outside world. Could I but have the daily Tribune[13] to read I should have less time for gloomy forbodings. Went to Capt. Pierce's[14] quarters & borrowed a book to read. And in its perusal a part of my melancholy flew away. I have just learned of the poisoning of one of our men belonging to the Elgin Co. He purchased an apple from some Missouri huxter and in a few minutes was taken violently ill. Antidotes for poison were administered and he was somewhat relieved but even now he is in a dangerous position.

Friday Oct 4th
Arose early, and drilled as usual. George Walker came over and we had a short visit. After noon I was passed out and saw some of the wounded from Springfield.[15] About 40 have come through, partly on foot, they being more able to travel than others. One had a toe shot off, another while upon his back loading his gun was shot in the nose, the ball passing down through his mouth and tongue and passing out between his chin and neck, a truly miraculous escape indeed. Another was hit on the top of his head, cutting a gash to the skull. In fact, every species of wound from the head to the toes,

were inflicted. Before returning I went about a mile up the railroad track where soapstone was found in considerable quantities. Nearly everyone in camp has a piece which they manufacture into pipes or anything else their imagination fancies. I shall send a few specamens home if I have an opportunity. The cars came in last night being the second train since we came through. The cars on this road are mostly used for carrying troops to Jefferson City and it is seldom one comes here. Oh! why cannot I hear from home. Just now the 15th Illinois[16] are marching to the cars to go to Jefferson City to join Fremont in his operations about that place. Going from my tent about 4 p.m. I heard a band playing the dead march and am going to the parade ground. I saw a detachment of about 200 from the Iowa slowly following one of their number to his grave. The corpse was wrapped in the American flag. The slow beating of the drum and marching of the procession together with the circumstances of his burial far from his home, unmourned by friends & in a distant and perhaps unknown grave. The scene was one that called up solemn reflections. Who knows but that some of us will soon be carried to an unknown grave, with no tears of friendship and affection to show that a link in some fireside has been broken. The soldier's grave! what a place for reflection. Friend Judson[17] came over in the evening and a pleasant hour was passed in our tent.

Saturday Oct. 5th

Wrote two letters, one to Melissa, then whittled a pipe for Lieut. [Albert M.] Hobbs, also some specamens for myself out of soap stone. Rained in the afternoon, and rest and reading was the order of the day. The rain has put out the fires & we are obliged with a cast iron cracker and a chunk of raw and greasy bacon for supper. To those who are in comfortable homes surrounded with everything that their luxurious taste desires, it would be no easy matter to reconcile their feelings to the hard fare of a soldier's life. But when the heart is warmed up and really engaged in any cause a person can do anything. Many at home thought me mad, or crazy to leave a comfortable home, a kind and loving wife, a lovely child to link me to my fireside—friends on whom I could depend, and all the care and responsibility of a husband and parent; to leave all these for the hardships of a soldier's life—its hard fare— fatigueing marches with the cold damp ground for my bed, and above all, the awful carnage of the battle field, to meet death in a thousand hideous forms. No one loves wife or family more than I. Yet my country has claims upon me strong as that of home and family. I had rather feed upon roots, pillow my head night after night upon the cold ground, rather than keep aloof from the struggle that now menaces our country, and her liberties. If I should live to return I shall walk upon the soil that gave me birth and exult in the thought that I am not unworthy of it. I shall mingle with my friends & feel that I do not dishonor them. I shall look upon my country and her own sacred sights and rejoice that I did not basely desert them. But should I be one of the victims I feel that to die is the irrevocable decree of Him who made us all. The silent village of the dead must sooner or later claim my body. Oh that my soul may be ready to meet death without dismay.

Sunday Oct. 6th 1861.
Immediately after guard mounting I went with George Walker to the creek and washed myself and clothes. I flatter myself that for the first time I succeeded admirably. I may perhaps be induced to get up a laundry shop. I passed most of the day out of camp, and during my scout gathered a quantity of hazel nuts. A mail arrived during the day and I was the lucky recipeant of a letter from home. Melissa remembers and I believe loves me. Her letter came so prompt and so opportunely and filled with affection and confidence. Never, never! can I forget or betray her. Our company being detailed for duty left me with nothing to do, but one of the first relief being sick I took his place and for the first time stood on guard. My first post was adjoining the Missouri 7th and from the sentinel adjoining my beat I learned many particulars of the battle of Springfield.[18] I also learned that on the first night of our encampment here a squad of 25 of his company were on a scout and towards day reached a house about 20 miles to the southwest of Rolla where they dismounted and fed their horses, the boys laying down among the bushes to get a little sleep. In about an hour a report came that 200 or 300 mounted secessionists were on their march to surround them. They were aroused and hurried away and did not miss two of their number for some distance. The two left behind were asleep and were surprised by 30 secessionists who fired into them on their attempting to flee. One was brought from his horse severely wounded and captured. The other, though severely hurt, having one arm broken and a finger shot from the other hand succeeded in making his escape and is now in the hospital, but in a critical situation. The last time I was on duty I had no one to talk with and was very sleepy. Thought however was busy and "home sweet home" was again in my mind and never did it appear so enchanting and lovely as now.

Monday Oct. 7th 1861.
Nothing beyond the usual routine of camp duties has transpired today. There is a small chance of my being transferred to Bowen's cavalry regiment[19] as Bowen's clerk if Judson does not succeed. I hope I may.

Tuesday Oct. 8th 1861

We buried one of our number today. He was a member of the Aurora cavalry[20] and was sick at St. Louis and came on without other sick [men] which accounts for his being separated from his company. Nearly all our regiment followed his body to the grave.

Wednesday Oct. 9th 1861

Was detailed nearly the whole fore noon to get wood. Our Captain [Charles D. Fish] went with us and pulling off his coat worked with a will, doing as much service as any of the men. This act will atone for much of his crossness. It has seemed of late that he could not do a favor with a good grace. But I suppose it is necessary that military matters be attended to in an arbitrary manner. We each received an additional blanket today. I now have three, which will make me comfortable. After dress parade three or four Missourians came into camp alledging that some of the boys had passed counterfeit money upon them. Their case was investigated and found strictly true and a Lieutenant in the Woodstock company[21] implicated and immediately arrested and is now under guard. I hope he will be severely and justly dealt with.

Thursday Oct. 10th

The regiment was drawn up and reviewed by Gen. Wyman. We were all so anxious to do well that we rather overdid the matter. And I think our appearance was not very creditable. The 13th Regiment with whom we have been on terms of closest intimacy commenced their march westward at 8 o'clock. Their destination I have not learned. It's supposed that they are to join Hunter[22] from Jefferson and march to Springfield and thus intercept any bands of rebels on their way to the South. About 9 a.m. it commenced raining, and has been a gloomy and cheerless day, in harmony with the cold and loveless world around. I pity the boys of the 13th on their wet and dreary march, but sickness nor storms do not interrupt the movement of armies. We had a most wretched supper. Not only was it inferior in quality but in quantity and nearly all but our cooks went to bed hungry. In the evening the matter was thoroughly canvassed in every tent and loud and bitter were the complaints of almost every man. Not only was the subject of rations discussed but also the usage from our superior officers. I need not add that Capt. Fish got his share of execrations for I must say he has of late been extremely overbearing and tiranical to some of the men, and the boys are fast learning to hate him, as they before loved and respected him. I have yet to learn that he has granted a single favor except to a certain few with a good grace.

Friday Oct. 11th.

The storm for some time brewing in the minds of the men burst in all its fury this morning. All who had been guilty of any fraud or meanness to the men were told distinctly of it and a general row was the consequence. Our officers during the day has been more courteous and gentlemanly to the men than ever before. The camp was all day startled by the most absurd rumors. One to the effect that Wyman was retreating and that Price[23] was upon us with an overwhelming force of the enemy was believed by very many and it was amusing to see the frightened expression on the face of many. A look at the map was sufficient however to dispell all fears on that account. With his position 120 miles distant and Fremont between us and him, how absurd to believe any rumor of this kind. Another to the effect that the Rail Road track was torn up 20 miles away was currently reported and believed. But during the night a train from St. Louis put a quietus to that absurd rumor. It is strange how easily the boys are gulled. One of the Lieutenants of the Woodstock company succeeded in passing quite an amount of counterfeit and worthless money upon the Missouri hucksters which come to camp with apples, chickens, pies, etc, and was arrested and put under guard. I am glad to see offenders occupying high positions thus summarily dealt with. They are no better than privates to go free from justice. Let Justice be done to all with impartiality. I realy pitied some of the poor women upon whom he foisted his worthless rags.

Saturday Oct. 12th 1861

Another mail was received today and I was much disappointed at not receiving something from home. I will read Melissa's old letters again; it will be some consolation. Oh! My poor Melissa; My absence may expose you to privations, but I trust to your own innate goodness, and to the protection of parents and brothers for thy safety. I cannot forget my little one—its bright eyes full of expression and fire. I seem to see it now toddling around requiring a mother's care and watchfulness to keep it from danger. I hear that the 13th are getting along finely and are far beyond the Gasconade. The Iowa 4th buried four of their men today. The result of dissipation and meddling with vicious women. Our Col. is very strict about the boys going outside of camp. No doubt that many of the boys now languishing on beds of sickness and death would now bless the same strictness in their commanders. It is hard to restrain the amorous propensities of many young men especially when brought in contact with vicious women. I notice a marked improvement in the conduct of our officers to their men. I shall be thankful for the volcanic explosion of temper manifested by the boys. We have now 16 in the hospital.

All but two are fast recovering. A man deserted from the Woodstock company last night. He stole a horse and succeeded in even passing the pickets stationed a few miles out. But they finally mistrusted and put after him and came near overhauling him when he left his horse and put in the brush and eluded his pursuers. Attended a prayer meeting in the evening at one of Young America's tents. Three parties of pickets were sent from our regiment today. One was taken from Co. E. Their destination and object I do not know, nor do they know themselves.

Sunday Oct. 13th

Attended divine services at 10 o'clock. Bro. Stonar of Newark gave us a good discourse, reproving the wild ones for their conduct & the vicious ones for their dissipation and pointed to the hospital with its corrupt and festering inmates as a witness of the sad results of excesses. In the afternoon while at the Colonel's tent to get a pass to wash he told me he would detail me at the fort to assist the engineers at 75 cents a day extra. The Oswego Co. & and the whole camp are again in a ferment. It appears that no officer in the regiment can get a commission until Walker's claims are recognized. The Co. was called up in line for another vote, but when it was discovered that Walker's friends were largely in the ascendency the vote was not permitted. The whole regiment is in a ferment about the matter. I hope our Governor will not permit our domineering officers to exercise tiranical sway over their equals in intellect and general information.

5 | Engineering Work in Rolla

October 14, 1861–November 3, 1861

Went early to the fort and by measurements established its exact center, also fixed stakes as a basix for future operations.

—Lyman G. Bennett, October 18, 1861

On October 14, 1861, Corporal Bennett's military life altered significantly: "While on drill I was sent for to go to the fort, and trudged up the hill on foot. Arrived at the place I soon ascertained what was required in the engineering line." Military officials would, from this time forward, utilize Bennett's surveying skills. At Rolla, Bennett worked first as a military engineer and later as a surveyor. While primarily involved with these duties, Bennett also spent time with the men in his regiment and continued to report on a series of controversies within the Thirty-Sixth Illinois Infantry.

Brigadier General Franz Sigel and his men occupied the Rolla railhead on June 14, 1861, and the town remained in Federal hands for the duration of the war.[1] On August 18, a train arrived transporting "four 32-pounder siege guns."[2] Soldiers painstakingly conveyed these to a hill "three-quarters of a mile south of the courthouse, a site which overlooked both the center of town where the railroad and government shops were located, and the roads in and out" of town.[3] A fort was now needed to house these big guns, and at the end of August, men from the Thirteenth Illinois Infantry began constructing Fort Wyman, named for their commander, John B. Wyman.[4] Colonel Grenville M. Dodge, the new commander at the town, intensified the construction by putting prisoners of war and refugees to work after the Thirteenth Illinois marched to southwest Missouri.[5]

Constructed as an earthen redoubt with gun emplacements at each corner, Fort Wyman reached a height of approximately ten feet with a six-foot ditch around the square-shaped site.[6] Working under the direction of Lieutenant M. LaRue Harrison, Bennett helped plan the construction of the blockhouses and the outer works of the fort.[7] The log blockhouses were outside the redoubt but were "connected to the interior of the fort by covered tunnels with entrances underneath the gun platforms."[8] Loopholes for riflemen dotted the blockhouses so men could concentrate their fire on the ditches around the redoubt.[9] As it turned out, the enemy never assaulted Fort Wyman, and it was repurposed at times as an overflow area for the county jail.[10]

Sprinkled throughout Bennett's writings is evidence of his curiosity and interest in collecting information. Although showing little awareness or interest in national politics, he regularly visited with soldiers about their involvement in military actions while also chatting with refugees who increasingly made their way to Rolla. These civilians, victims of "Neighborhood violence and intimidation" as well as the movements of armies, fled from southwest Missouri and some of the Arkansas counties bordering Missouri.[11] Army officers issued food rations to the refugees with the military eventually bearing much of the responsibility for tending to their care.[12] Although many of these displaced persons were Unionists, others were pacifists, and some were hostile toward the federal government.[13]

In this set of diary entries, Bennett wrote during a time when Rolla was the primary relief center for hundreds of refugee families.[14] Military enlistment benefited when many men joined fellow refugee John S. Phelps's Independent Missouri Regiment, while others entered the Twenty-Fourth Missouri Infantry and several Missouri cavalry battalions.[15] Once the army reoccupied Springfield in early 1862 (this time permanently), many refugees decamped from Rolla and moved to Springfield.[16]

Monday Oct. 14th.

While on drill I was sent for to go to the fort, and trudged up the hill on foot. Arrived at the place I soon ascertained what was required in the engineering line. After looking around a while I went out to a Co. of Missourians who arrived yesterday. Never in my life did I see a worse looking set of ragmuffins. Two thirds were without shoes and some without hats. All were dressed in that universal garb of butternut breeches, if the rags which took the place of that important portion of the human wardrobe was deserving of that title. Scarcely one but that both knees and posterior were without covering, exposed to sun and storm. A more lank, long haired, long bearded, ragged and dirty set of beings I never beheld. Their shirts, once white were worn until they were the color of tobacco juice. They numbered seventy men and had been organized as home guards since spring. As such they had been robbed and hunted like wild animals by the Arkansas secessionists. They came from Douglas County on the borders of Arkansas, and at home were constantly in danger of their lives. The secession desperadoes from that state were constantly committing depradations by driving off their hogs & cattle, stealing their horses and in some instances hanging the citizens for no other offence than defending themselves. One man had his ears & other members cut off and then stabbed to the heart. For a month this company had been driven into the woods and dared not venture to their homes only by stealth. Finally when their sufferings were past endurance they banded together and on Thursday the 3rd instant they marched against 300 of their oppressors. A party of 12 Union men being detached from the rest encountered 200 secessionists at Bryant's fork of White River about 2 ½ miles south of Very Cruze the County seat of Douglas county.[17] The fight lasted about 15 minutes when the secesh fled leaving 12 dead on the field & 10 wounded. But one of the Union men were touched with a ball and that only slightly on his arm. I saw the scratch myself, now nearly well. They were unable to pursue the enemy for want of ammunition.[18] The whole body of secessionists left for Arkansas soon after but the Union men fearing a more formidable invasion started for Rolla to enroll themselves in the service of Uncle Sam. Two days after the fight on Bryant's fork another party of secesh were met and one killed and three wounded without any of the Union men being injured. On their way to Rolla a party were in the house of Mr. Bates, a Union man. While there a Lieut. in McBride's regiment[19] with four secessionists rode up and mistaking them for secessionists commenced a conversation in which the plans of the secessionists were fully revealed. Capt. Adams, after hearing him through and getting all the information he wanted called upon the Lieut. to surrender, which he did and he is now a prisoner at Rolla and will be sent to St. Louis for safe keeping by the next train. One of their number was a man 75 years old. His name was Solomon Collins.[20] Their hats were nearly all gone to seed, and he had not a shoe to his name, and the remainder of his raiment hung in jingles, perfectly glazed with dirt. He had fought with Jackson at New Orleans and was now emulating with the young men to fight again. He had been hunted like an animal and forced to live in the bush away from his home. One of his sons had been caught and hung by the secesh for nothing more than loving his country.

Sunday Oct. 15th.
Pro ceeded to the Harrisons'[21] tent and drafted a plan of the block house and surrounding works and after dinner proceeded to the fort and made

preparations for taking the levels inside. I went among my butternut friends of yesterday and while there one of their friends arrived from their homes. All thronged around him after news. He left 4 days after the others and only succeeded in getting away by pretending to be a secessionist and on his way as a spy. He reported the secesh back again at their old tricks of pillaging and destruction. One had been hung and one taken prisoner. Our party of scouts came in tonight. They were stationed 12 miles out on the Springfield road. Among the items of news was a reported fight[22] at Lebanon between Wyman and a large body of secessionists. Wyman's cavalry 400 strong under Major Bowen,[23] being in the advance encountered 1500 secessionists & charged upon them, killing and wounding 25 or 30 and taking a number prisoners. But one of the cavalry was killed. This news came through the hands of Missourians and may be exaggerated, for there can be no dependence be placed in the butternuts. That there has been a skirmish there can be no doubt. As I write (9 p.m.) I hear the whistle of a train from St. Louis. I will surely get a letter tomorrow. Melissa cannot delay writing longer. Last night a corporal while in the bushes near one of the guards heard the countersign as it was given to the relief. Going to his quarters he persuaded five of his friends to accompany him on a plundering expedition. All six came in loaded with chickens, geese, and cabbage which they fondly anticipated would make them a fine feast on the morrow. But unluckily the affair came to the ears of the Col. He had them arrested and court martialed, which resulted in the corporals being reduced to the ranks and all sentenced to ten days imprisonment and accordingly all were marched under guard to the jail and securely locked up. It is said some of them cried upon finding themselves within the precincts of a prison. Justice has been done, and I hope for the honor of our regiment and the love of them behind us every similar offence will be strictly punished. Our presence in Mo. makes stealing no less a crime than when at home.

Wednesday Oct. 16 1861

I went to the fort this morning & by the assistance of Mr. Harrison took the levels inside the fort, also laid out the ground for the Magazine and also leveled the counter [extra spaces in typescript]. The afternoon was employed in calculating earth work and drafting plans. Lieut. Chappel[24] of the Elgin Co. died about noon. This is the second death at Camp Rolla. His remains will be sent home for burial. Lt. Walker of the Oswego Co. presented himself[25] to headquarters and reported himself to his company for duty. His friends greeted him warmly. But a hurricane, or an earthquake could not have more disturbed his enemies than his appearance here. The Colonel was beside himself with rage and pitched into all within reach. Clutching a pick he rushed upon the boys loitering around headquarters and scattered them like geese to their quarters. A seargent of the Woodstock company[26] being slower than the rest was soundly kicked. Col. Greusal had better not resort to such measures when in a passion, for he may find his match. Not every one will submit to be kicked like a dog. I learn the Woodstock boys are in a rage. A thousand rumors in regard to Walker are aflote in camp. Col. Greusal is a stubborn man and will not soon succumb to others. Our Co. were on guard again today and as some of the boys were sick I stood on guard from 1 a.m. till 3. All was still as death save the constant cry of the cricket & Katadid. As I trod my lonely beat I thought of my home & friends. My whole life came up in review. Oh! what a time for speculation and prayer. My loved ones were presented to God and His mercy and loving kindness invoked in their behalf.

Thursday Oct. 17th.

Went to the fort in the morning and established the grade for the block house and set the men to work. After noon I cleaned my gun & then platted the dimensions of the ditch and whole works surrounding the block house. It is fun to see the Oswego Co.[27] in dress parade. Both Walker and Murrell [Orville B. Merrill] stand side by side in the 1st Lieut's place as if it took two to discharge the duties of the office. I find Mr. Harrison to be a man and a gentleman. His tent is a pallace by the side of my own. I have written to Melissa and Drake today. Wyman's battle is confirmed but I hear no particulars.

Friday Oct. 18th.

Went early to the fort and by measurements established its exact center, also fixed stakes as a basix for future operations. From the point where I stood I could look down upon the camp of the 36th and see all that was going on. I saw the escort bear the body of Lieut. Chappel to the R. R. depot. The men were formed in two lines extending from camp to the depot, between which Co. A marched to the music of the burial march. It was a solemn scene. Capt. Baldwin, I understand, goes with the body to Elgin.[28] Returned to camp at 11 a.m. and found a letter from Melissa, being the 2nd I have received since leaving home. There has nothing occurred in my soldier's life that has done me half as much good. A letter from my old friend McDowell[29] comes just in time and is particularly refreshing. I have at last something authentic from Wyman's command, from some prisoners released from the Springfield Hospital who witnessed the fight. Wrote a letter to the Free Press afternoon and in the evening attended a prayer meeting. I am feeling more contented and happy than at any time before since I enlisted.

Saturday Oct. 19th.

Attended to my duties at the fort and returned at noon. Started after dinner for the fort and on the way was informed that the secession prisoners captured at Wet Glaze were coming into town. With a crowd of men, women, and boys I went to the headquarters on a lope and arrived just as the head of the column of cavalry was filing by, followed by a dozen or more wagons filled with prisoners from a Lieutenant in Capt. Montgomeries company[30] I learned the following particulars of Wyman's march and the exploit at Wet Glaze.[31] The day the 13th[32] left Rolla they travelled 24 miles through mud and a driving storm and crossed the Gasconade on Saturday night the 12 inst[33] they encamped 4 miles this side of Wet Glaze and 24 miles from Lebanon. A fireing between pickets during the night apprised our boys that the enemy was near & it being ascertained that a considerable force was a few miles ahead Gen. Wyman dispatched four companies of cavalry under Major Wright[34] and four companies of infantry who set out early in quest of secesh. In the mean time about 400 secessionists under Col. or Major Turner[35] took up a position on the side of a hill at Wet Glaze overlooking the road which passed through the ravine and partly up the side of the hill. The country around was hilly and cut up by ravines, the sides of which were covered with bushes and jack oak and was peculiarly well adapted for the secession mode of bush fighting. The enemy had formed and were awaiting the approach of our force when several ambulances with wounded from Springfield approached & were ordered to halt before crossing the ravine, with the threat that the last man would be shot if a wheel was moved. In this position they were detained three hours, and frequently jeered by the secessionists & informed they would soon have another load of wounded to carry along with them. The result proved that this remark was more ludicrous than brutal. This condition of things lasted two or three hours when suddenly two companies of Federal cavalry under Capts. Montgomery and Switzler,[36] who were in the advance of Wyman's detachment came rousing over the hill partly on the left flank and rear of the enemy and when within one hundred paces plunged a destructive fire into their ranks, which scattered the secesh like chaff. After one volley the order was given to charge and the boys pitched in, each singling out his man and sabering him to the ground but when out of reach of their sabers they drew their pistols with which they did more execution than any other way. The fight was over in a few minutes, the rebels flying precipitately up the ravine through the brush in a perfect rout. They were taken so completely by surprise that they hardly had time to return a few straggling shots. It was a dash—a shout and a gleam of death from our side and a wild and frightful scamper for life among the rebels. The ambulances and cavalry soon met and three rousing cheers were given with a will which made the woods and mountains ring. The wounded party met the cavalry with tears of joy and welcomed them as deliverers. Our force engaged numbered only 93 men. One man belonging to Capt. Wood'[s] Co.[37] was shot in the breast and was the only man touched on our side. Sixty three of the enemy were found dead upon the field and thirteen wounded. It is supposed that many more were more or less injured. About forty were captured. The boys said that the secesh would fire at our men & if charged upon would throw down their guns and piteously beg for mercy. One of our boys was fired upon, the ball grazing his head. He turned upon the one that fired, who threw down his gun and implored for mercy. The trooper commanded him to stand still, then, loading his pistol deliberately shot him down. One darkey belonging to montgomery's Co. fearlessly advanced ahead of his comrades and blazed away at the secesh until the order was given to charge when he plunged in among them, killing one and capturing two. He also killed the first man that fell.[38] The prisoners were marched under a strong guard to the fort and put in the charge of Col. Phelps.[39] They are a sorry looking set and some of them look forlorn enough, others are cheerful, and profess to have been deceived by wire pullers and knaves. They were provided with tents and rations and cared for as well as our men, being no more vigilantly guarded than our own boys. They were dressed in the universal butternut garb but were not as ragged as the company of Union men who came in a few days ago. What their fate will be I know not. Some of them were officers among which was a Captain & two or three Lieutenants. Being employed about the fort I have free and uninterrupted access to the prisoners, and will question them tomorrow in regard to events in their region of country. About 30 of the prisoners were captured at Lynn Creek, where Wyman's force were at the time the escort left. Geo. Walker sent me a letter but as he was too late for the fight he gave me only the particulars of his march and that a store belonging to a noted secesh was broken into and plundered by the boys. This was done in retaliation for a similar robbery upon the store of a Union man a few days before. In the evening while writing to Melissa the mail arrived and I was the happy recipient of a letter from brother Guy.[40] I did not retire until nearly 11 and then could not sleep for musing upon the events of the day and thinking of the loved ones at home. About 8 p.m. the report of a gun near our guard caused a rush to that part of the camp. The officers however promptly ordered the boys back to their quarters. It was found that the alarm was caused by a fracas among the cavalry boys in which one received a pistol shot in his leg.

Sunday Oct. 20th.

Oh what a bright and beautiful day this has been from early dawn till dewy eve, no cloud has obscured the sun's bright rays. I believe there is a difference between this climate and that of Illinois. An abundance of peaches and other fruit proves this. Last night two sentinels of the Iowa 4th approached each other at the terminus of their respective beats and commenced maneuvering with their pieces, in good humor, when the gun of one accidently [went off], killing the other instantly. I believe no such carelessness has been displayed in our camp and hope no accidents of this kind may occur. Attended divine service at 3 p.m. Do not like our Chaplain[41] much, not a word did he preach, adopted to boys far away from home and friends & many far from God. He appears to be selfish, and too foppish to suit me. Visited the Hospital, and attended prayer meeting in the evening. We were reviewed in the morning, all our clothes, blankets, &c were inspected.

Monday Oct. 21

My business, as usual, called me to the fort. During the day rumors of the enemies' advance was freely circulated, and though none could tell the source from whence they came yet many of the boys were disposed to believe that a detachment of 8000 from Price's army were not 20 miles distant. This story was modified during the day materially, nevertheless Col. Phelps thought it best to unlimber the guns of the fort to be ready at any moment. Accordingly I detailed some of my boys and we soon had the guns ready for shooting. About forty were detailed from the 36th to work on the fort today.

Tuesday Oct. 22nd.

A seargent major from the regular army came to the fort and examined the guns, firing one, which startled the boys in camp and for a time many supposed the secesh were surely upon us. In the evening I went into the cavalry camp and was much interested in hearing them tell their adventures. I learned the cause of the rumors of yesterday. It appears that about 20 of Price's brigands who live 15 or 20 miles to the west have recently come home, and are skulking around in the neighborhood where their families live. Last night a party of our pickets in that neighborhood were fired upon. Their fire was returned, with what result has not been learned. None of our men were hurt. Today Capt. Wood has scoured that section through but has not succeeded in hunting them out of their holes. Today 4 men with a load of chickens came to camp to sell their truck. A boy was with them and coming to the fire to warm himself the shrewd questions put to him revealed the fact that the 4 hucksters were bloody secessionists. The butternut gentry were soon nabbed and locked up in the Rolla jail.

Wednesday Oct. 23rd.

Late last night some of Woods' cavalry brought in 4 more secesh, being a part of Price's men who have returned and are skulking about their homes. One of them was wounded in the face with a charge of buckshot, one passing through his mouth and another near his eye. This morning all the cavalry at Rolla except one company started for Wyman's regiment. They are a bold, hardy set of dare devils & though mostly Missourians are enough for four times their own number of secessionists. Wrote two letters, one to the Free Press[42] & one to Thomas Simpson. Passed a pleasant evening with Lieut. Stonix.[43] He is a gentleman and a Christian. Were all our officers like him we would have no Walker affair to disturb the peace and harmony of the Regiment. I did not get to sleep until near midnight. Thought was busy, wandering o'er life's pathway from infancy to the present time, lingering nowhere but in the heart of that loved being. (Wife). And that shadow of herself. (Child). Thoughts, not sad but sober drove sleep far from my eyes until wearied nature at length sank to quiet rest.

Thursday Oct. 24th.

Our work progresses slowly at the fort. The whole excavation is rock, mostly soap stone, but large boulders of flint occur all through the ditches and it is almost impossible to excavate it. One of the Block houses is commenced and in a few days will be completed. Our force of laborers have been largely increased and I am obliged to work nearly the whole time to keep them agoing. I was somewhat disappointed in getting no mail today but hope for better luck next time. Wrote until 10 o'clock & retired to rest somewhat fatigued.

Friday Oct. 25th.

Today is the birthday of my own one year old baby. One year ago today I little thought I should love that babe so well. Here I am with a wife & child and all the means of happiness and can not see them. Well, who is to blaim but myself. I will not repine but how much would I give to now be with my own loved one if twere only for a day. Today thirteen Union men came in from Wright County near the head of the Gasconade about 65 miles distant. In that region the secesh are in large numbers and commit all kinds of depredations, driving off cattle and horses and even stripping the beds of union men. It is also reported that the underclothes have been taken from women and made into shirts for the secesh. Four prisoners were brought in today from near the Gasconade. They were caught in the act of robbing a house and taken prisoners. One attempted to fly but was shot down by a charge of buck shot in his face, gouging out an eye and otherwise spoiling his physiog.

They are confined in the Rolla jail. It rained some in the morning and but half the boys came to their work.

Saturday Oct. 26th.

The secession prisoners were put to work today. They worked well and if all employees upon the work do as well as they it will soon be completed. I heard no complaints from them except one who went to Col. Phelps and asked if it was the custom to work prisoners of war. The Col. pointed to a shovel and told him to go to work which he did without another word. I was on guard from 1 to 3 a.m. One guard from the Elgin company was found asleep on his post. This is the second offense and I fear it will go hard with him. Col. Greusal threatens to shoot him. I hope this will not be resorted to and that something besides death will be resorted to for an example to the other boys.

Sunday Oct. 27th.

Was relieved from the guard at 9 a.m. and immediately proceeded to Mr. Harrison's tent and wrote several hours to friends, and one to Melissa. At 3 p.m. I went to church but was too sleepy to pay attention to the sermon. In the evening went to the creek to wash and procure water for cooks use. Weather cool but pleasant.

Monday Oct. 28th.

Attended strictly to my work and was kept busy, on account to the additional number of hands, the secesh prisoners being employed on Saturday. On my return to supper I found the camp, or rather our portion of it in a perfect ferment. It appears the cooks were caught in the act of selling a portion of our rations and dividing the money between them. This has been repeatedly done and though the Captain [Charles D. Fish] has often been apprised of it, he has never taken any measure against it. The boys are firmly resolved that this system of wholesale plunder shall not be continued longer and are anxious that we go into mess cooking or that the cooks be changed. The Captain will not permit either, as he thereby will be minus his little rations. Our cooks and officers are now in league, they to have the proceeds of what can be plundered and the officers live out of the boys' rations & get the money for their own. The cooks save the choicest of everything for the officers while the slush and refuse is given to the boys. In the evening H. Wagoner,[44] D. Cromwell[45] & myself were selected as a committee to await upon the Captain and acquaint him of the wishes of the company. So tomorrow morning I shall expect to witness an explosion of some kind for I well know the Captain will do nothing that will in anyway deviate from his arbitrary and overbearing disposition. It is strange how soon a man will change and lose all sympathy & feeling for his former friends. A little brief authority will make an entire change with some men. I understand that Capt. Fish when at home was not considered in intelligence, information, and sound judgement quite up to the average & was quite as apt to ask favors as anyone, without receiving insults in return. Such men seem to forget that they are no more than men & may yet see the day when he may beg for the friendship he now spurns. Judge Semple Orr,[46] a noted Union man from near Springfield was in town today, and will start tomorrow for his home, from which he has been an exile for three months. [A] large number of refugees are going with him and expect to find their homes free from the cutthroat secesh. They are well armed I understand. From all the information gathered from that quarter the enemy are leaving in haste from the sweeping march of Fremont.

Tuesday Oct. 29th.
Immediately after roll call "We the honorable Committee" presented ourselves humbly before his august presence, the Hon. Capt. Fish formerly of Little Rock &c, &c. and very cooly and civily made known to him the wishes of the company. To have our present cooking arrangements suspended. The Hon. Capt. politely informed us "to go to Hell," that such a change could not be made & should not be made. A few words more and the subject was dropped. At night when I returned from camp I found no secession of our troubles. A petition was in circulation and signed by 85 of our boys for the removal of our present cooks. I do not believe it will have the least influence with our Captain.[47]

Wednesday Oct. 30th.
Well, the war in relation to our cooks has in no way abaited. The boys are however bound to have a reformation. I observe however that our "grub" is in better order and the cooks less overbearing and abusive. The fight has been productive of some benefit at least. The work of engineering and keeping accounts of the laborers keeps me pretty busy. Mr. Harrison, my superior, I find to be a perfect gentleman. Our intercourse is most friendly and intimate. There is nothing of the overbearing aristocrat about him. Would I could say as much of many others.

Thursday Oct. 31st.
Today all the forces of this place were reviewed and work was suspended at the fort. The whole review and inspection occupied nearly all day and was very creditable to us. In the afternoon I made out the pay rolls of the extra help on the fort and it was not until 11 at night that I completed them.

During the afternoon our pickets brought in two span of horses as contraband. I did not learn the particulars of their capture or where they came from. One team is given to our company so we will not be without means of transportation if we are ordered to march. I also learned from our pickets that in several directions have been frequently fired upon lately, but so far without injury to any of our men.

Friday Nov. 1st.

The weather is getting decidedly cool and for several days our full quota of clothing have been put in requisition. The men in camp huddle about the fires and those at work are obliged to be busily employed to keep warm. For several days we have had the same lowering cold weather accompanied by a northwest wind. Posted as I am on the highest knob in the country I feel every blast that passes by. While eating my dinner a man was observed running from the direction of the 36th and passing within forty rods of the fort. At about 100 rods distant down in the ravine were several Missouri ponies grazing in the brush. Our pedestrian caught one of the horses and made off. Capt. Coleman[48] and several of his men started in pursuit and were rapidly overhauling the horse cramper when he jumped off his steed and put into the thick brush and managed to escape. If Capt. Coleman's men had come within rifle range he never would have lived to cramp another horse. It is supposed he was a spy on his way to give notice to a secesh camp 36 miles distant of a project on foot to capture them. After the hands had been discharged I remained some little time posting books and crediting their time. As I started for camp I met Col Greusal at the head of our cavalry, who came filing by in good style the clatter of hoofs over the steep and stony road and the rattling of sabers in their scabbards appears more like war than anything I have yet seen. The boys were lustily cheered as they passed and were in high spirits at the prospect of a brush with the secesh. I started on, and soon met Co. B, and a few rods behind them 60 men from Co. E. O how I wished to be with them but there was no chance for me. The Iowa Regt. sent also 200 men making near 600 men all told. I learned that our pickets have reported a large body of secesh at or near Salem, collecting for some purpose not understood. Some conjecture an attact upon Rolla, others the waylaying of the Specie train and robbing it on it's way from St. Louis to pay the boys. I hope our boys will have a brush and test their mettle. We are also under orders at the camp to sleep upon our arms, prepared for any emergency, consequently every gun is loaded. The Major is considerably excited and thinks there is more probability of a fight here than where the boys have gone. It must have been 6 p.m. when the boys started, and if they march far it must be through as thick darkness as can well be found, for the night is intensely dark.

Saturday Nov. 2nd.
Last night it was found that two well diggers who had been employed the whole summer about the Union's encampments dousing for water were no more or less than secession spies. On their persons was found an account of every movement of our troops, our numbers and position accurately described. I have no doubt that every day professed Union men but at heart rabid secessionists, visit our camp and give all the information they can glean to their secession leaders. Our portion of the camp is nearly deserted, very few are left. This morning I went to headquarters with a requisition for lumber, but Capt. Small would not recognize the requisition. It being ascertained that our pay rolls were full of mistakes and at Smith's request I was excused from the fort and was busy nearly all the forenoon correcting them. After dinner proceeded to the fort & took charge of the work there. Mr. H being absent most of the time I also formed the acquaintance of Capt. Rich[49] & had a long & interesting conversation with him. Found that he was acquainted with Doc Canniff in Kansas. I wish I could get a position on Col. Phelps' staff. I would willingly leave the 36th. Two more secesh prisoners were set on the works today. Orders for the detention of every one who comes to Rolla within ten days have been issued. Consequently several men and teams are detained here. Strong picket guards are sent in every direction. The Major is bound to not be surprised. I learned that our boys went 12 miles last night and camped at midnight.

Sunday Nov. 3rd. 1861.
Received two letters from home at 11 p.m. and this morning answered Melissa's in time for it to be mailed. Was glad to once more receive assurance that I was remembered at home. I must write immediately to my mother. She is ever the same kind and loving mother as of yore. Attended church at 3 p.m. and then washed and spruced up and had a general good time with the boys. In the evening a messenger came in from the boys & reported that they were 50 miles from Rolla and still pushing on. No enemy in sight & I did not learn if there was any prospect of meeting secesh. The boys were in good spirits and eager to push on. The Iowa 4th buried another of their number today, and the body of another dead soldier was taken to the cars to be buried at his home. The burial of the Iowa soldier was not completed until after dark. The funeral durge as it swelled upon the evening air touched a tender chord, and set in motion a train of thought upon the uncertainty of life, and the immense sacrifice of human beings in this unhuman conflict. Oh! the desolated firesides and woe that over spreads like the black ball of death our whole land.

6 | Surveying Work in Rolla

November 4, 1861–December 15, 1861

We went down one of the most miserable looking ravines I have yet seen. And what is stranger still people were living there.

—Lyman G. Bennett, November 8, 1861

On November 4, 1861, Corporal Bennett received a request "to do some surveying for Col. Dodge." A year older than Bennett, Grenville M. Dodge worked as an engineer and railroad surveyor in Illinois, Iowa, and Nebraska before the war. Commissioned as colonel of the Fourth Iowa Infantry, Dodge soon exhibited a knack for war and went on to become a major general.[1] Throughout the conflict, Dodge showed a strong interest in maps that clearly reflected his engineering background.[2] Both Confederate and Federal officers tried to acquire maps, however and whenever possible.[3] Postal route, railroad, and section maps were deemed particularly valuable, but even those lacked certain features needed by military commanders.[4]

According to Dodge, most topographical engineers were, like Bennett, detailed for that specific service from their regiments, and because of that, they often missed out on promotions in spite of their important work.[5] Many Civil War topographical engineers were "self-taught amateurs" who were artistic, liked to keep personal accounts, and had literary interests.[6] Bennett had many of these qualities, although he certainly never achieved the fame of other cartographers such as Jed Hotchkiss, D. H. Strother, or Ambrose Bierce.[7]

In only a few months of service, Bennett had probably, and unsurprisingly, received no training in military mapmaking. At this point in time, Bennett did not work in the way that topographical engineers often did during the Civil War. Accompanied by an aide, a "typical Civil War mapmaker would ride with a drawing board resting on the pommel of his saddle."[8] Sketching quickly, the mapmaker investigated roads and paths often using a prismatic compass, a portable instrument used for measurements.[9] A simple aneroid barometer could be used to roughly measure changes in altitude.[10] Bennett knew how to survey land, and he employed these skills when he worked for Dodge.

In the time span of forty-two days covered in this chapter, Bennett spent sixteen of them surveying. Working roughly counterclockwise, he and his group surveyed the area around Rolla for a distance of five to seven miles outward. His tools probably consisted of a theodolite and a surveyor's compass in addition to the surveyor's chain and notebook that he later mentioned.[11] Stretching for sixty-six feet, the surveyor's chain had one hundred links with a brass ring for every ten links.[12] Bennett worked with a team of soldiers detailed for the survey, and they probably worked like this: "The rear man stood by the starting stake with one end of the chain, while the front man, carrying the other end and a set of tally pegs, walked toward the mark, unrolling the chain as he went. . . . At the end of 22 yards [66 feet], a tally peg was inserted, the rear chainman came up, and the process was repeated . . . 80 chains made a mile. . . . The axmen who accompanied the chainmen chopped away trees and bushes . . . the surveyor checked the position by taking compass bearings on a tall tree or an exposed hilltop, then blazed the tree or marked the hill, and entered the details in his notebook."[13]

Although no doubt accurate, surveying was a slower method of military mapping. In addition to his sixteen days of field work, Bennett spent another thirteen "platting" the maps themselves. Unfortunately, none of Bennett's maps of the Rolla area have survived. According to Bennett, on November 26, "Col. Dodge expresses himself highly pleased with the map, or as much as I have completed," but one senses that Dodge knew little about the process of military mapmaking during this period of his wartime career. Later in the war, commanders came to see the value of a timely, reasonably accurate map that could be produced quickly.[14] On the other hand, the army was in a static position at Rolla which may have shaped the method that Bennett employed.

Later in the war, Bennett described a faster method in his January 12, 1865, diary entry:

> Let me here describe the manner in which a topographical engineer makes his surveys[.] Having ascertained some well defined starting point such as a section corner, In his field book he notes all the objects he wishes to be shown on his map such as hills, fields[,] houses, streams[,] roads, ravines &c &c and notes their distances

in his field books. He then follows some known line either a rode, a section line noting the intersection and distance and course of all the objects on his rout[e]. One leaf of his field book may include an 80 acre tract of a quarter section and when the surveyed corners are standing, it is but an easy job to pass around the section and through it if necessary, and get an almost mathematical[ly] correct map of the tract and when the several 80 acre tracts are put together the map is completed. A person wants considerable practice and a good knowledge of distances. In this manner a number of miles may be traveled and a wide scope of country surveyed in a day.

Throughout this set of diary entries, Bennett showed that he had the eye of a topographical engineer, and Dodge revealed precisely what a military commander needed to know as he "wanted the whole country surveyed and a correct topographical map of the whole country made showing roads, valleys, hills, woods, clearings, springs, streams, houses, &c, &c." In order to plan and time marches, officers had to know where men, wagons, and transport animals could actually travel.[15] The steepness of roads, elevations, the location of fords, and tree cover were of vital importance.[16] Many of the diary entries in this chapter served as a type of map memoir that topographical engineers created as a written record to supplement their drawings.[17] It is likely that Bennett used his diary as a source for adding notes to the maps of the Rolla area as they include detailed descriptions of topography, water sources, types of trees and rocks, and other data requested by Dodge. Bennett's one surviving map set (covering March–July 1862) included notes of this type. Civilians often supplied commanders with critical information about the road network and directions. Bennett's fascinating, and often negative, comments about civilians that he encountered during these months probably related to efforts to collect their names and map the locations of farms and houses.[18]

Interestingly, Bennett documented two attempts by officers in his regiment to stop his survey. His immediate commanding officer, Captain Charles D. Fish, was apparently jealous, and Colonel Nicholas Greusel was angry because Bennett wrote a negative newspaper article about his leadership. Colonel Dodge apparently overrode both of these efforts to halt the survey. Besides surveying, Bennett wrote enthusiastically about a spelunking expedition, documented a disturbingly violent punishment inflicted by a company commander, and intermixed the whole with frequent weather reports.

Corporal Bennett, though, increasingly missed his family, and on December 7, 1861, he wrote to Major General Henry W. Halleck, the department commander, explained his inability to complete his survey to his satisfaction due to a lack of proper instruments, and requested a furlough home where he had proper instruments and a work area. Halleck did not grant the request, but on December 12 instead ordered him to St. Louis to complete his map of the Rolla area, a situation that Bennett confessed "somewhat surprised me for I either expected an order home or else politely informed to remain at Rolla & do my platting."[19] Bennett packed his possessions and left for St. Louis four days later.

Monday Nov. 4th. 1861.

About 9 last night the report of a gun, among the camp guard, startled some. Upon enquiry it was found that one of the guards discovered a stump in the darkness and mistook it for a man and haled it three times and failing to get an answer blazed away at it. The affair was somewhat ludicrous. The guard however did his duty if he honestly supposed it to be a man. Co. C will do its duty every time whether on guard or in battle. While eating my breakfast and as usual expecting to go to the fort Maj. Barry[20] and Mr. Harrison came to me and requested me to do some surveying for Col. Dodge.[21] I at once assented, expecting but a few hours work, and went to headquarters where I found that the Col. wanted the whole country surveyed and a correct topographical map of the whole country made showing roads, valleys, hills, woods, clearings, springs, streams, houses, &c, &c. Of course I accepted the job which will keep me busy for at least a month. I had not time to ask about extra pay but am confident if I do a good job I shall be liberaly paid. I went to the shop and had my tally pins made and at 11 o'clock commenced the survey. Four from my own company were detailed at my request to assist me. I commenced at the Rolla Court House and followed up, or rather down the Springfield road. At first the road passed over about the highest ground, crooking here and there to avoid hills and ravines, but about four miles out it decended into the valley of Beaver creek, and followed the sinuosities of the valley. We found some very fine springs of pure cold water, also wild grapes in abundance. This road is much traveled but little worked. But few families lived along the road and we passed but one farm that looked like live and with a framed house. The other buildings were all of logs and some were the most miserable tenements for human beings I

Carte de visite of an unidentified topographical engineer holding a transit and tripod. In his other hand, the young engineer is holding a plumb-bob, and a level or ruler is in his breast pocket. Editor's collection.

ever saw. The valley as we proceeded became narrow and deep, with rocky sides, and covered with timber and sometimes dense underbrush. The people, as we passed, looked half frightened out of their wits, not knowing what was up. We ran five miles and a half and quit rather later than we should and found we would not get back to camp until after dark. We returned on the R. R. track. At one place large stone abutments were built up forty feet high for a bridge to span a valley and small rivulet. Such a crooked R. R. I never imagined could be run by R. R. cars. In places the track was excavated through solid rock forty & fifty feet deep. Where under the sun business can be found to sustain a railroad in this God forsaken land I can not tell. We reached camp about 7 p.m. tired and hungry, but our cooks had a good supper prepared to which we did ample justice. Received the Oswego Free Press[22] and wrote a hurried note to Melissa to send my draughting tools.

Tuesday Nov. 5th

This morning started to my work at ½ after 7 and ran on the Houston road. Stopped at the fort two hours to assist Mr. Harrison in adjusting his levels[.] We passed over much the same country as yesterday. At noon we reached the pickets and prepared our meal at their fire and after a good dinner and reading the paper we resumed our work. At 3 ½ miles we decended into the valley and found a small stream of pure water, but about a mile beyond it had entirely disappeared in the ground and was only visible at times during the remainder of our line. We found an abundance of walnuts, hickory nuts, and sweet grapes. We passed but two houses during the day on the road that were inhabited. Oh, this is a hard country. Once in a very long while a little patch between the bluffs is tilled, but everything from the half naked dirty children and linsy woolsey tattered mother down to the pigs & poultry proclaim the barrenness of the land. The trees even are gnarled and scrubby & scarcely a tree can be found that will make decent timber. I can not find a hickory tree larger than a person's leg, and yet all have nuts upon them. Springs of cold and clear water gurgle up in all the ravines but often disappear among the rocks which are every where about. The surface rocks are generally sandstone. Lime stone is abundant & of the finest quality. Flint exists in a greater or less degree. The timber was scarce along our line & many of the hills were naked except tall grass. We ran 5 ½ miles and returned to camp at sun down. The weather has been warm & pleasant, and the sky clear.

Wednesday Nov. 6th. 1861

Today my survey was on the Salem road. We were at our work at ½ after 7 o'clock and ran six miles and a half. The country was less timbered than my previous surveys and a portion of the way might be termed prairie. A large part was Oak openings and the whole country was more level and better for cultivating than any I have yet seen. I however passed but three farms, and they presented rather a dilapidated Appearance, being decidedly of the Missouri order. One man I found who had resided on his farm fifteen years and had a large peach orchard of 1000 or 1500 trees. He stated that he sold an immense quantity & not less than 2000 bushels were wasted. I went into the orchard and under some of the trees the ground was completely covered with pits, and the dried fruit. Large quantities still clung to the trees and were completely dried and cured. We all ate what we wished but they were far behind the ripe fruit. This man also had about two acres of cotton. It was the first I had ever seen and I picked a number of boles and will send it home if I have an opportunity. At noon we went more than half a mile off from the road to a creek for water to help along our dry dinner. While on our way a deer started up from the grass but a few rods from us and went bounding into the woods. Just as we were ready to return to camp a butternut Missourian came along on foot. He came upon us rather suddenly and was considerably frightened, not knowing but that we would be the death of him. He protested that he was a union man driven from his home and was on his way to Rolla for work. We obliged him to go with us and on our arrival at Rolla I took him to headquarters. Col. Dodge after questioning him a while told me to take him to the Guard house, give him his supper, and bring him over in the morning which I did. On arriving at camp I found Ted Joslyn back from his Springfield trip and all the boys in safely. I went to the Aurora cavalry[23] quarters and for an hour listened to a recital of the incidents of the trip. They went as an escort to a part of the 13th[24] who were recovered sufficient from sickness to be able for duty & to protect a large quantity of army supplies. A number of fugitives who had been compelled to leave their homes at the time of Seigals [Sigel's] retreat were also on the train and returned to their desolate homes. The secesh were apprised of the movement of the train and were mostly away from their homes & their seedy wives, were nearly frightened to death at the appearance of our boys and protested solemnly that they had no husbands "but were poor lone widders." The situation of many told that they had not been "widders" long. Very many of the secesh when our boys came in sight took to the woods and could sometimes be seen a mile off streaking it over the hills for dear life. Frequently our boys would put after them and succeeded in capturing 4 or 5 and 6 horses. They went within 20 miles of Springfield and the train being within Fremont's pickets they came home, being tired and sore. They suffered some for the want of provisions, but endured all the hardships and privation like men and without one word of complaint.

Thursday Nov. 7th

Today I confined my survey to two or three cross roads and did not go more than three miles from Rolla and ran but five mile of line. We struck the Springfield road two miles from town about the middle of the afternoon and then returned to camp. The day was very warm & we could not work with any degree of comfort. On our way back we found a quantity of persimmons which we gathered and ate. I do not dislike to eat them, I had heard so much about them that I had quite a desire to see and eat of the fruit. Last night the Mo. 12th[25] came from St. Louis and today are encamped two miles out on the Springfield road and intend to march through soon.

Friday Nov. 8th.

This morning I commence on the ridge road at the point where the cross road from the Houston road intersects it and prosecuted the survey down a ravine with the Springfield road at a point about ¼ a mile above the termination of Monday's survey. We commenced at that point & pushed down the Springfield road ¾ of a mile to the main fork of Beaver creek, from thence we returned via the Railroad, and surveyed a road leading down a valley where the Iowa boys had been cutting timber and making shingles. We went down one of the most miserable looking ravines I have yet seen. And what is stranger still people were living there. One "Juems Wooley" had a "claring thar" & said he had lived "thar 15 years" and that this miserable hole was much better than a large part of Missouri. If this is the best part of the country the Devil may take the remainder. We returned to camp after sundown with pockets and haversacks full of hickory nuts. We found our boys back from their expedition,[26] all feeling in good spirits and in a mood to tell their adventures. The first night out they marched 12 miles and at 1 o'clock at night encamped, laying on the ground. The next day they marched within five miles of Licking, at Crow creek where they encamped until Monday when they proceeded to Licking, finding the place almost deserted. The Union men having been stripped and cleaned out by the secesh who were in the majority in that section. Our men quartered in the empty houses and during their stay took everything that could be found in the shape of sheep, horses, mules, cattle, and wagons. They returned with from 10,000 to 15,000 dollars worth of property. The cavalry were in the advance and did all the skirmishing. They went as far as Houston and tore down a secession flag floating from the Court House and ran up the stars and stripes. Col Grensel [Greusel] issued a proclamation to the effect that if the flag was torn down he would return in 10 days and burn down the houses of secessionists. I learn today that the flag has been torn down and that the Col. will go there next week & burn the whole secesh part of town to the ground. Fourteen prisoners were taken and confined in the fort and will be set to work with the rest. Among the prisoners one Captain, one Deputy Inspector General, one Quarter Master General[,] one Sergeant Major and one Orderly Sergeant. They all have taken an active part in the war and been in the battles of Springfield & Lexington. It looks to me rather hard and cruel to set old gray headed men to work on fortifications with no pay but their living. I should hate to see as old a man as my father a prisoner and compelled to work at hard labor. But when we consider the enormity of their crimes, humanity in their cases ceases to be a virtue. Many are the wails of distress, gone up from famished and persecuted Union families who have been barbariously robbed and plundered by them, even to their last morsel, their last bed, & all but the clothes from their backs. It is a pity that the world is poluted with such beings, but he who will rebell against the best government upon earth, is but little else than a barbarian to which the most horrid crimes would be familiar. This has been a beautiful day, weather warm and pleasant & I have worked in my shirt sleeves.

Saturday Nov. 9th.
I went early to Col. Dodge's quarters for another blank book to note surveys in and while there learned that the paymasters would settle with the 36th Regt. and concluded not to work today but to arrange the details of hands that in future our surveys would not be subject to interruption. After dinner Co. E was formed in line and each in turn walked up and received our little article of pay. My wages amounted to $89.00 aside from extra pay, which will come in a day or two. I received $35 in U. S. drafts and $40 in gold which I exchanged for two $20 drafts and went up to the fort and wrote to Melissa enclosing $40 to her which I doubt not will be of some service to her & me. On my return from the fort I found friend Harrison writing a label for Col. Greusel's trophies, but having considerable business on his hands I turned in and assisted him. While at work I heard our Orderly report me as absent, which somewhat vexed me as he had no business to do so. I found that the Regimental officers were on my side and gave me to understand that I was right and no harm should happen to me. In the evening I was attacked in a most abusive and insulting manner in regard to my being detailed. I could not hold in and think I gave him as good as I received. None but narrow minds and the utterly depraved and selfish will ever envy and persecute others for their good fortune. My being detailed is a double favor to me for I not only get good pay but am out of the control of Capt. Fish.

Monday Nov. 11th.
Mr. Harrison having business on his hands, I went to the fort this morning and took charge of the work in his place. Nothing of note except for release

of one of the prisoners. The mother and sister of one of the prisoners came on horseback to see him. To see the mother's tears and the sister's cry and implore that brother renounce his rebellious scheme, to swear allegiance to the United States and return home was an affecting sight. But that stubborn boy, with the utmost composure and firmness avowed he would never take the oath as long as he lived, and that if he worked for Uncle Sam he would even wear gray, he would never violate his oath to the Confederate States. After that there occurred nothing worth recording except attending services at the fort for the benefit of the secesh. I also wrote to Gen. [George?] Walker, Mr. Humphrey, and dined with Capt. [Josephus G.] Rich of the Missouri Regiment. He is a gentleman and as free as water. His wife had just arrived and was with him at his tent.

Tuesday Nov. 12th.

I awoke with a severe headache & knew I was in for the sick headache and went to Doc Young[27] for an antidote. He gave me a dose of morphene which stopped the headache but so completely I was overcome so that I could not arise without fainting. Towards night I grew better and hope to be able to proceed with my survey tomorrow. The weather has been really warm and I sweat properly as I lay in my tent.

Wednesday Nov. 13th.

I have felt quite well today and ran seven miles on the North Salem road. The country on this line is much the same as that heretofore surveyed. We kept up on the ridge & hence found but little water. In the afternoon we went half a mile off the road for water and found some at a house which looked clean and decent, being the first one I have seen in Mo. Although the woman & children were dressed in homespun yet they all were clean & tidy and the house was also in order. When about to return we came across a man in rags, on his way to Rolla. He had been hunted all summer by the secessionists & lived in the bushes. His hair was long, ditto, his beard. And the lower portion of his pants were entirely gone, and his moccasins were worse than none. He came to camp with us and proceeded to the fort, where he knew many who were with Col. Phelps. I received a long letter from home. Oh how I was rejoiced to hear from Melissa again and to be assured that I was loved. Never have I received a letter that did me so much good. So much good advice and Christian resignation reminded me of days gone by. Not one word of reprimand or reproof, but all was love and resignation. Long after I had laid down to rest my memory dwelled with pleasure upon past home scenes and the joys of my early married life and I hope to dwell again in the midst of home joys & to taste again a wife's love & devotion. These give me strength, these give me hope, and my mind is full of fond memories. It is a great comfort to know I have a happy home and a good and loving wife, and could I believe she was not so I would never wish to return, but hope for death to terminate a life of misery and woe.

Thursday Nov. 14th.

Last night upon my return I found Co. B of the cavalry in a terrible excitement and the whole camp in a ferment. It appears a private of the Co. had been on a drunken spree for a day or two over in the town, and was not present to answer to his name at roll call. After getting sober he came to camp and reported to the Col. where he had been and what he had done. The Col. reprimanded him telling him as this was his first offense he would not punish him, but to keep sober in the future. Upon going to his quarters he reported himself to Capt. Smith[28] who immediately gagged him by tying his hands behind him and then tied a rope around his head and through his mouth. After which he jerked the gag from behind until the poor fellows mouth was badly cut and bled profusely. The string under his tongue was cut off & he could not speak. Not content with this he kicked him in a most brutal manner until the soldier could not stand. At this point the company interferred and rescued their comrade from further violence. When the extent of his injuries were known, a more indignant set of men I never saw, and they all made a rush for Capt. Smith with the intention of killing him. But the Col. put a guard over him and finally facilitated his escape from camp. This morning while on my way to work on the Northern Springfield road I saw Capt. Smith at a private house some distance from camp. His early exit from thence was all that saved his life. The road today was mostly in Beaver creek valley which I found more thickly settled than any road hitherto surveyed. But such objects in the shape of human beings can not well be found. Butternuts cannot express what they were. Copper bottoms would be far more appropriate. Men, women, children all were most thoroughly smoked and dressed in the most primitive style. We found an abundance of pure springs, walnuts, butternuts, and grapes. We ran only about 4 ½ miles & returned early to camp.

Friday Nov. 15th.

After a cold, frosty night, the sun arose clear and beautiful & again I sallied forth to prosecute the survey. I took a by road pushing off from the Salem road and after following over hill and dale through brush and prairies in a wild and unfrequented country I have yet seen. A little after noon we reached a little one horse town on the R Road called Dillon. The few inhabitants I saw looked wild and frightened and gazed at our little party from behind house corners and stumps, and were in a position for instant flight the instant we made a move to approach them. The grog seller of the place was not quite as fearful of us and did not leave his den as we entered it. Some of the boys took this opportunity to wet their throats with a smack of 80 rod whiskey. The seargent of the picket guard stationed here came up and had many questions to ask in relation to our business, after which he treated my boys to another dram. I learned from him that he had taken Capt. Smith from Rolla in disguise the night before and when the train for St. Louis came along succeeded in getting him on undiscovered and in safety. He advised him never to come back to his company again, for as shure as he did so he would be a dead man. We surveyed the main road from Springfield, from Dillon towards Rolla to within 1 ½ miles of camp. During this day's survey I went to a house on the way to enquire the proprietor's name. None but women and children were there. When I enquired the owner's name the women were frightened and supposed we were around to impress her husband into the service of the United States, and notwithstanding I assured her that we could get men enough without drafting, she still could not understand why we wished her husband's name. The children also shared their mother's fright and came wildly staring at us. The oldest, a girl of perhaps 16, had but a single article of dress upon her, that a dress, with as many windows to admit sunlight and air to her person as an Illinois two story house. Her hair I would almost take my oath had never felt the impress of a comb. The bushy locks were thickly interspersed with straw, feathers, & a number of Spanish and bur dock burs. How I wish I could have had an artist present, to take the thing "true to life." "Those matted hair," that eyes sparkling from amidst thick scabs of dirt and matter. Her nose—well there I can't describe nor want to for even now the thought is sickening. I went with Mr. Harrison to the depot and waited until 9 in the evening for the cars. On the arrival of the train a large number of secession prisoners were taken from the train and escorted under guard to the fort. These prisoners were captured by Capt. Jenks[29] of the Aurora cavalry. Mr. Harrison's boy was also on the train & told me that my drafting instruments were along. I had almost forgotten one little incident on our return from work we met a boy with an ox team. He halted and offered to show his pass, taking us for pickets. We found he had apples with him. He gave us all we could eat and we permitted him to proceed without showing his pass, which he thought was a great favor on our part. He said "he got his pass a right smart while ago but thought it good yet."

Saturday Nov 16th.

I received a letter from Ed Lyon[30] which gave me much pleasure but the pleasure it gave me was more than counterbalanced by the receipt of an order from Col Greusel forbidding me to proceed with my survey. On enquiry I found that Wm. F. Sutherland[31] had shown him my article in the Free Press where I had criticised his conduct toward Walker. The Col.'s anger was aroused and he threatened to disgrace me to the ranks and to stop my prosecuting my survey any further. I immediately wrote to Col. Dodge the particulars & cleaned my gun & prepared myself of company duty. I felt bad. I could not help it for the prospect of doing well at this survey I found suddenly blasted. Though surrounded by thousands I felt indeed alone, and without friends save those that could be bought. As I looked around me I beheld the cold and selfish expression on the faces of each & felt an inward loathing for the most of mankind. I long for that *loved spot, my home*. No, I am not altogether friendless and alone. There is a still small voice which reaches my inner heart and whispers that God is near and that I have a friend in Him. As I was preparing for dress parade a note from Col. Greusel was handed me, containing assurances of friendship and esteem, and that a new detail was made out for me to renew my survey. Of course I felt better and was grateful to Col. Greusel, and I hope that in the future I will not write about anything except my own business and the news. I found my messmates mutually glad with me, which is an assurance that the boys are my friends.

Sunday Nov. 17th.

Arose with a slight headache & proceeded to wash and change my clothes. Learning that the 13th Reg. was encamped a mile & a half away on the Springfield road I walked to their camp and had a pleasant visit with my old friends. Geo. Walker had however left the camp and I missed of seeing him. Returned just in time for meeting, after which I wrote to Melissa a long letter, and *such a letter*! Oh it is good to have one true & faithful to which *heart secrets* can be entrusted. I have told them all to her, and I hope for mercy from God, may may she forgive the many wicked thoughts and unkind actions I have at times shown her. George Walker came to our camp and stayed until night, and a right good time we had. I retired to my straw pillow with more serious thoughts than ever before. I could not, neither sought I to banish the thoughts of my family from my mind. In thought I was far away to my home. I thought of the many hundred miles that separated me from those I loved. I recalled with delight the scenes of other days. I remember my once cheerful home which I have forsaken & which perhaps is cheerful no longer in consequence of my absence. I fancied I could see my wife once more, but with a heavy heart & she sighed and wondered what had become of me, who should have been her stay and support. I thought, and grew sad as I thought, until tears came to my relief.

Monday Nov. 18th 1861.

The morning was so foggy that I scarce could see a half dozen yards from me. It being impossible to survey, I went to the town for repairs to my compass after which I went to the fort until noon & assisted Harrison in making a bunk for sleeping. After dinner I proceeded on the survey of the Rolla & Dillon road which I completed. [Missing: The next two pages are too faint to copy.]

Tuesday Nov. 19th 1861

The dark fleecy cloud sweeping up from the southwest betoken an approaching storm, and I decided not to survey but platt up the notes already gathered and I proceeded to the fort for that purpose. The wind however blew the dust and dirt upon my papers in such quantities that I could not work and as we designed in the end to move our tent into the fort, we did so now and before the storm came on we were snugly ensconced in our new quarters. We had a drenching shower which was much needed as the roads were very dusty. In the evening I attended prayer meeting and never before did I feel as if I wanted to tell my feelings and throw myself wholly into the arms of God as then. O that I was a good Christian and by my daily walk & conversation could be brought near to Christ. After returning to my tent I had a long argument with Todd,[32] our drummer, on temperance. Todd likes his dram & sprees occasionally & stoutly argued against temperance societies and reform. After awhile he became more conciliatory and the subject being changed somewhat Todd, myself and three others entered into a compact not to swear or use any bywords in the future. The erring ones to be mildly reproved by the others. In short, we were to do all we could to become gentlemen in our conversation & actions. God give us success for some who entered into the compact are horably profane. My good friend [Hiram] Wagner Wagoner [Wagner] is an exception however.

Wednesday Nov. 20th. 1861.

It was cold and windy today but I early commenced my platting and made good progress until noon when I heard the expression outside of "how do you do, General" and going out I found Gen. Siegel [Sigel][33] in the fort and giving directions to Mr. Harrison in relation to the prisoners he had brought in. It appears that when his army reached Lebanon word was brought that a small detachment of "Home Guards" at that place had been captured by a superior force of Freeman's[34] band of secession marauders, who was hurrying their prisoners into Texas County. Siegel sent a detachment of cavalry in pursuit, who overhauled them after a long and tedious ride & killed two of the enemy and captured two, one their Captain Bohannon. But Freeman being near with an overwhelming force, the cavalry returned without the home guards & with only their two prisoners. Siegel however brought along several noted secessionists from their homes as hostages for the safety of the Home Guard prisoners who are confined in the fort the same as other prisoners only not obliged to work.[35] Among the hostages is a Lieutenant Colonel Somers from the Southern army. Siegel is a small wiry Dutchman and talks as broken as any Dutchman. He has not the sluggish appearance of the general average of lager beer dutch but is quick & active in his movements and conversation. His black, fiery, restless eye attracts attention the first of anything about him. There was nothing about his dress or haughty bearing to indicate he was a general. Many 2nd Lieutenants in the 36th dress far ahead of him. Still there is a something about him which tells the stranger that he is no common man, and that he will not disgrace the position he occupies. Upon the whole, I think Gen. Siegel richly deserves the laurels he has won in the bloody battle fields of Carthage[36] and Wilson's Creek[37] and that a broad niche is in store for him in our country's history. In the afternoon friend Roseman[38] from the Morris Co. and myself went directly from the fort to the camp of the 13th Illinois Siegel's camp and part of Gen. Asboth's[39] camp. Never have I seen so many armed men before. Several Regiments were on parade but I saw none that could equal the 36th & the 13th. Indeed Gen. Asboth said today that no troops in the service equaled the 36th in the manual of arms. This compliment from so high a source means something more than empty compliments. On the parade ground of the 13th I saw a dead horse and on enquiry found that on battallion drill a mock skirmish had occurred between the cavalry and infantry when one of the guns of the latter accidently went off & killed the trooper's horse, the ball striking not inches from the rider. It seems such accidents are pure carelessness and deserve punishment.

Thursday Nov. 21st.
Cold and windy still and the wind plays all kinds of fantastic tricks with our tents, loosening the guy ropes and thrashing the ground with the ropes ends and finally tieing it into knots hard to unravel. Sit down to work & when you fancy yourself in security from the blast first one & then another

rope gives way & ere I am aware one whole side of the tent is flapping in the wind and rudely slapping me in the face. The cold increased in the afternoon & I was obliged to suspend work at my plat, and pitched another tent in a less exposed position and removed our beds & cooking utensils to it. I went to the 36th['s] camp and moved part of my household goods to the fort with the intention of taking up my abode there until my surveying & mapping operations are completed. Harrison, his boy, and myself slept very comfortable in one bed, in fact I slept better & more warm than any time since coming to camp.

Friday Nov. 22nd. 1861.

The wind and cold has increased in intensity and today the prisoners were not set to work. I could not work myself at the plat but contented myself by staying and fixing up our new home for living. I sadly feel the want of money. I should not have sent so much home but I expected long before to receive my pay for extra work. I can do but little toward procuring supplies for our comfort and convenience. Among the prisoners brought in last night were Capt. Keshon & ten of his neighbors. The nature of the charges against them I know not. The old Capt. is [a] very well informed man and resides 22 miles from Rolla. Has been an old sea Capt. and professes to be a Union man and says the charges trumped up against him were by a horse thief whom he has been active in bringing to justice, and he in order to screen himself and avoid The State prison has preferred charges of treason against Kershon and his neighbors. I find many of our officers interested in his behalf & believe he will not stay here long. The present distracted state of mind is peculiarly favorable to horse thieves & robbers. By pretending to be Union men the worst outrages are practiced against any one they have an antipathy for, whether Rebel or Union men. Wrote a letter to Melissa and at late bed time retired to a good night's rest.

Saturday Nov. 23rd. 1861.

The guard from the 36th brought a letter from Melissa which gave me much pleasure and I read & reread it over & over again. O these tender messages from *home*. I learned of the safe arrival of my remittance and that two of my debts were paid, & so many avenues of uneasiness and foreboding effectualy stopped. Receiving this letter in the morning I had but little time to muse upon its contents, but little time to wander far away to the past, recalling fond recollections of old scenes, and but little time to leap forward to the future & become perplexed in conjectures regarding my final fate. The cold had increased in intensity & I found it impossible to work upon my map. Going to head quarters & representing the case to Col. Dodge, he very kindly furnished me a warm room and facilities for working. No work for the prisoners today. Two teams were sent to the fort to work & Lieut. Harrison directed one to draw wood and another water. The water man refused to do as he was ordered, when Harrison spoke to Col. Phelps about it the Col. walked up to the teamster and taking his whip gave him a sound thrashing. The driver finally promised to do as ordered, but got on to his mules and was about starting for town when Col. [John S.] Phelps drew his revolver and threatened to shoot him down if he did not start immediately to his work. The Irish driver went to his work without more complaints. Col. Phelps is a brick in his way & when once aroused will venture anything to accomplish his objective. I platted a large amount of my survey and at night returned to the fort with Col. Phelps and had an interesting argument with him in regard to slavery. Wrote a letter to Guy [Bennett], after which I took lessons in the game of chess. When I get home if Melissa will learn the game we will amuse ourselves at it occasionally.

Sunday Nov. 24th.

It was a cold and cheerless morning & I was reluctant about getting up but when I finally did make my advent from the tent I found it snowing a little but not enough to whiten the ground. This has been the first snow of the season here. There was but a little dash of it however and by 9 a.m. the weather began to moderate & the wind go down. I passed the forenoon in my tent reading and writing up my diary which was four days behind. At 3 p.m. two prisoners were brought in from Crawford County forty miles distant from whence they have come since morning. The names of the prisoners were Hinch and Hardee. It appears that during last night they came to where a family of fugitives were camped for the night. The man's name was Thomas Green[40] from Springfield Mo., and on his way to Illinois for safety. The two highwaymen made rather free with one of Green's daughters, threatening to shoot Green, and finally took 3 saddles, 2 blankets, an overcoat and also Green's money. Two small girls, frightened nearly out of their wits ran near a mile and found another camp where three or four families had stopped for the night, and giving the alarm four men fully armed started for the rescue, and arrived in time to catch the two thieves, a tussel ensued and the villains were overpowered, one being thrown down and stamped in the ashes. They begged loudly for mercy and offered all the money they had, some $200.00 to be allowed to go free, but no begging or coaxing could induce their captors to release them. The chopfallen villains are fearful that they will be shot, as they richly deserve. The gentlemen also stated that in the same neighborhood a fugitive was shot on Friday night by secessionists and that his body was still unburied as they came along. Never was a country in a worse condition than Missouri at the present time. One year more of guerilla warfare will leave it without inhabitant[s] save predatory bands and not a vestige of improvement or sign of civilization left. After noon Harrison and myself went to Siegel & Wyman's encampments. I saw George Walker and found him much better than when I saw him last. Harrison found a brother in law in Col. Knoblesdorf's[41] Northwestern rifle regiment. I was introduced to the Col. and found him a conceited Dutchman, unworthy, in my opinion of the position of a non commissioned office. On our return to the fort we gathered a few persimmons and on our arrival found Capt. Kershon & part of his companions released on parole to return tomorrow morning at 10 a.m. The weather is much more moderate than in the morning but still cold.

Monday Nov. 25th.

Dishwashing and kitchen duties are new to me and detained me from my work until after 8 a.m. after which I resumed my mapping. As I was at my work the constant rattle of government wagons remind me of Lake Street and I could hardly persuade myself that I was in the worst of all places, Rolla. It is really astonishing to see the vast amount of supplies needed for the large army here, and all day long the town of Rolla is alive with men and teams in motion. George Walker visited me this afternoon and at eve went to the fort with me and thence to camp. A vast amount of uncertainty exists in regard to future movements. I learn that the rebels are returning in large numbers and occupying the country vacated by our troops, and that their pickets are 15 miles this side of Lebanon. I cannot comprehend the recent movement of our force upon Rolla unless it be to draw Price & McCullough[42] into a trap. We shall see in a few days what is to be our fate. It is really astonishing to see the vast number of fugitive Union men fleeing from Southwestern Missouri to Illinois for safety. The roads are literally alive with teams & hundreds of passes are issued each day by the Provost Marshal for them to pass the pickets. Four prisoners were released on parole after taking the oath of allegiance. Towards evening a member of Co. C was buried. Measles was the cause of his death. This is the third death in the regiment. Though a funeral procession is no uncommon sight yet I cannot see one without feeling sad.

Tuesday Nov. 26th. 1861.

All day I platted and nearly completed the notes I had collected. At night I went to the 13th and got George Walker detailed to assist in my survey.

It was after dark when I arrived at the fort. Col. Dodge expresses himself highly pleased with the map, or as much as I have completed. I think Col. Dodge is a man and a gentleman. He is a favorite with his men. I was introduced to Gen. Wyman and had a long talk with him. I believe him to be a gentleman in spite of his bandy looking nose. The fort guard has a hard time of it their quarters being in the ditch. Weather cold but pleasant.

Wednesday Nov. 27th.

Rather late in getting off to my work, walked four miles and surveyed four & then walked from Dillon to Rolla—six miles. Found a man direct from Wet glaze [Wet Glaze, MO], who told me there was no doubt but that Price or McCullach was marching for Rolla as he had seen large bodies of southern troops and that Union men were hurrying away as fast as possible. If this [is] true it looks as if we would soon have a fight on our hands. Let it come. There are lots of boys here who are even anxious for a fight and I believe we have a sufficient number for nearly any emergency.

Thursday Nov. 28th.

My survey today was almost due north from Rolla, from the fort it looks to be level in this direction & I was surprised at coming suddenly upon a deep valley at half a mile distant from town. Down this valley we plunged and followed it for four miles, then we left this valley and ran over ridges and across ravines until we accomplished six miles. George Walker assisted me today & has been detailed for ten days. His health is not good and he was nearly exhausted upon our return. We propose running toward and visiting the mammoth Cave of Missouri. The boys all are anxious to go & I have no doubt we will enjoy ourselves. It is a warm and beautiful evening & about the finest sight I have seen is a night view around me. South of us and near a mile away the camp fires of Col. Phelps' Regiment lights the trees around them & the Heavens above. South west of me, along down the Beaver Creek valley and following the sinuosities of its course for miles, the camp fires of Wyman, Siegel, and Asboth's brigades present a bright and beautiful appearance and on the clouds above the various camps are all mapped by the bright fires below. I can trace every camp by the bright reflection in the Heavens. The 36th and 4th Iowa are almost under my feet at the foot of the hill & I can almost fancy what the boys are talking about as they sit around the fires or gather in groups inside their tents. I wish my friends at home were here for a day or even an hour to enjoy the scene with me. The fort guard gathered about our fire and for all I know sat there until morning telling their yarns. One old sober deacon looking corporal named Baughman[43] excelled all the others on tough yarns. Talking about moskeetoes he remarked that down in Arkansas they were so thick that a man by holding his arm extended in the air a minute and then suddenly withdrawing it would leave a hole in the air the size of his arm. Old Baughman is a brick in his way. Among his stories he said that when he first settled at Springfield his neighbors (which like angels' visits were few & far between) were mostly North Carolinians. For his first year's provisions he raised a patch of buckwheat & taking it to a mill owned and run by a Carolinian to get it ground, the miller knew nothing about buckwheat for bread but for the novelty of the thing purchased quite a quantity for his own use. The miller's wife, equally ignorant in regard to it undertook to make light bread and after two or three trials and failures threw it away, and the miller gave away his supply on hand, declaring that Baughman was a scoundrel and a fool to use buckwheat for bread. While in the way of telling Missouri yarns I must tell a good one upon Harrison. He went to the house of a Missourian early one morning for milk, and found the woman out, "pailing the cow" as she called it. Harrison, while waiting for his milk entered into conversation with the good dame and finally asked her if her cow was a good one. "Mighty good" was her reply, "she don't give a right peart flow now but I reckon she gives a right smart sprinkle."[44] One of the 13th tells of calling into a Missouri house for a drink of water, and found the family at dinner. Two girls 12 or 15 years old were dipping bread into the molasses dish and playing grab the best they knew how. One of them, failing to get her share, cried at the top of her voice "Mam! Sal dips twice into the deep to my once in the shaller, and you know lasses is scarce now." I might mention many other odd phrases of these wild Missourians but the above is a sample of their whole conversation. It amuses me vastly.

Friday Nov. 29th. 1861

About 8 a.m. myself and party started on our survey & excursion & by 9 was at our starting point 4 miles from Rolla down the Springfield road. We passed through all the camps which extend down the valley for 4 miles. The morning was cold & cloudy & when we arrived at our work it began to snow and continued for 15 minutes to fall quite briskly nearly whitening the ground. Our road led up a ravine branching off from the main valley which we followed for a short distance & then struck off to one side up the bluff and on the ridge. The road was very crooked and the woods dense. We pushed on until about 1:00 o'clock and came to the camp of the timber cutters. They had just finished their dinner but friend Reel got us up a good dinner which was eaten with a gusto. After dinner it was decided to go to the cave that day and if necessary remain in it over night. Accordingly we started out after being reinforced by a number of Reel's men. We trudged on 6 long weary miles over ridge, through woods, and finally down a deep valley & small creek. Geo. Walker was taken sick on the way & our progress was necessarily slow. Finally Geo. grew so bad that we urged him to stop on the way and not go farther but he would not listen to such a proposition and pushed on with us, though suffering at every step. After following the valley to within ¾ths of a mile of the Gasconade, we then struck off to the right up a narrow ravine and over a narrow path, difficult to travel for half a mile when suddenly emerging from the bushes the mouth of the cave yawned open before us.[45] A small stream was running from it, over a gravelly bed. The bottom of the cave for 400 feet is gravelly and wet from the waters of the creek. At that point two passages branch off from the main passage and the bottom rises quite steep for 50 feet almost like climbing stairs. Here we rested & prepared for night's adventures, for it was near sundown. George Walker could proceed no farther and laying our overcoats on the damp ground & leaving him as comfortable as possible, with lighted candles we took up our line of march. We entered the right hand passage which led to a magnificent chamber or series of chambers, the roof of which was almost wholly covered with magnificent stilaetites which like icicles hung suspended from the roof. Occasionally one extended to the bottom of the cave and resembled a pillar. These pillars were very numerous and of various sizes from a man's arm to three feet in thickness. Very many had commenced forming from the bottom & stood like stumps or telegraph poles here and there. One I noticed resembled the pictures of the leaning tower of Pisa. Another branched and resembled the following diagram [drawing here]. Some of them were honeycombed, some nearly transparent and others white. One very beautiful specimen hung from the roof resembling a rose bud [drawing]. I cannot describe all the beautiful things I saw. The rooms were wet from constant dripping from the wall and in some places the bottom was quite muddy. This passage was not very long and a half hour served to explore it. We returned to the main passage with a few not very beautiful specimen. Finding Walker unable to go with us we set out up the stream making slow progress up the crooked and craggy passage. The rough rock projected from each side forming kind of a shelf, over which our pathway led. The small rivulet trickled over the rocks a few feet below us in a narrow bed worn by the action of the water for it. We followed in this manner for nearly a mile when the passage grew so small that we were obliged to stoop & walk in a cramped position. Small chambers were frequently found, where we would stop & rest & then push ahead again groping, stooping, & occasionally crawling on our hands & knees, finally the passage grew so small that for several hundred feet we were obliged to crawl snake fashion on our bellies. Still we pushed on, the sweat pouring out in great drops & the warm, close air weakening us every minute. The passage presenting no new features we finally decided to return. Not until we had blown out our lights and looked at the darkness a while and firing a pistol and listening to the stunning report and the loud echoes reverberating through the hollow passage. Shortly after 8 we reached the entrance in safety but completely exhausted. The unwholesome air and want of proper facilities for sleeping quickly determined us to leave and groping our way through the bushes we went on to the next which likewise was deserted. But our wearied bodies would go no further and we took possession of the tenement & building a rousing fire we soon were comfortably warm. George being sick we put him to bed, but the rest lay on the floor, occasionally renewing the fire & keeping us as comfortable as possible. The night was cold & the hard floor fatigued more than rested us. I slept but little that night & felt that my visit to the cave was a sore one.

Saturday Nov. 30th 1861.

Stiff, cold, & hungry, just as day began to peep we were up & away from our anything but hospitable hotel. The morning was without a cloud & the sun arose & smiled upon us as if in mockery at our miserable situation. We slowly wended our way through the woods and reached camp at breakfast time, which with a two hour rest, reinvigorated us in a measure, and we preceeded to our work. It was ascertained that the occupants of the two houses we called at for lodging were the wives of secessionists now in the Southern army. They saw so large a crowd of soldiers & so late in the day, supposed we had evil intentions toward them and fled to the house of a Union neighbor. We apologized to the mistress of the house where [we] stayed for our unceremonious occupation of her house pursued our way. This day we ran a new road toward Rolla of about five miles & a more foot sore & weary set of boys I never saw. We felt like some poor sinners I have heard about. "Weak & wounded, sick & sore."

Sunday, Dec. 1st. 1861

It was a clear but cold day & as I had neglected my washing I was obliged to wash my clothes before I could change my garb & it was afternoon before I was able to leave the fort. I went to the camp, had a good visit with the boys, wrote a letter to Melissa, which comprised the business of the day. About 50 of the prisoners were sent to St. Louis and about 25 released on parole upon swearing allegiance & pledging themselves not to bear arms against the government. A few sick ones were taken to the hospital thus emptying Fort Wyman of all its prisoners. I am glad they are gone, they are a poor, degraded set. I am informed but three or four of the whole lot can read or write.

Monday Dec. 2nd. 1861

Arose early and found it snowing quite fast and before noon the whole ground was whitened with 3 or 4 inches of snow. I proceeded to headquarters and proceeded with my platting. Winter is really upon us. I have felt it's chilling breath more than any time before. Oh my poor wife, wilt thou, O God of heaven protect her from the cold, provide for her wants and be a father to her. I can endure the keen northern blast, I can suffer in the cold and not murmur but the loved of my heart must not suffer from the cold.

Tuesday Dec. 3rd.

It seems that trouble will never end in our Regiment. A quarrel has arisen between Col G. and Mr. Buck[46] the Q.M. and the Col. has dismissed him & appointed Jack [John] Van Pelt in his place. A very bitter quarrel is raging between the Col. & Buck & I understand Col. G. intends to resign. I care but little about the matter. With the old proverb I believe that when rogues fall out honest men get their dues. One phase of the difficulty is owing to the sutler selling liquor to the soldiers. The Col. wishes to abate the nuisance & I hope his efforts will be seconded by every decent man in the Regiment & let whiskey [Edward S.] Joslyn go to the shade as he must eventually if he continues to soak as he has of late. Platted most of the day. Weather cold, snow still on.

Wednesday Dec. 4th. 1861

Platted as usual. Received my butter from home & took it upstairs of the Post quarters. News came in today of a fight at Salem.[47] Major [William D.] Bowen went in that neighborhood last Saturday. Early this morning a house in the outskirts of town where some of our men were posted was surrounded by Freeman and Turner's secesh bands and while the inmates were asleep fired in the windows & killed and wounded some of our men. Our men rallied around the Court House and after a brief fight drove the secesh away, killing a number. The messenger did not know the extent of the injuries on each side. We look for more news tomorrow morning. A reinforcement has been sent to [William D.] Bowen's relief. I hope Freeman will be wiped out as he has been the terror of the whole country South of Rolla. Particulars tomorrow.

Wednesday Dec. 4th. [Confusingly, Bennett repeated the date.]

Platted all day. Near night went to the express office and found a keg of butter for me. Charges $3.55 which I paid and removed the butter to Col. Dodge's quarters.

Thursday Dec. 5th.

Platted most of the day. Sold 32 pounds of butter at 20 cts a pound. Mrs. Grensel[48] bought 15 and Mr. Buck 17 lbs. Had a long talk with Lieut. Walker in regard to his affairs. He is feeling in good spirits. Went to the 13th at night and notified Walker of my purpose to survey tomorrow. O God remember and care for the loved at home.

Friday Dec. 6th.

Surveyed 6 miles of the Vienna road which runs nearly Northwest from Rolla. A description of the country would only be a repetition of what I have already said of other sections of the country. One thing remarkable was the discovery of a *school*. I had supposed there was none in Missouri but we found one today & the natives said the teacher was a "right smart feller" from Tennessee. George Walker nearly gave out and was quite sick on our return. He can not stand it to go with me longer. I wish he could get the privilege of going home. Weather warm.

Saturday Dec. 7th.

My stockings and shoes were outside the tent last night & were completely soaked with the rain which came down in torrents last night. I had a muddy walk to my work. Before commencing at the plat I wrote a letter to Gen. Halleck[49] for permission to go home. I have little hopes of being successful but then it is worth trying for. About noon Quarter Master Buck was arrested and put under guard and all his books and papers seized. It is alleged that he is $10,000 behind in his accounts. I have had my suspicions that he was plundering all the while & this impression is general throughout the camp. And he finds but few sympathizers. Would that all the plunderers & wire pullers attached to the army were dealt with as they deserve. I have not yet learned the full particulars of the fight at Salem. Our loss was 2 killed & 8 wounded & that of the enemy sixteen killed, twenty wounded & 10 prisoners.[50] No rain today but the streets were muddy. This has been the only rain of any consequence since our arrival here. Seven paymasters & $1,000,000 of money arrived last night to pay the troops here. I received a letter from Melissa which did me much good. The assurance of her unalterable love is worth more than a mint of money to me. I had begun to be somewhat uneasy about my child but this letter assured me that she was better again. God of Mercy! protect and care for my little one, and all the dear ones at home. Grant that we soon shall see each other, and that thy love & protection shall never fail.

Rolla Missouri
Dec 7" 1861

Major Gen. Halleck
Dear Sir
    I have been detailed by Col. Dodge commander of this post to make a survey and topographical map of the country surrounding Rolla. The survey is now completed and a rude sketch partly made. I find it impossible to do a credible job here, and ask permission to go home for a few days, where I have instruments and facilities for platting which I cannot obtain here. I am confident I can expedite the business and do a better job there. Hoping you will grant my request. I subscribe myself

Very Respectfuly
Your Obt Servt
Lyman G. Bennett
Corporel Co E 36" Regt Ill.

Bennett Compiled Service Record, Thirty-Sixth Illinois, National Archives.

Sunday Dec. 8th. 1861

The sun arose bright as ever in tropic clime & shed it's genial influence over the hills, valleys, & rocks, giving new life & joy to our camps, and making the heart glad. Tis a day calculated to draw the mind to God and to Heaven, in thanks for His goodness and mercy. I wrote a letter to Melissa, mended a rent in my pants, and at noon went to the 36th & attended divine services. But little of the quiet of a home Sabbath prevails here. From the valleys comes the music of many bands and the unceasing rub a dub dub of the drums. On the parade grounds the game of ball is played with as much zest as if this was not God's holy day. Squads of horsemen gallop over the hills. The shrill report of guns all around tells that the same carelessness every where prevails. Though I do not keep the sabbath as I should yet I cannot forget when Sunday comes. The report is circulated that Freeman is captured and his force routed with a loss of 150 killed. This is most too good news to be true. I hope it is so, however. The shadows of evening are drawing around and the approaching twilight casts it's shadows over the mind. I have heard a band playing a funeral dirge, making the third burial train that I have witnessed today. The soldiers more have ceased from their battles, laid down their arms and have gone to their long home above. Not a day passes but the silent village of the dead on yonder hill receives from 1 to 3 or more inhabitants. From an army of 15 or 20,000 numerous deaths must be expected. I can never see a funeral cortege without feelings of sadness, and a prayer of thanks to God for life and health. This evening Mr. Harrison, James, and myself have had a rich treat among the poets. Mr. Harrison has a taste for poetry, is conversant with most authors and quite a poet himself. Here is something he has written after the style of Edgar A. Poe which has as much truth as poetry about it: "Tattered flag in ribbons flying, Prostrate arms in puddles lying, Cautious guard and mud holes shying. Wraps him closer than before. Wakeful Corporal is peeping, When the guard relief is sleeping, But his officer is steeping, Brandy sling behind the door." Here is another in the same style which the recent rains suggested. Long before daylight I was awakened by the patter of rain on the canvas, to which my head lay in close proximity and found Harrison writing and the following is the result:

> "In my tent as I lay napping,
> I was wakened by a tapping,
> And the sound was like the rapping
> Of an angel at my door.
> Then it faster kept on coming,
> With a ceaseless, tireless humming,
> Keeping up a wierd like drumming
> Louder, clearer, than before
> But I thought, perhaps I'm dreaming,
> Of a visitation seeming,
> So I raised me up, and leaning
> On my elbow toward the door
> Only heard the canvas flopping,
> Never ceasing, never stopping,
> Rain drops constantly dropping
> Down upon my earthen floor."[51]

I have not thought of writing poetry for years. Were I in that line of business my present life would furnish abundant themes for poetry.

Monday Dec. 9th.

Surveyed a branch of the Vienna road over six miles from Rolla. For the first time saw a fine farm in Missouri, owned by Mr. Payne. A widening of the valley of Spring Creek left sufficient land for a good farm and Mr. Payne had about 200 acres fenced & under cultivation. He also had a number of slaves & from all appearances was quite wealthy. His house and barns though behind the average of Illinois houses, yet had an air of comfort about them which I have not yet seen in Missouri. Pickets on the road told me that two of the fords of the Gasconade were occupied by our troops. The defeat of Freeman is again confirmed. I hope it is so yet fear the truth will not confirm it. It has been a warm and pleasant day, almost like summer. Not a particle of frost or weather cold enough to wear coats.

Tuesday Dec. 10th. 1861.
A walk of four miles brought us to our work, and then we surveyed but two miles and returned to Rolla a new route across the valleys and hills without reference to any road. Came across a poor dilapidated shanty without floor and almost without sides or roof. As we came in sight three or four children were found outdoors at play. One little urchin, I judge 2 or 3 years old being entirely naked. We came upon them unawares. As soon as we were discovered all scud for the house. A naked child outdoors at play in the winter time is something I never saw before. While on our way to work we met near a score of women, and I will not pretend to say how many men on horseback, riding to town & on our return we met them on their way to their homes. Horse back and ox cart are the modes of transportation here. Warm, but cloudy until toward night when the wind changed to the north & it became colder. Received a letter today from Mother.[52]

Wednesday Dec 11th.
We had a cold night and a clear, cold morning. The bright sun however made the day pleasant and we got along with the survey very well, passing over ridges & entering no valleys on our route. At times, when on top of a high hill the view of woods and valley was splendid. We passed no house on our route & save the road no signs of civilization. We returned through the camp of the 13th, found Walker some better than when I saw him last. The 36th marched through the various camps and showed themselves to good advantage. Our regiment are a fine body of men not excelled by any troops here. Bowen's command returned today from their chase of Freeman whom they chased into Arkansas, capturing about 50 of his men. I will gather the particulars as soon as possible.[53]

Thursday Dec. 12th 1861.
Did not intend to survey today but on arriving at headquarters I found the boys all ready for work, and concluded we would complete the Beaver

Valley road. We found this a rough unfrequented road down a deep valley overhung with high and almost perpendicular rocks and bluffs. It was very densely timbered and I observed a few fine cedar trees, the first evergreens I have seen in Mo. I received a letter from Major Gen. Halleck ordering me to report to St. Louis for completing the map on which I am now engaged. The order somewhat surprised me for I either expected an order home or else politely informed to remain at Rolla & do my platting. I am in hopes I shall be able in the end to go home, Harrison thinks I am homesick. Perhaps I am, I own to a strong desire to see my wife, child, and dear ones. I shall say nothing to Capt. [Charles D.] Fish in regard to this order. He & I are too much at loggerheads for me ever to crouch & perform the menial part. I hope that I shall be able to secure a permanent position in the engineer department.

Friday Dec. 13th.

Worked all day faithfully at my platting. The table where I work is small and two of Dodge's clerks, a provost marshall, a lawyer, and myself work upon it. This is entirely too circumscribed for me and I shall never be able to do a good job & am therefore glad of the opportunity of going to St. Louis. Lieut. Walker and a lawyer from the Iowa 4th were all day engaged in getting up some charges against Col. Greusel. I believe them to be true and if Walker can prove them Col. Greusel must lose the position he now holds. Col. Greusel on the strength of these charges was arrested and now has his limits assigned him. This was announced on dress parade in such a manner as to produce an effect upon the men & some companies at the command of their officers cheered the Col. and groaned for Walker. I hear some bewailing the affair and fear Greusel will lose his place & at the same time they say the Col. is right. If this is the case he has nothing to fear from a court marshal & if he is wrong he deserves all the penalty the court sees fit to inflict. I hope right prevails[.]

Saturday Dec 14th.

Worked at platting today under my usual disadvantages. Walker & his lawyer as yesterday monopolized the table. I received a letter today from Col Thom in regard to pay & was ordered to make out the pay rolls of myself & help and bring it to St. Louis and it would be paid.[54] The day has been pleasant indeed, not much like a December day. Rolla presents a busy scene just now. Scores of teams are busy hauling provisions & forage for the different encampments. Groups of horsemen dash along through the streets and parade ground. Soldiers & citizens are everywhere met and the different guards and patrols lounge lazily about but watch closely events that are passing around. Several Artillery companies came and drilled on our parade this afternoon but did not fire their pieces as I hoped they would. Took dinner with Walker at his boarding house & was treated to an oyster stew. This is something so much better than usual that I appreciated it highly. Harrison & his boy both sick tonight.

Sunday Dec. 15th 1861.
Changed my clothes and went over to Col. Phelps' regiment. Capt. [John S.] Coleman was very courteous and more than ever strengthened my good opinion of him. There were four dead bodies lying in the camp. Measles are prevailing to an alarming extent in all the camps & appears to be particularly fatal to Col. Phelps' men. There are very many deaths in the various camps now and almost any time of day a burial takes place, and is so common that one scarcely notices it. My pay rolls are all made out & I hope to get my own & assistants' pay for surveying. The weather is very warm, almost like summer, & thus far we have had an extraordinary winter. I have learned some particulars of the Salem expedition. After Bowens' surprise, and the flight of Freeman from Salem, a reinforcement from here was dispatched to his assistants who went in pursuit of Freeman and came so near overhauling him that he left the valley down which he was retreating for the mountains. Captains [Theodore A.] Switzler and [Bacon] Montgomery conjectured from the comformation of the country that this was but a ruse to get them off of their course & that Freeman would again strike the valley further on. Accordingly our force kept down the valley until it was conjectured they were ahead of the rebels. When striking across the country they came to an open place after night & discovered the enemy entering the opening from another direction. The order was given and our men formed in silence and at the word of command emptied their carbines among them. The secesh was so surprised that they fled in every direction without firing a gun. A large force of the enemy being in the neighborhood it was thought best to go no farther, and our whole command came back to Salem. The loss of the rebels was not known but it was supposed to be serious. Rumors of a movement from here are rife and it is but reasonable to suppose that a move will soon be made. To remain in camp will but foment quarrels and demoralize the army. The evenings are as bright & pleasant as the days & the moon now in her full never shone more bright or presented a more smiling face. I love to contemplate the moon. Its calm beauty reminds me of the calm, sweet, and almost divine face, many miles away. Melissa, do you think of me tonight? Yes and every night! Tomorrow I shall probably be in the busy city, of the Mississippi valley. May it be but one step to you.

I have just heard that Kimball Smith of Co. I is dead. Poor boy, he has left this world of sorrow and will nevermore feel the coldness and scorn of this world. His warfare is now ended and his spirit is at rest from its campaigning. I also learn that Lieut. Clark[55] of our own company is dangerously ill and may not recover.

7 | St. Louis

December 16, 1861–January 12, 1862

This morning I felt like a motherless chicken, and scarcely knew or cared what I did or wanted. I ate no breakfast, nor could I [have] eaten any thing, were it prepared by wife or mother. The day was about as dull and gloomy as my spirits.

—Lyman G. Bennett, December 22, 1861

Several days after arriving in St. Louis, Bennett dispatched his completed diary to Lieutenant M. LaRue Harrison, who delivered it to Mrs. Melissa Bennett while on a journey to visit his critically ill wife, Rebecca. Bennett himself increasingly complained of ill health, which led to his hospital admittance on December 23. With scant appetite and suffering from jaundice, Bennett experienced a brief hospital stay.

Not surprisingly, given its status as a transportation hub, St. Louis also became a center for medical care. By late 1861, several hospitals opened to serve the large military population anchored by Benton Barracks with its twenty thousand soldiers.[1] St. Louis hospitals were greatly aided by the Western Sanitary Commission, founded on September 5, 1861, at the suggestion of Dorothea Dix.[2] Designed to serve soldiers in the trans-Mississippi, the Western Sanitary Commission (WSC) was the sister organization to the US Sanitary Commission.[3] Sometime after the battle of Lexington, Missouri, the WSC opened City General Hospital, that, based on Bennett's description was probably where he recuperated. Staffed by the army, City General was "a five-story marble-fronted structure on the corner of Broadway and Chestnut streets."[4] The WSC's primary efforts were directed at providing medical supplies and food to St. Louis hospitals.[5]

Homesick and ill, Bennett occupied his time by reading, resting, observing fellow patients, writing poetry, reflecting on spiritual matters, and gazing out the window. His spirits improved, however, with every letter received either from home or from a fellow soldier. The arrival of the Biddle sisters raised the morale of every patient at the hospital. Nieces of Nicholas Biddle, the former director of the controversial Second Bank of the United States many years before, the Biddle sisters may have been working under the auspices of the St. Louis Ladies' Union Aid Society.[6] Why Philadelphians Catharine C. Biddle, Hannah Stokes Biddle, Elizabeth Newbold Biddle, and their widowed sister, Mary Vandervoort, were in St. Louis at the time is a mystery. One of their uncles, Thomas Biddle, a War of 1812 veteran, had resided in St. Louis until his death in 1831, which resulted from a duel over an argument that concerned his brother Nicholas's handling of the bank. Congressman Spencer D. Pettis and Biddle fired at each other from a distance of only five feet, and each man mortally wounded their enemy as a result.[7] Catharine started philanthropic work in 1856 along with her two younger sisters, Hannah and Elizabeth, and all nursed in Philadelphia hospitals during the war.[8] One of the sisters, according to Bennett in his December 29 entry, was considered "our *guardian angel*." Lyman became so well acquainted with the Biddle sisters that he eventually was invited to dine at their house in what became a memorable occasion.

Monday Dec. 16th 1861.

Arose rather earlier than usual & after eating my breakfast and hurriedly collecting a few necessary articles for use in St. Louis I bade Mr. [M. LaRue] Harrison & James good by and proceeded to Col. [Grenville M.] Dodge's headquarters and finished up the map already commenced. The cars did not leave until 10 ½ a.m. I had but little time to see my comrades but all bade me God speed & with many kind wishes from them I left Rolla with something more like railroad speed than when we came here. I had thought that Rolla was as near to the fag end of creation & any place could well be, but after a ride by daylight I am led to believe that Missouri is about alike throughout its whole extent. Copper bottoms, buckskin pies, secesh & niggers are about all that can be found in Missouri. I had a severe headache all day & when we reached St. Louis at 7 p.m. I felt more like dying than doing anything else. I went to the St. Charles hotel & calling for lodging was soon ensconsed in a 4th story bed which to my weary bones & aching head was a luxury I scarce have enjoyed for four months. Capt. [Theodore A.] Switzler and a dozen guards from Major [Clark] Wright's battalion came on the train from Rolla with 18 of the Salem prisoners. Their destination was the arsenal where they will be put for safe keeping for awhile. Col. Phelps' wife[9] was also on the train. She is a fat, jolly dame & with her sun bonnet and half homespun ways looks but little like a congressman's wife. It is said that after Springfield was abandoned by the federal forces, McCulloch visited Col. Phelps' wife & had a long talk with her. She told him that in the move our army had made it was the design to cut off his supplies at Pocahontas, in Arkansas. McCulloch gave credence to the story & the next day left Springfield in the utmost haste. This has been a sick day for me. O that my dear wife was with me to soothe my aching head.

Tuesday Dec. 17th.

After breakfast at the St. Charles which was about the first clean meal I have had since leaving home I proceeded to headquarters & reported myself ready for duty. Gen. Cullum,[10] Col. [George] Thom, and all the engineer core [corps] were absent and Col. Kelton[11] hardly knew what to do but finally ordered me to report again tomorrow morning and gave me tickets for subsistence, lodging &c. I improved the day in strolling about the city & in the afternoon undertook to walk to Benton Barracks,[12] little dreaming how far they were situated from the city. I trudged about two or three miles and then enquired how far the Barracks were, and found I had two miles yet to go. Feeling somewhat fatigued I got into a street car & returned. My fare is poor enough but somewhat better than I generally get in camp. I cannot afford to idle around the city & tomorrow must go to work or I shall spend every cent I have got & be no wiser in the end. The city is the place to ensnare soldiers and wring from them all their money.

Wednesday Dec. 18th. 1861.

Agreeable to orders I was early at the desk of Col. Kelton ready for duty. I found by reading the morning papers that Col. Thom & Gen. Cullum were at Cairo & would probably be absent a week. Col. Kelton sent me to Col. Totten's[13] room to confer with him in regard to the matter, but he being absent I asked the privilege of going home, which he refused. Determined to be idle no longer I proceeded to the engineers' room & finding paper for the purpose I went to work without consulting any one. I was fearful if I did not go to work I would be ordered to Rolla, which I was unwilling to do. I hope I shall be able to go home after the job is done, & I also hope to get a transfer into the engineers' department. It would suit me much better than where I am now. I have a splendid room to work in and everything handy as I could wish. I have sent a box by express to Melissa & if I am able to go home will take something more. I do not like my sleeping apartment, everything looks slovenly & dirty and being slept in by hundreds before me I am fearful of getting lice. I hope for better quarters as soon as Gen. Cullum returns.

Thursday Dec. 19th. 1861.

Worked all day at my platting & was interrupted by no one. In the afternoon being down in the Adjutant Genl's room I saw a familiar face & on enquiry found it to be Eugene Lake,[14] who resided at Winona in 1857 and boarded at the same place with me. He went to my room & we passed a pleasant hour in reviewing the events of the ever memorable winter of '57. He now belongs to an Ohio artillery battery in Asboth's brigade, now stationed at Rolla. He has been sick and away from his brigade & now wishes a pass to go to Rolla. In the evening went to a comic theatre. A troupe of darkies and half dressed white girls were the principal performers. Dancing, in which legs figured largely was the principal feature of the evening.

Friday Dec. 20th 1861.

Platted all day and am tired tonight. Saw a procession of Masons pass along the street. Found that Frank Briggs,[15] formerly of Oswego was in Gen. Halleck's body guard. I hope to have a talk with him tonight or in the morning. The weather is quite cold today. I fear our warm summer days are over & that now we will have severe cold. Have not heard from Rolla or home yet. A letter from Melissa would do me good now. Incessant work keeps the blues away but at times I feel as I could fly to the refuge of my own home and friends.

[The last sheet in the book:]

Dear Melissa,

    Mr. Harrison will take this to you. It is not full quite, but I thought I would not neglect this opportunity to forward it. In it you will find all that has transpired with me since I left you. Mr. Harrison is a gentleman in every sense of the word & if he calls I wish you to treat him as such.

Ever Yours,
Lyman

Saturday Dec 21 1861

At a time so near the close of the year it may be thought strange in me to commence a diary, and I should not have done so, had not an excellent oportunity occurred to send it home by friend Harrison who, much to my surprise, sent a note from the St Charles that he was in the city & on his way home. He had received a telegram from Aurora telling him of the critical state of his wife's health and that she lay at the point of death.[16] He had only time to reach the cars, and was only able to get a free pass from Co[l] Dodge

& without a single explanatory word from him. For a time Col Kelton was not disposed to grant a furlough but through Col Thom's influence one was finaly granted for twelve days.

Mr Harrison has agreed to call at my house, and I send Diary & other little notions by him. Would to God I could get the privelege of going home too. Nothing now would give me more pleasure. *Going home*! how sweet the words to me.

I have been quite unwell to day and am now under the influence of a dose of pills. I trust soon to be able to go on with my work with some degree of comfort. As I now am I can no more than half work, and it is with difficulty I work at all. Col Thom came back to day and took my pay rolls &c for work at Rolla. He appears to be a gentleman and I trust my connection with him will not be disagreeable to either. May God incline his heart to assist me in getting home.

I cannot commence this new diary, without *one little word* in remembrance of the dear ones at home and a repetition of that oft made prayer, that God in His infinite mercy will care for them and grant a speedy union of husband and wife and of father and child.

Cloudless has been the risings & settings of the sun, and all day long that glorious orb went shimmering on its way with not a cloud to dim its transcendent lov[e]liness. Not so this morning for no trace of sun was seen through the dark pall of clouds, and as evening comes on apace, the snow comes thick and fast, and is wrap[p]ing the dirty city in its mantle of purity.

Sunday Dec 22 1861

My new acquaintance and friend Mo [?] invited me to sleep with him at the quarters of the Kane Co Cavalry. This company numbers about 80 men under Captain Dodson and are mostly from Kane Co with a few from Dupage.[17] They are a fine company, and Gen Halleck has selected them for his body guard. To this Company Frank Briggs belongs and I find them a whole souled company of men.

This morning I felt like a motherless chicken, and scarcely knew or cared what I did or wanted. I ate no breakfast, nor could I have eaten any thing, were it prepared by wife or mother. The day was about as dull and gloomy as my spirits. The dust and snow in the street were about equaly divided and it would have required the nicest mathematical skill to determine which predominated *land* or *water*. The ringing of church bells reminded me that I was yet in a christian land, and kindled a desire to once again go where God was worshiped[.]

While slowly plodding down the street & while passing one of the army hospitals what should I see but the well known countenance of John Martin. It was realy him standing on guard at the door, and the meeting was like a ray of sunshine from a threatening & stormy sky. On my return from church I went with John over the building which presented such a contrast to the filthy hospitals at Rolla, and with my preconceived ideas of life at the hospital, that I at once resolved should I not get better to seek admittance to this one. The beds were clean no disagreeable stench which usualy prevails in sick rooms could be detected. The nurses seemed to feel the importance of their charge and did all they could for the comfort of the sick. This was a large five story building and I learned that 500 sick were within its walls. John had been here since the middle of October and was now well and getting fat as a buck. The physicians and men in charge all liked him, and have put him on extra duty at $.25 pr day extra pay and hence he has no particular desire to join his regiment.

I passed the remander of the day in reading the newspapers, and long before dark went to bed in my quarters of last night. My skin gets more yellow, and I am getting to be less and less a well man.

Monday Dec. 23rd

I did not leave my bed very early. Without an apetite and without a breakfast I proceeded to Col Thom's room and stated my inability to work and my desire to go to the hospital untill I should get better. Col Thom immediately went to Col Kelton with my request and I received an order to be placed in the hospital. A clerk received me and registering my name took me to John Martin's room. Here a servant brought clean clothes and water, and after a thorough washing and changing my clothes a bed was furnished and I found myself regularly installed as one of Uncle Sam's patients. I am allowed only half diet consisting of only tea and toast. I have plenty of reading matter and contrive to pass my time quite pleasantly. Still no letters for me or word from home. I begin to feel cruely neglected. I cannot say that I am better or worse to night. Hope to be about soon.

Tuesday Dec 24th

The sad quiet of the sick room is in marked contrast with the din and bustle in the streets. Busy crowds are in shop & town preparing for the festivities of Christmas, and other hollidays. The poor soldier away from home, there is no glad festival for him. His only consolation is in recalling what has been, and yet more anxiously pray for the return of peace, and the days of yore. Heaven grant that this may be my lot, and that a speedy union with my family will soon occur. I have written to Melissa and trust to get a quick reply.

With nothing that my hands can do, and with scenes of sickness and death around, the heart grows sick and sad and yearns for some slight memento from the loved and absent. To day I learn that 1200 prisoners have arrived and put in confinement in some large building[.] Everybody feels well at so large a haul of rebels[.]

Wednesday Dec 25

I find a communication from Rolla in one of the city papers, and in it is a notice of the death of Peter Scryver[18] one of the musicians of Co. E. He has had a severe cold and cough for some tim[e] but when I left was not considered at all dangerous. He messed in the same tent with me, and his death is coming nearer my own threshold than the generality of deaths. The other was also a member of our Co. and has been sick for a long time. At one time he was out of the hospital, but a relapse was brought upon him, and he now is numbered with the dead. Oh, the dread horrors which war engenders. Death holds high carnival over the land, homes and hearts are left desolate and forlorn[.]

Have written to father[19] to day. Am no better of my ja[u]ndice.[20] Weather pleasant but cold. The river still open and boats running[.]

Thursday Dec 26

A blister[21] covering my whole stomach was drawn to day and the effect thus far has been to half craze me. I can not stir without the most intense suffering. It may do me good in the end but now it seems as if Promethian fires would be a comfort and a luxury to one severe blister. More than the object of my enlistment has now been accomplished. I have slept on the ground, I have seen the worst features of camp life and have got enough of it and am now ready for home. I begin to sigh in secret, aye and in public too for the bread and milk[,] the savory viands and downy couch of civilization. There are nine patients in my room beside myself. One from Ohio is dangerous and I fear will never more know the delights of home. His mother is here and daily & hourly stands by his couch to minister to his slightest wants. Give me a wife for love but a mother for care[.]

Friday Dec 27

I have felt somewhat better to day, and have read most of the time, examined the newspapers and read my bible. There is much of consolation in that Book of all Books. Mark the following "Him that cometh to me I will in no wise cast out.["][22] This is a selfish world and when we call upon the most High we are apt to crave His blessing. His aid and assistance only for ourselves & our own selfish wants and wishes. There are four sisters by the name of Biddle from Phelidelphia now stopping in the city who make it their whole business to visit the hospitals and minister to the sick & dying[.] Three are unmarried & the fourth is a widow[.] They are sisters of Nicholas Biddle of bank notoriety and are reported welthy[.][23] One talked long and ernestly with me this afternoon, left me books to read, and before she left me I had learned to regard her as all the others do as more of a sister and friend than a stranger. How many sick and sorrowing soldiers have cause to bless the fair angels of mercy, that cheers and braces them up or smoothes their dying pillow. Here is a field where many a Florence Nightengale might have free scope to exercise their philanthropy[.] Still no news from home. I will not yet despair, for I know Melissa's fidelity *and love.*

Saturday Dec 28

It was a long time last night before I could get to sleep. My stomach was sore where the blister was drawn, and the least contact with clothes would so iritate it that I was almost in perfect misery. A dull heavy feeling in my chest also helped to drive sleep far from my eyes. As I lay upon my bed, the gas light dimly flickering, seemed only to reveal the gloom around and render each sound more distinct and impressive, and I even fancied that that dull light seemed to sharpen the perceptions of the mind, for never before do I remember being so nervous and sensibly affected by every little noise[.] The man in the cot next to mine is afflicted with a cough, which I could not well compare to anything else than a *perpetual motion*, ᶠᵒᵍ ʰᵒʳⁿ ᵒʳ ᵗʰᵉ ʰᵒᵃʳˢᵉ ᵗᵒⁿᵉˢ ᵒᶠ ᵃ ᵗʰᵘⁿᵈᵉʳ ˢᵗᵒʳᵐ and thought twould never end. Twas was not a small hacking cough, escaping from just beyond the lips, but ᵒⁿᵉ ᵗʰᵃᵗ ˢᵘʳᵍᵉᵈ ᵈᵉᵉᵖ ᵃⁿᵈ ᵘⁿᶠᵃᵗʰᵒᵐᵃᵇˡᵉ ᵗʰᵃᵗ ˢᵘʳᵍᵉˢ up from the lowest deabths & ʷʳᵉⁿᶜʰⁱⁿᵍ ᵉᵛᵉʳʸ ˢᵖʳⁱⁿᵍ of the mortal system, wrenching not only knee pans but toe joints and nails. That cough would long ago [have] wrecked any common [illegible crossed out word] ᶜʳᵃᶠᵗ sailing on its huge ocean ᵗᵘᵐᵘˡᵗᵘᵒᵘˢ billows. That cough was enough to supply a regiment with each a liberal allowance & then [illegible] an or surplus [?] for a [illegible]. On the other side was a great lubberly fellow, who thinks more of candy and his gout than anything else. It is ludicrous to sit and listen each day to the recital of his various ailments. It would form the subject for a chapter longer than I can insert here. Well, he is terribly given to *snoaring*, and such *deep unearthly snores* comeing from the unfathomed deabths of his cavernous lungs, and rushing like a hurry-cane through shattered and distended nostrils.

> The shadowy nymphs of night
> Would scatter in affright,

> And would never venture more
> Should they hear but *this one* snore.

Or

> The ancient mother porker
> With her frightened progeny
> With tails and ears erect
> Would quickly leave thirsty

and let this fellow do the snoaring for their whole race, and he could *do* it too. There's no such thing as *fail* in the notes he gave us that night.

> It was a snore so deep, sonorous
> As to shake the ceiling over us
> To scare all the sleepers round
> And *make Heaven and* earth resound.

Another, in a distant comes, and laying near the cot of a dutchman, was all night long talking in his sleep. His dreams were vocal and it would require the most skillful reporter no note the various topics he introduced during that long, and to me almost sleepless night. At one time he would be at work on the farm and busy in the management of his team[.] Again, engaged in the fierce brawl, an[d] woe to the victim of his harsh upbraiding[.]

Very few in that chamber of [illegible word] sick attended to their own business or showed a willingness not to disturb others with their sickness. The poor dutchman in the ~~corning~~ corner was nearly frightened out [of] his wits, and when he heard an extra snore "Wilder fiercer than before" could see him raise up in his bed and casting a malignant glance around, cry "Say you dare. What for be you dodats no shleeps none at all[.]" Sthop you [i]t, no goot ish me[.]"

Thus might I go the whole sound amo[rest of word illegible] the wheezy breezy, groaning moaning sighing dying rueful visaged inmates of this hospital, and perhaps each would have some peculiar characteristic of their own which would attract his [?] notice.

This has been a lucky day for me, for early in the day three of Melissa's *best* letters (They all are good) were handed me. News from home just now seems to come by the cart full, and jumped into my lap just when needed most. I had no more than read them and written an answer, when two more were handed me. One from Geo Walker and one from Mr Harrison.

I will not at this time dilate upon the many excellencies of my dear wife's letters[.] I will only add, that a few repetitions of the same kind of medacine will thoroughly eradicate all traces of disease and rid the hospital of one of its patients[.] I am better to day, materialy better and no thanks to pills or physicians prescriptions either[.] A good wife's love and devotion is worth more than all[.]

It has been a day for the admission of visitors, and women particularly have come in droves. At such a time make known but your slightest wish and tis gratified[.] Those who have been longest here know best how tis done and take advantage of their charitable credulity to fill themselves with apples and the forbidden luxuries of the hospital[.] Our *friend* the *roarer* or *snorer* laid in for a fresh supply of candy which cost him not a cent. Ill [I'll] warrant he is hereafter a fixture here, while candy & [less?]

Sunday Dec 29"
Though the black clouds of civil war is all around us, blasting and deforming the fair face of Nature, Leaving only in its wide and desolate path broken and smouldering ruins: strewing earth with the mangled and bloated corpses of its victims: yet our sunny day our bright mornings and clear twilight evenings all tell us that God has not left us yet: for each succeeding day tells in language sweet of the smile of God upon a sin cursed earth. How well would I like upon this beauteous sabbath morn to inhale the pure air and listen to the life giving words of Christ's gospel. But this previlege is is denied me and must make my self happy here. _____ [hand-drawn line]

It is pleasing to know that there is much of *love*: *true* and *heart* felt, yet left in the world. Dwight Follett[24] from Ohio, with as patriotic impulses as ever actuated man, left a home, where peace smiled, plenty abounded and nought of earthly bliss was wanting. All these he left for the defence of his country and the untried hardships of the camp and battle field[.]

A few weeks he bore up bravely, did all that duty required, from a patriot and soldier. But disease had marked him for its prey, and we find him in the hospital, languishing upon a bed of pain. Away off yonder, speeds the sad tidings to that happy home. Without stopping to inquire the *name* of the disease, without careing to *know* the *extent* of its ravages, that mother hastens to the cot of her boy. She only knows that he is *alone*, without friends to cheer or care for him. She finds her boy very low—almost at the gates of death. For six long and *dreary* weeks has she sat and watched by the side of her darling boy. Each visitor to his sick room, ominously shakes the head and whispers "*he must die.*" The ph[y]sician looks into his pallid, almost ghostly countenance and sees no ray of hope. Not so that mother! Her strong faith that her God is a *hearer* and an *answerer* of prayer, tells her that yet her son shall live. And from that distant home, comes the father's word of cheer. He, too, with unshaken and abiding faith in the God of earth and Heaven, has p[r]ayed that the sick one's life may yet be spared and believes not that death will yet claim his victim, away from his parents' home.

How strong—how abiding that mother's love—Awake from sleep at morn and by the sick couch she stands. Come when you will from early morning to frosty evening—she still is there and in the middle watches of the night noislessly and oft she steals by the side of her sleeping son. Next to God a mother's love is unfailing. Yesterday when all but hope had died away, one little cheering ray of life was discerned in his glazed and fixed eye. To day he is better still. Oh! tis good to see that mother's heart again thrill with gladness. With an unshaken trust in God she believes *her boy will yet be well*.

Although feeling much better in health and even in spirits, than any other day since my incarceration within these almost prison walls, and although the bright appearance of everything out of doors was well calculated to make me contented with myself and surroundings, yet as evening drew near, and I had written all I knew to write about, and read everything readable that I could lay my hands upon,—a feeling of discontent and utter lonliness began to creep over me[.]

A longing for *somthing*, I knew not what came softly stealing up into my thoughts. As usual, I thought of my home—my cherished idols there:—but the chilling thought, that distance as well as time intervened between them and me only served to heighten the gloomy thoughts which now came fast crowding round.

> While laying listless on my bed, the sweet words
> of Eulalia came to mind
> "Oh, there are many brilliant spots,
> "To gild life's lonliest hours."[25]

Just then, who should step into the room but the "angel" of the hospital Miss _____ Biddle[.] We all regard her as our *guardian angel*, old maid though she is. A smile—a word of cheer, she has for each, and seems to enter with the greatest zest, and most absorbing interest, into all our little stories of trial and sufferings. The victims of disease & death may murmur much at their hard lot, and fail to see one single feble glittering ray of hope—aye, plunge headlong down the deep abyss of dispare, yet God with us is kind withal—for has He not sent these angel visitants to cheer us with words of love and friendship? Yes the words of cheer coming from strangers as it were, through a bright gleam into the chamber of sickness, that with its halo lingers around our couch after they have long been gone.

At length it came my turn to be talked too. She asked me of my wife.—my little one.—and at once their loved forms rose up before me in all their transcendent lovliness—Friends, home, all came in unbidden and yet welcome. She told me of religious services in the c[h]apel below and bade me go. She gave me books to read, and a thousand things she said which I cannot write.

I can believe Miss Biddle's motives and [are] prompted from none other than a sense of christian duty. She cannot certainly seek a mate from among the miserable and motely crew of lame sick, halt & blind which are jumbled together here. No sooner is the dread toxin [tocsin] of war sounded over our land, and its accompanying woes and horrors let loose among us, than we see her like Florence Nightengale[.]

> Swiftly like an angel go
> To aid a brother in his wo[e].
> _____ [hand drawn line]
> That heart must be a heart of stone,
> Unthankful for such kindness shown.
> [hand drawn line]

Monday Dec 30

This day has been much like yesterday, clear, bright and beautiful. What winter was ever as this has been I have not seen it, and my memory stretches over, more than 25 long years.—But the wether must not furnish a never ending theme to talk and write about. Scenes in the hospital are getting stale and threadbare. What *can* I write about? Oh, I have it. I am gaining finely, and it requires but a little stretch of the imagination to fancy that I am sloughing off my yellow skin. I am about as willing to say "*good by old brass*" as any one. Wonder if any smutty mulatto girl would fancy me for a mate? No one can conceive a life more dull than in a hospital. Kindness of nurses[,] frequency of visitors, nothing can relieve the dull monotony of the hours as they slowly drag along. Much of the time I have sat at the window, watching the moving multitude in the street below—& here are

_____ [hand drawn line]
Some of the Things I saw.

What can be seen from my window
    Shall be the theme of my song
What scenes are transpiring below me
    In the crowd that are dashing along
A huge marble pile is before me
    Far up its dome reaching high
And the swallows that build their nest there
    Have their homes in the sky.

Far beyond cornice and chimney
    Over roofs, spires, and pilasters too
Through a niche in the wall I'me [I'm] given
    One glimpse of the skies azure blue.
But why tell of birds, skies, and sunlight
    And pleasure derived from the view

I fear the great God that made them
    With cities has little to do.

There hobbles the ragged street beggar
    Here dashes the belle of the town
Broadcloth and kersey commingle
    The baby—the banker—the clown.
There are dinahs fair, sleek and shining.
    Seedy boys retailing the news
All the fag-ends of creation
    But none like me has the blues[.]

Tuesday Dec 31st 1861

This is the last day of the year 1861. Thus day succeeds day, and the new year comes close upon the footsteps of the old. A retrospect of the past year finds much of disappointment and much that causes pain written out against me. Much that calls for repentence and sorrow, and much that I have done that were better to have been undone. I can not live over the life that I have lived nor can I undo what has been done. But God helping me I will try and improve upon the lessons of the past year and live more as I should live.

    My health has been better to day and after dinner Dr Barns[26] permitted me to go out for an hour or two. Went to the post office, mailed a letter to Melissa & received one from H. [Hiram] Wagoner [Wagner]. From thence

"Camp of 36 Illinois Regt Rolla Mo." Notice the upside-down flag.

went to the levee and found the river full of floating ice, but not in sufficient quantities to stop navigation. To fill up the time I sketched a view of the camp of the 36" Regiment which will be found on the preceeding page. My walk somewhat fatigued me.
[drawing of buildings follows this entry]

Wednesday Jan 1st 1862

Before I was up, or even fairly awake my ears were saluted with "Wish you all a happy new year," from Irish John, who is one of the day nurses in the 3d ward. And after the usual exchange of New Year greetings, we proceeded to put our room to rights, and try to pass as "happy" a new year as the circumstances would allow. But from some cause my exuberant feelings of yesterday were gone, and I felt a more than usual depression of spirits, perhaps partly owing [to] a sore mouth caused by the medacine I have been taking[.] It is getting so bad that I can eat with difficulty. My teeth trouble me much by aching[.] I have all my life been free from this affliction but now it seems I am called upon to bear my share of this vexations affliction.

To while away the time I have drawn an eagle & banner on the opposite page. I have another picture engraven on my heart which I wish I could transfer on paper, with every lineament as well defined as it is there.

Tis a picture of a little *Home circle*[.] Of a young—beautiful and trusting wife—so fair so guileless, so loveing.—And a little third spirit, fresh from heavenly hands and in both are smiles, which lends a holy charm around. That little angel spirit is nestled closly in its mother's arms.

Melissa—my wife—my beloved! and Minnie—my child—You are graven on my heart of hearts, and no time or distance can efface you from my memory. Father in Heaven, shall I not see this picture soon, and be with wife & child again?

Columbia's *Banner! O long* may it wave

It may not be inappropriate at the commencement of the New year to form new resolves of rules of action for the future[.]

    First—I am resolved not to swear
    Second " " not to get angry
    Third " " to lead a temperance life
    Fourth " " Wrong no person
    Fifth do my duty to my country & fellowmen
    Sixth fear God and love and serve Him as I ought.

"Columbia's Banner!"

It is my firm purpose now to live in accordance with these rules! May God help me to do it[.]

Thursday Jan 2nd 1862

Passed rather an uncomfortable night my sore mouth and a generall bad feeling oppresses me nights, and I sleep but little[.] I am also subject to night sweats which tend to weaken me. In the day time I feel better and on the whole think I am on the mend. Was permitted to go out in the morning. My first trip was to the P.O. but found nothing for me—thence went to Col Thom's room. He expressed no hurry in the prosecution of my work, and told me to stick to the hospital untill well. On my return part of a five cent investment in apples was taken from me by the guard at the door. A slight fall of rain & sleet fell all day and freezing on the sidewalk. I saw from my window boys, with skates doing as good sliding as if on a pond of ice.

We are about to lose our snoaring friend. He is pronounced well by the physician and tomorrow starts for Tipton to join his regiment. We will miss our quandum friend for none have given more audable tokens of his presence than he. I trust hereafter I can be allowed to sleep in peace.

One of my room mates has the eurysipclass[27] and notwithstanding all the doctors has done, at times his face will swell. Poultices and other applications are put on to keep the swelling down but no sooner does the doctor & patient begin to rejoice at a permanent cure than the obstinate face begins to swell again. The last application has completely transformed him into a darkey, and very many who do not know him readily take him for a big buck niggar. He is a quiet good soul and rairly complains of his aches or his usage. I think he may be safely booked as a fixture here for some time to come.

Friday Jan 3rd

Muddy streets, and icy sidewalks presents a dreary sight from my window. Pedestrians can scarcely pass at all and the poor horses attached to the street cars reek with foam and sweat. I wrote a letter for a poor sick soldier to his mother. There is not much improvement in the condition of the sick in my room. A poor Bohemian has been confined a long time with the rheumatism, and a portion of the time so bad as scarcely to be able to walk. Several blisters have been drawn on his back, which though very painful for a time would give the poor fellow relief.—At length he began to be sick of blisters and when asked if he would have another shook his head and said "blister no good." This is about the extent of his english. After using lineament several days without any apparent benefit, he this morning said to the doctor, "plaster good" thus signifying that he wished another blister drawn. A large one was prepared and his whole back covered with it. By some means the plaster slipped down, and when at night an examination was made his posterior was found completely blistered. The poor fellow can neither sit or lay down. No paddling could more effectually skinned him than the blister. The boys can not restrain their laughter, though they pitty him at the same time. ————————————[hand drawn line]

Mrs Follet[28] was taken sick with the measles and obliged to quit the bedside of her dying son. For a day or two he has been slowly failing. And last night about 8 P.M. he breathed his last. But a minute before he asked for water and no one supposed that death stood so near. He died without a struggle. Poor boy! he has suffered a thousand death[s] on the bed of pain to which he has so long been confined. I pitty his poor mother, and trust when life with her is done she will meet her loved one in Heaven above. His remains are to be taken to Ohio for burial. The corpse dressed in military uniform looks well. I have written a long letter to Melissa, and expected to hear from her, and was much disappointed at the announcment "No letters for you Bennett." This being a day for the admission of visitors, we were much cheered by the words of kindness from from good Samaritan sisters, which called upon us. [written in toward the right margin November 12 1862]

Saturday Jan. 4

Have not been at all well to day. Have read or written but little. A head ache and a pain in my side, has left me in no mood for reading or enjoyment[.] But little change in the atmosphere which overhead is dark and lowering and underneath it is as bad as rain and mud can make it. In my room is a boy named B. F. Wells[29] but 16 years of age. He is a drummer in Col Birge's sharp shooters,[30] and is a fine looking and noble fellow. He has won the friendship of us all, and his honest and childish way has enlisted the sympathy of visitors and many are the apples and choice delicasies which have been smuggled under skirts and muffs[.]

He has no mother or father, is an orphan, and has two brothers on the Potomac in the 1st Minnesota Regt.[31] Little Benny received a letter from his sister in St. Charles, Kane Co. Illinois and as he can neither read or write, I read his letter for him. It will be a pleasure for me to do what I can for him or any of my companions. God grant that he may soon recover his health and that sweet blue eye recover its beam with health & hope[.]

Sunday Jan 5

Here is a gem published in yesterday's Democrat which so nearly express my own sentiments that I copy a portion of it

What I live for.
I live for those who *love* me,
  Whose hearts are *kind* and *true*:
For the heaven that smiles above me,
  And awaits my spirit too:
For all human ties that bind me.
For the task by God assigned me.
For the bright hopes left behind me.
  And the good that I can do.

There are four other verses equaly as good as this one, but I have not time or room to copy.

After retiring to bed last night two letters were brought me. One from brother Guy and one from Melissa. Her sorrow and sympathy for me in my sickness touched a tender chord, and scalding tears though not of sorrow, came to my relief[.] Oh the strong & endearing love of a wife[.] Be my life long or short, that life shall be for her and my God. I am much better to day than yesterday and feel that I shall soon be able to work again.

It has been a stormy day and a portion of the time it snowed very fast, and before night street, pavement and all was white with a pure sheen of snow. The storm king has been slow to leave his drear domain among the glaciers of the north and the ice wrapped poles, but now his reign has commenced and we may expect to feel the rigor of his power.

Monday Jan 6 1862

About 9. or 10. P.M. I was awakened from a sound sleep and Wilson Judson[32] was announced below and desirous to see me[.] I was not slow to dress, and to find him[.] And we had a short but joyous meeting[.] I learned that he and others from his regiment had been detached for six months on recruiting service in the State of Illinois[.] The particular locality where he was to operat[e] he did not know, but hoped that it would be near his home. I am glad that he has a position to which he is eminently adapted and which I trust will suit him as I know it will. If there is a young man in the service whom I wish success & who richly deserves it, that man is Wilson Judson. He is honest, temperate, moral and every way a model soldier and man.

This morning Judson & myself went to Benton Barracks. They are situated near four miles from the hospital and we took the horse cars in preference to walking. We found no difficulty in gaining admittance and egress from the grounds. The stalls for cattle &c on the fairgrounds are now used for stables for cavalry horses[.] The quarters for the men looked comfortable and clean. All around were Regiments, companies and detachments on drill.

We were told that 20,000 men were stationed there. Why is it that so remote [illegible word] any scene of action and where soldiers are not needed, so many are kept idle. Oh, my country where are thy leaders to take thee to victory over the hosts of traitors and rebels. Nothing but incompetency and corruption is at work on every hand and the consequence is humiliation and defeat in every important battle. Will a republican people submit to this much longer? And as if ~~was~~ our humiliation was not yet fully complete, our best generals are forced to quit their posts to be filled by some traitor or coward. Sigel, the brave & victorious Sigle, has been forced to resign[.] He is the only one aside from the lamented Lyon[33] that has shown himself in every way up to the times and competent to the task before him. How long will such men as Halleck usurp positions of power and responsibility and abuse the confidence reposed in them to the ruin of our country and her brightest ornaments. A move to hurl such men from place and power would be hailed with joy by millions of patriot hearts[.] I have written five letters to day and read considerable, and feel less fatigued than many other days when I have sat still. Mrs Follet starts for her home in Ohio. She has watched by her son—soothed his dying hours and heart broken, returns to a desolate home.

Tuesday Jan. 7

Weather more moderate than yesterday, and the streets getting muddy again. Except a pain in my side I am feeling very well to day. I am now given full diet and sent below with a large crowd of convalescent ones into the common dining hall situated in the basement. No sooner is the signal given when one grand rush is made and a general jam at the door ensues. The fair is not very extra but enough and I have no fear of starving.

Wednesday Jan 8

So muddy are the streets & lowering the skies that I have no wish to go out. I have composed the following poetry, but find it no easy task.

> ### The Soldier's Dream
>
> Night has spread its sable mantle,
>   Wraped the camp in its dark pall;
> Not one star was seen to twinkle.
>   Darkness brooded over all.
> So dark so gloomy and appalling.
>   Not 'ene was heard the zephyrs swell,
> Or any sound, save the sentry calling
>   The hour of night "and all is well."

And all that day, from early morning,
    Till night shades gathered in the west,
The soldier wearied with his marching.
    On the damp ground seeks for rest.
What, though war clouds oer him lower
    With threatning wrath?—he heeds it not.
What though his head with pain is throbbing
    In sleep his ills are all forgot.

But now a scene to him most cheering.
    To his enraptured sight appears;
Again his own loved home appearing
    With its scenes of happier years.
Dreams not he of banners-gleaming;
    Other memories clustering come.
He is dreaming—fondly dreaming,
    Of the loved ones left at home.

He meetes his wife, so faithful, trusting,
    Clasps her to his trembling breast.
One cry of joy, with love unfailing,
    Clings she to her place of rest.
And there's his child, sweet prattling thing.
    A spirit fresh from heavenly hands
Its little arms around him cling.
    Glad welcomes are at his command.

Soldier! waken from your dreaming,
    Bid adieu to child, and wife;
The twanging drum and trumpet peeling
    Calls you to the deadly strife.
God of battles! Stay this contest.
    Break, O break its lurid gleam;
Each home circle fill with gladness;
    May this not be all a dream.
St Louis Mo Jan 8" 1862

Thursday Jan 9 1862

Was quite unwell during the night & though feeling better to day, am in no mood to stir about much. Received a letter from friend Dicky at Rolla, who gives a graphic description of what is going on there.

Both Capt [Samuel C.] Camp and O. B. Merrill are under arrest, and an order from Genl Curtis[34] and Gen Wyman have placed Lieut [William] Walker in command of Co I. Of course the sore heads in the Regt, and the company writhed and wriggled and threatened and blustered about. But Col Joslyn, now in command promptly put a quietus to all this. The day after Walker's restoration to his position, Co I was on guard Walker being officer of the day, and [William F.] Sotherland [Sutherland] officer of the guard. During the night, a few of the sore heads and among them Sotherland, made an effagy, and hung it up in a tree in the company's quarters. On one side was written, "I have *got my* posish," and on the other "*O what an ass.*" This remained till late in the morning, attracting large crowds to see it. At length Col Joslyn and Major Barry came around and inquired into the matter[.] But Sotherland denied any knowledge of it and protested he knew not how it came there. Major [Alonzo H.] Barry asked him if he was not officer of the guard and how such a thing should escape his notice, and finaly gave him to understand that his lies could not quite go down with them, and gave him to understand that a repetition of any more of his insults would get him into one of the biggest kind of little scrapes[.] Of course Bill wilted and since has carried himself more circumspect. But after all there is not the first glimmering spark of manhood in him, and we may look for nothing but the most consummate meanness from him. Camp & Grensals trial is set for the 8" inst. I look forward with interest to the result, for I cannot but believe that wrong acts will come back in some shape to hurt the wrong doer.

In talking with some who are the ever ready apologists of Camp[,] They tell me that they know as well as I that underhanded games, and wire pulling have been used to get rid of Walker, but that he was a *fool*, and deserved to be set adrift.—and when you talk to them of the abstract principle of right—they will with a flush of indignation, ask me—"Who is there that acts right in all things? No man! As to strict honesty in war, it is impossible. It is a system of sharp dealing all around. Every man goes in for himself—and all is fair in war[.]"

Now I beg to demur at such reasoning[.] At home, these same men profess honesty and fair dealing. They take their family to church sundays, and profess the principle of the Golden Rule "As ye wish that men should do to you, do ye even so to them" [35]—I believe this *can* be lived up to, and *must* be lived up to, to reach Heaven and immortal joys after death[.] Again—right never comes out of wrong—never has from the beginning & never will to the end—and the man who consoles himself with the idea that *trickery* is *right*, has learned his morality from some book that I am unacquainted with. Trickery and wrong never prospers. It may be thought wonderful smart at the time, but such smartness does not pay in the long run. These wire pulling officers may loose their position, or they may not, but one thing they will loose, and that is integrity.

Friday Jan. 10 1862

My health is improving again, and the idea of release from confinement is cheering. The appearance of the outer world from my window is more pleasant. The mud is either drying or freezing up. I long to be out again to breathe once more the pure air of heaven. At Miss Biddle's request I wrote a copy of the "Soldier's Dream" for her, and took the occasion to thank her for the more than sisterly kindness shown to a poor sick soldier. She has given me an invitation to dine at her house tomorrow and I shall only be too glad to escape, if but for a little while, from my confinement here to not fail to importune the doctor for a pass, to enable me to share the proffered hospitalities of a friend. I am reading the life of Josephine and find it very interesting. I do not lack for reading, but it is not at all times such as I would select. Received a letter from Melissa. Each day but strengthens my love for her. A life's devotion will be but a poor recompense for her truth, fidelity & love.

Saturday Jan 11 1862

The little scraps of poetry I have composed to beguile the tedious hours, has given *some* of my roommates rather exaggerated ideas of my powers. One clever soul by the name of Peeslee from Indiana from sundry hints intimated that he wanted to use my "*poetic*" bump for himself. And to day he slily told me he had a sweet heart "out in Indiana["] and wanted me to "*git up some right peart verses to send to her.*" I could hardly restrain my usibles as he made known his wants. I thought of Barnum and for a time was half inclined to "*Blow her up a little,*" and then "*touch her feelings*[.]" But the idea of imposing upon a poor honest hoosher dissuaded me from this project and I told him honestly that it was a difficult matter for me to "versify" & that I seldom did it only and then only to drive away *ennui*. Let a person show any particular [talent] I care not how bungling he may be, and there will always be credulous ones enough to magnify this particular attainment into a mountain. To illustrate, not a day passes but numbers come to my room, many of them strangers to look at the sketches I have made, and go away saying "*they never saw the the likes*[.]" Wrote a long letter to Melissa and then donning my "regimentals," sallied out—first to the post office—then to Col Thom's office then to the river: and near noon I reached the residence of the Misses Biddles. I had heard they were welthy but did not dream of a large three story pallace of marble, or of furnature the most magnificent. But such it truely was[.] I had fancied a neat but unostentatious residence with furnature to correspond—for there is nothing in their appearance at the hospital in their dress in their manners to lead one to suppose that their style of liveing would be beyond what their every day appearance would indicate: But such it was: and had it not been for the warmth of friendly feeling evinced by the ladies I should hardly have known what to have done with myself. One cannot help but be at home in their society—Books in endless numbers were on the tables and on the library shelves. Pictures rich and tasty hung around, and a steroscope with views of a thousand different European scenes afforded rich entertainment for me. I was then more than an hour before dinner, and a more happy social hour I have seldom passed. They have traveled Europe over and are familiar with every scene travelers and historians have told us about. And why was I a poor bush bread and unpretending Corporal, the guest of those who strictly belong to the higher or more aristocratic circles. Then being in this circle I believe is more their *fortune* than their *desire*, and is not attended so with that insolence, pomp, pride and sense of superearity, which is generaly a characteristic trate of those whose wealth, or circumstances have placed them in these upper circles.

Well dinner was announced—and that too was on a par and corresponded with the whole arrangement of the house—The turkey, the steak the potatoe—the bread, the pie, cake, apples, all was the result of superior cooking, and of a style to suit the most epicurean. To one so long accustomed to pork, beans, hard bread, and hard fare generally it was a *feast*. Melissa may tell of her New Year's & Christmas dinners, and I shall not say these were not good—even better than my dinner to day[.] But this was none the less a feast of good things to me.—[hand drawn line]

Returned to the hospital shortly after dinner and passed the remainder of the day in reading, writing &c. Was agreeably surprised to find a quantity of paper, envelops and five postage stamps under my pillow. Can guess where they came from. Heaven bless the good donor.

Sunday Jan 12 1862

Read and wrote most of the day. At 3 P.M. attended divine services in the chappel of the hospital. A goodly number of the soldiers were in attendance and several ladies from the city—The minister was an old fashioned good old man, and his discourse dull and prosy enough. In the evening was visited by my good friend Miss Biddle, who when she learned that I was to leave

"My Childhood Home"

tomorrow gave me a fine pocket testament as a token of remembrance. I also found some paper, envelopes, and postage stamps on my bed. How can I ever be grateful enough for the many kindness I have received from their hands[.] When sick and among strangers I have found friends, and when my poor heart was almost overwhelmed with despair and gloomy forebodings, God has comforted and cheered me. I find Frank Briggs an inmate of the hospital and a number of other members of the Kane Co Cavalry. It seemes that sickness prevails to an alarming extent among the soldiers[.]

8 | Work in St. Louis

January 13, 1862–January 30, 1862

But the deeds of the past are forgotten in the mighty whirl of the present.— This should teach us that after all, there is nothing stable and lasting but God, and his word to the believer is ever new, and always interesting.

—Lyman G. Bennett, January 19, 1862

On January 13, 1862, Corporal Bennett was released from the hospital, and he spent much of the remainder of the month completing two maps of Rolla and its surrounding area. The present entries illustrate his continued interest in Bible reading and reflections on religious matters. He was highly critical, though, in his January 19 entry of a Congregationalist worship service that he deemed as having too "much pomp and parade" for his taste. In these months as well, financial problems led to a fortuitous loan from one of the Biddle sisters that allowed him to pay his hotel bill. Two of the most humorous passages of his writings are in this chapter: a reflection on St. Louis mud, and an account of his arrest for failure to carry a proper military pass. Bennett's continued homesickness, his wife's loneliness, and severe tooth pain led him to ask for a surgeon's "certificate of inability to serve in the field" that was granted and quickly led to approval of a ten-day furlough. After an absence of about four months, Bennett was homeward bound.

Monday Jan 13
Awoke from a good night's rest, and feeling well in body and in spirits. At length the much wished for day of deliverance had arrived, and with a light heart I made all needful preparations, and bidding my comrad[e]s good by, was again a well and a free man. And yet it was not without a shade of sadness that I said good by to the kind friends whom I had found there. John[,] our day nurse, is a noble kind hearted fellow and I felt like parting from a brother as I shook his great friendly hand, and so it was with stewards, and

the lady nurses too[.] I could ask for no better usage or no better friends that I have found during my three weeks confinement. Proceeded to Col [George] Thom's room and was soon busy as a bee over my map, and I scarcely paused in my work untill night—not even to notice two regiments of soldiers who marched by our quarters to embark for Cairo. Col Thom agrees to pay me $.75 pr day & I select my own place to board. Being among strangers I did not know where else to go except the hotels. Went to the St Charles and was agreeably surprised to find friend [M. LaRue] Harrison just on his return to Rolla. We had a good visit and at the invitation of Col Budd joined a Union Club[.] Harrison & I were room mates, and we talked much of Illinois and times in general.

Tuesday Jan. 14

Bade friend Harrison good bye and went to my work, and was busy more than an hour before any of the others came to their work. Have done a good day's work and in the evening written to Melissa[.] The wether is cold and the ground froze as hard as a rock. Tis said that yesterday was the coldest day on record in St Louis[.] I have seen many colder days in Illinois[.] The river is not frozen over, yet tis full of floating ice and unless the wether moderates it must soon freeze over.

Wednesday *Jan 15*

Worked all day long as busy as a bee—I commence near an hour earlier than the others and work an hour later. I would not do it only I wish to get my money. I am desperately in need of some. Went to the Planters house after supper and read the newspapers. In their news room is papers from all sections of the country and one can choose what paper he pleases and read as long as he wants to. Wrote a letter to R. McDowell.[1] Do not have time for sketching and writing as much as when in the Hospital. Evening and day cold but clear. I am glad I am so well situated. In camp I should be sick or die this cold wether.

Thursday Jan 16

Very soon after commencing work, a letter was brought me from Melissa. Dear girl! how lonly and sad she feels at times. May a kind Heavenly Father cheer her and protect her.—and if by letter or any other possible means I can make her feel happy I shall do it. Worked hard at my map untill about 3 ½ P.M. when Col Thom wanted me to assist him in collecting and arranging his instruments and platts. Some of them were gone which did not suit him much. The men about head quarters are in the habit of taking just what they want and never returning it[.] Several days ago I commenced a sketch from memory of the first home I have any recollection of. I have not finished it untill to day [see p. 107]. I was but 10 or 12 years old when a new one was built and the old log home torn down and made into fire wood. I can remember just how the old house look[ed]. The roof was of slabs with here and there a crack just large enough for the snow to sift through and many a cold winter morning have I got from a warm bed and on my way down stairs left my tracks in snow that had sifted through during the night. I remember the old sweet briar at the end of the house its long shoots reaching to the chamber window—and under its ample foliage, chickens, turkeys and the whole poultry race upon the farm would flee for protection from rain & storm. And the cherry trees and cherry pickings. Bushels of the red and juicy fruit have I gathered. The huge old fashioned fire place, though not represented here, is fully delineated upon my memory. And when the huge logs were piled up, and the Christmas and New Year's gatherings came around and the laugh, the story, the joke, the apples and cider passed around the happy circle.—the slab roof and rough log walls were quite forgotten. O! those were happy days. And often through the subsequent journey of life, when surrounded by all the elegant furnature that wealth could command, I have remembered the old log house and the happy *hours*, aye—*years* passed within *its hospital* walls.

> "How dear to my heart are the scenes of my childhood
> And its fond recollections which comes to my view"

Friday Jan. 17 1862

The wether has been more moderate than for several days past. At 3 ½ P.M. went to the hospital and found but few of my comrades left. A few days produces many changes there, the well ones being discharged and new ones admitted. Two of the Biddle sisters were there, but my favorite was at home. I sent a little note of sympathy and if I hear she is no better will call, for I cannot quite forget her kindness to me. Wrote a letter to Melissa and after supper strayed to the Pacific R.R. depot and nearly the first persons I saw were the members of the band of the 36 Regt. I went with them to their hotel, and learned many things in regard to my old comrades. The 36 left Rolla Tuesday morning for Springfield, the boys all feeling well at the change. The whole force at Rolla have marched or are going to, and that post occupied by new troops. I trust that something will now be accomplished in Missouri and that we will not again endure the disgrace of a retrograde movement just at a time when the enemy are in our power. I cannot say that I have a very strong desire to be with them this cold wether. O that I had some money with which to send some slight token of remembrance to Melissa[.] Perhaps I can borrow of some friend.

Saturday Jan 18

The weather has become mild, the frost has disappeared and an interminable sea of mud spreads over our streets and sidewalks. Sent a package of letters, papers, &c by the Band boys to Melissa. To raise money now to get a few things for her is impossible. The paper to day has the following from Rolla—————[hand drawn line]

"A brigade composed of 4 regiments of infantry and two batteries, under the command of Gen Osterhaus[2] have moved west from this place. The troops were in excellent spirits, and were as follows 36" Ills, 35 Ill, 25 & 44 Missouri. *"The splendid appearance of the 36" Ills, 1,200 strong, in their march out of town received the unqualified and unanimous admiration of the spectators.* This is no unmerited compliment. The 36 cannot be equaled by anything in the service, and give them the oportunity and they will sustain the[ir] character as *fighters*. Gen [Samuel R.] Curtis who w[a]s the command[er] at Rolla has released from arrest Col Greusel, Capt [Samuel C.] Camp, O. [Orville] B. Munell [Merrill] & S. [Isaac] M. [N.] Buck and ordered them to duty. The paper states that all the charges were withdrawn. There [is] something about this I cannot understand[.] A review of artilery took place about noon[.] They passed under the windows of the office when I was at work. I was told sixty pieces of cannon were in the column. To each gun carriage and each amunition wagon six horses were harnessed and as the whole went lumbering along notwithstanding the muddy streets & unfavorable wether, they presented a fine sight[.]

Had a wretched teeth ache in the afternoon and I could scarcely do anything.

Sunday Jan 19

Mud, fog, and anything but agreeable wether[.] Attended church and listened to an excellent sermon. The house was full, and the full tones of the Organ added to the interest of the services.————— [hand drawn line]

In reading my testament to day my attention was called to John V.39"

"*Search the scriptures*: for in them ye think ye have eternal life; and they are they which testify of me." How many are they, even among professed Christians who search the scriptures as they ought[.] They open it with preconceived, opinions and old prejudices of their own, and they read it only for the purposes of confirming or proving in their own peculiar views. They do not set as high an estimate upon it as they should upon the sacred oracle of God.—There is no writing now in existence as ancient as some portions of the bible. They have withstood the varied changes of thousands of years. Profane history is but a record of the doings of men, & from it we learn that man in all ages has been about the same[.] But the deeds of the past are forgotten in the mighty whirl of the present.—This should teach us that after all, there is nothing stable and lasting but God, and his word to the believer is ever new, and always interesting. The pen which wrote it, though held by human fingers was dictated by God; and every line bears the impress of Divinity—Go among strangers enter the family circle—and it can soon be told who "searches the scriptures[.]" If you find upon the center table, a bible in gilt with no finger marks upon its pages or none of its elegance tarnished—that Bible is not read, and is only there for show. Or if on an obscure shelf in the library, you find a volumn covered with dust, which for months and years have accumulated upon be assured, *that* Bible is not read.

But the *"Old Family Bible"*—that source of Heavenly consolation—every page of which is a life gushing fountain, and which is read and reread, and pondered, and prayed over.—As you turn it over leaf by leaf, every page will tell you, that it has been read. Here, are pencil marks upon the margin indicating a passage of peculiar interest—of consolation—of promise. It can not be wrong to mark a bible thus. Here you will find unmistakable traces of tears. Here a drop of candle greas[e] shows thy night and day, that Old Book is used. Portions are well worn, but not obliterated [?]—like the lone lines in the spelling book, each line has been studied—little fingers (not always the cleanest) have traced each word untill they become like familiar faces, easily distinguished. So with passages in the bible[.] They are like familiar faces, or old friends quickly recognized, and ever welcome[.]

In looking over my testament (the gift of Miss Biddle) I find sweet passages, marked with pencil, all through, and on nearly every page. Each passage marked is something of encouragement, to the weak and desponding—some bright promise to the faithful, and all attesting to the long suffering—the infinite goodness and mercy of God. Many thanks for these pencil marks. I will never erase one. *When* and *how* can I ever repay the fair doner for her previous gift.

For two days past a regiment of cavalry from Ohio have been encamped on the opposite side of the river in Illinois, and unable to cross for the ice. Today the ferry boat was again in operation and the regiment crossed, and marched by the head quarters and my office with flying colors and lively music. They were a fine appearing body of men, and if the oportunity is only given them, will do good service.[3] Attended church in the evening long ways back from the river. It belonged to the congregationalists and had every appearance of being patronized by the aristocracy. There was so much pomp and parade, introduced in the exercises, that I could not help but contrast it with the simple and unostentatious worship of God by the early Christians—and by the brave and hardy pioneers of new and backwoods countries, where a log cabin, or a barn sufficient for a sanctuary for preaching the gospel of one who was always poor, a companion of the lowley and who had not [any]where to lay his head[.]

The music though excellent, was better adapted for a theatre and none but the performers themselves could unite in singing to the music selected by the choir. I know I should not criticise so closely the services of professed worshipers of God, But it sounded to me more like sounding brass & tinkling silver than pure heart worship[.]

Monday Jan 20 1862

The wether is warm—ice fast disappearing from the river—but the dense smoke of the city and the lowering clouds which hang over us—together with the almost omnipresent mud of the streets, renders it very disagreeable. No hash pudding—or chemical compound, could be more thoroughly mixed than the mud & water is, by the thousand vehicals which in the course of the day traverse the streets. There is a long unwritten chapter of amusing scenes in a[cci]dents and mishaps connected with St Louis mud. For instance the Irish drayman, who all his life has reveled in dirt, is in his element when plashing through the filthy poriage. His face, his hands, and his clothes from slouched hat, down to the heavy cowhides which cover his feet, speak more are more eloquent of mud than of humanity. As he drives furiously along what cares he if he does plash great patches of mud upon starched shirt bosoms or unsoiled kids, or what if a broadcloth is rendered like Joseph's coat, of many colors. He cares not and rather laughs at the "*goak.*" It does appear to me that this unpleasant state of affairs, increased rather than diminished the number of lady pedestrians, for at all times of day, and night too, muddy skirts or muddy ankles are visible[.] Look out when I will, and one or more women are at the crossings on tip toe and with a huge mass of silks, crinoline, linnen and women geer of every name and description gathered up, at times to an immodest hight. It is a glorious oportunity for studying, not exactly Physiology—but *ankle ology*, or even a *higher* order of ology still. But then we men had rather witness such exhibitions of neatness, than the heedless sloven, who cary a greater amount of mud attatched to their clothes than their whole amount of clothes[.] I have seen dresses literaly bedaubed inside and out with the half liquid surface of

St Louis and instead of precautions to prevent its accumulation, no effort was made to prevent the broad expanse of silk or calico from performing the office of street sweeper. Went to the hospital but saw but few of the old patients & my companions. Changes are constantly occurring, a type of the transitory nature of life and all things in this world.

Tuesday Jan 21

I awoke in a quandary, have been all day in a quandary, and in a disturbed state of mind. And why? At the St Charles it is the rule to pay all bills weekly. And five dollars is now due from me, in the shape of an eating and sleeping bill. Let us examine the state of my finances. After a thorough examination of pockets and other receptacle[s] of this world's filthy lucre, the sum *"teetotum"* of bankable and other disposable funds is as follows

```
Two 3 ct Postage stamps—$.06
 "   1      "        "    .02
Total————           $.08
```

What is to be done? Five dollars must be paid, and I hardly dare ask Col Thom for an advance untill my work is done, & they that pay even then are called good paymasters[.] Let us go out in the street—over to the hospital—for I have received more friendship there than any where else in St Louis. But who is this! "Good evening Miss Biddle[.] I am realy glad to see you[.]" "How do you do Mr Bennett[.] You are not well are you. You look forlorn enough—What is the matter." O—nothing—in particular, or rather everything that poverty is heir to. Debts, dinner, duns—and the blues[.] "Is that all! Well try that (*placing a five dollar bill in my hand,*) and see if it won't relieve you—Why you look better already and here is another dollar. take that. You are improving[.] Guess you will soon be well. If not call again and you can have the dose doubled."

Well I did feel better—well in fact. But how she could tell this by my countenance was a mystery to me, for in spite of every effort on my part, the *man* gave away to the *woman*, and tears glistened in my eyes, and in tiny rivulets ran down my whiskered cheek. O! the luxury of tears! How they water the dry desert of the human heart. And where they fall, upon the *steril[e] soil* of *man's nature*, plants of gratitude, affection, good resolutions & holy purposes, spring up, bud and blossom[.]

Well, it is night, my board bill is *paid* two silver half dollars are in my pocket[.] I have returned from a visit at the Miss Biddles, and if ever fervent prayer went up from a greatful heart, and was registered in Heaven, that prayer, is now on God's records above. Others than Melissa's name, have been presented to the Infinite and rich blessings implored in their behalf.

Wednesday Jan 22
Completed one copy of my map, which was approved by Col Thom—Wrote a letter to Melissa, and after supper strol[l]ed up to the Planters House, to read the Chicago Tribune, & other papers. Saw a notice in a city paper that all soldiers in the city without passes would be arrested by the patroll guard, and taken to the military prison[.] So many soldiers find their way from Benton Barracks to the city, and find their way to bad houses[,] drinking saloons, &c and when they do finaly return to their quarters are either drunk or diseased that it has become an intolerable nuisance, and hence an order to arrest all without passes. One regiment when ordered to march could find but half their men, the others being away. In view of these stringent measures I thought it best to return to my hotel & the next morning provide myself with a pass. Stepping out from the planters to cary this resolution into effect, I was nabbed by a patroll which just then was passing and without ceremony was ordered to fall in & march to the quarters of assigned for stray military birds. Remonstrances, and protestations were worse than useless, nothing but a *pass* would avail me, and go I *must*. The gleam of a score of bayonets around me showed escape impossible, and with as good a grace it was possible to assume under the circumstances, was marched off to the prison. On the way, saloons & suspected houses were searched and a goodly number of half drunken & rowdy soldiers added to the procession. As we reached the prisoner's quarters, an old theatre, that had been abandoned, name and Regiment recorded, and then introduced to the main, or prisoner's room. Twenty or thirty were already there, had taken possession of the stage and were performing the first act of the night's entertainment[.] Each performer played the part that suited him best & in his own way, and no fault findings or criticisms were instituted provided each did all he could to accomplish the main object of the performance, viz—make all the noise and confusion possible for human beings to make. Scene first—The tin screens for the foot lights of the stage were tied to the heels of a score or less or performers, and then break into a double quick, march, or gallop around the room the tins banging like so many gongs, the company shouting laughing and whooping "a la indian" untill the whole theatre shook to its foundation[.] Scene Second—The boards of which the scenery was made, were collected, & split broken and mauled up untill the whole floor presented the appearance of a chip yard. A few long strips were used as arms, and a private emulous of official renown undertook the task of drilling these

soldiers with their pine board arms—The Co was ranged in line on one side and, the officer in front. At the order "ground arms" the boards came to the floor with a crash, which made everything ring, and which reverberated around the large room like the explosion of a powder magazine. In wheeling woe to the unfortunate being who stood within reach of the swinging pine firelock. Hats if not heads, were sure to fall. At Last the Officer imprudently gave the order to charge bayonets. Instantly pine boards were leveled at the only enemy which appeared in front, which unfortunately was the officer himself. The charge was irresistable and down came Mr officer sprawling on the dirty floor. The blunt ends of the pine bayonets had not penetrated very deep and a few slight bruises were the only wounds received. About midnight a window was raised underneath a stair case and a shirt torn in strips and one end fastened inside & the other hanging outside. By every imaginable device the guard's attention was drawn away, and three of the prisoners escaped in safety, by sliding down the rope to the ground. A fourth was more unlucky for scarcely had he cleared the window when the rope broke, and chug he went on the hard ground below, falling about 12 or 15 feet. We heard the poor fellow grunt, & utter a few muttered oaths, and then go limping off, the extent of his injuries are unknown. Fortunately I had a book with me, which I read through about 2" o'oclock and was so absorbed in its perusal that I scarcely noticed what was going on around me. After finishing my reading and I began to realize that I was a prisoner my thoughts began to be of a more lugubrious character. I did did not however swear as most of the others did nor did I indulge in unpleasant murmurings, such as was heard all around me, but slowly paced the floor backwards and forwards, ocupied so[e]ly with my own sad thoughts. It seemed as if the dull hours would never terminate and that that night was protracted into weeks. I thought of my "*home duties*" Melissa with her raven tresses, her tearful eyes and choking sobs, at our last parting at camp Hammond, which made me repent ever being a soldier. My smiling babe, the picture of love and innocence were also there. I also thought of my past life, my misdeeds and in short a thousand other things, good[,] bad and indifferent, untill it seemed an infinity of time was consumed, and still no signs of coming day. To sleep was impossible, for with an aching head and tortured imagination, the dirty and hard floor was an uninviting place to seek for rest & repose. Daylight came at last, but with it no signs of release. The same, heavy & measured tread of sentrys outside & in continued, but no indications of returning freedom. Some of the boys, unnoticed by the guard tore a portion of the staging down and quite a number made their escape. I resolved however ~~that I would~~ not to release myself in that way if I staid there untill doomsday. The guard missing some of his prisoners soon discovered the rat hole from which the boys were escaping and put a quietus upon it. The last fellow out hearing a movement of the sentinel and thinking himself discovered, made a break into the street and "leged it" as if his only hope of salvation depended, upon the way he "*dug in*[.]" The last that was seen of him he was dodging behind a corner and disappeared in a cloud of mud and gravel stones raised by his flying for cowhides.

I finaly wrote a note for Col Thom to come & release me but none of the guards were humane enough to carry it. About 11 A.M. a Lieut from Camp Benton in quest of some of his missing men, had the kindness to carry my note to the Col and in a few minutes he appeared & released me. If ever man was grateful, and glad to get his liberty again, it was me. Thus ended my first and may it be my last experience as a prisoner[.]

Thursday Jan 23d

From, the theatre I went to the Adjutant General's office & procured a pass, to secure myself against future arrest, and then to dinner 18 hours fast had given me an appetite like that required buisy waters [waiters] & the best viands of the St Charles to satisfy[.] Worked but little [in the] afternoon for my head ached and I was about sick & about four PM, went to bed[.]

Friday Jan 24

Last evening after awaking from my afternoon nap my head was aching as if it would split. Teeth, face, throat & head were also in the same pitiable condition. I however crawled down to supper and then back to bed again, feeling worse than before. About ten P.M. I could endure the pain no longer and sent to an apothecary for something to ease me[.] A bottle of neuralgia[4] medacine, applied to my head, teeth & face soon quieted me and I went to sleep. I have seldom suffered as I did for a few hours last night. This morning felt much better but thought best to keep quiet untill after dinner. Resumed work in the afternoon and though not right well did a good half day's work. Weather plesant. Some what muddy during the day, but nothing to what we have had. A rich secesh merchant has been banished from Missouri by Gen Hallack which creates some excitement.

Saturday Jan 25 1862

Worked faithfully all day, and the progress made with Map No 2, reminds me that my stay in St Louis is drawing to a close[.] Have devised a plan for going home as soon as I am through here, and have written to effect to Melissa—Think there is no doubt of my success. Friend Briggs comes nobly to my assistance and I am almost confident of success. The weather is pleasant and the streets seem more than usualy thronged. The number of soldiers loitering around has materialy diminished since the stringent order of Gen [Henry W.] Halleck has been put in force. This is well, for dens of vice have lately been filled with soldiers.

Sunday Jan 26 1862

After a good wash and changing cloth[e]s I went to the office & wrote & read untill church time. Liked the sermon very much. I think the minister's name is Brooks. Though not much of an orator, yet he has a faculty of drawing the attention of his audience and [the next page has an etching of "St. Louis from the River" pasted in] keeping them awake—And then his sermons are so applicable to the condition and wants of his hearers that I cannot fail to like them.—

The wether is mild and pleasant, not much like what a northerner would call winter.

Have read my testament through to day, but will not cease reading it in the future for there are portions which never grows old or which can be read too often. Its bright promises, and words of cheer to the desponding heart is enough to endear it to those who believe in Jesus as their savior. Have also read a book called *Imitation of Christ*[5] written many hundred years ago by a monk or catholic priest. I did not know that in those days of superstition and bigotry so much genuine christianity existed as is expressed in the little book I have been reading. Went to church again in the evening. I do like the preaching very much. It is not the same prosey rehash of the same subject each time, but something new & interesting and with energy and life throughout the whole.

Monday Jan 27

Awoke in the night by the sound of thunder and found a regular thunder shower in full blast[.] And on going into the street this morning found, the mud and water scenery of a few days ago spread out before me, and now again the streets are to be one vast mud hole from now to the end of time for ought I know.

During the forenoon a regiment of Iowa boys marched by in plattoons— followed by a half mile or more procession of wagons, with camp equipage. Some of the boys were literaly plastered over with mud somewhere between ⅛ and 2 inches in thickness. In one of the baggage wagons were two women—one with a child in her arms, and the rain, pelting woman, child and all. How foolish for a woman and a small crying infant to accompany an army[.] Have worked faithfuly and nearly completed Map No 2—and what is best of all received a letter from Melissa with good news from home. If fortune favors me, in another week I shall be at home and with family & friends once more. [next page is an etching of the St. Louis Courthouse—pasted in]

Tuesday Jan 28
About noon my map was completed, but as Col Thom is sick and not attending to the business of engineers I do not know when I shall be able to get my discharge or my pay[.] Saw some soldiers from Rolla and learned that the 4" Iowa and nearly all the troops there had moved to the west. The roads were very muddy and traveling nearly impossible[.] Also met parson Lyon in the street and was promised a call from him. Col McPherson[6] tells me to remain patiently in the office untill Col Thom gets better. When that event will occur, "mercy only knows".

Took a good long stroll over the city after dinner, passing by the McDowell's college[7] with its 1100 secession students, under the tuition of some of Uncle Samuel's sons. Called at the hospital and saw many of my old room mates. The rain prevented my extending my peregrinations as far as I wished. The heavens seem draped in mourning, and weeping bitter tears upon the earth. Can it be that mans depravity caused the very Heavens to weep [?]

Wednesday Jan 29" 1862
My teeth are a source of great affliction. They ache a little in the morning—the pain increases as the day advances just enough to sweeten my temper and keep me in good humor. After dinner they ache *some*—and during the afternoon a *good deal* and in the evening a *good deal more*—At bed time they set in for an all night's ache, and after aching the next day, and the succeeding night and day, then commences a spell of the tooth ache, such as no previous experience can give an idea of. Were it confined to one, I would not care, for I think I could stand all the ache that could be concentrated in a single tooth[.] But when every one, aches in concert—then comes the tug of war. Should I pull those that ache I would be as toothless as an infant. Am not sure but it would be far better to be so.

After raining constantly for a few days it finaly set in for a rainy day. The skies were clothed in funeral black, and the water poured down as if lake Superior had been turned over pan cake fashion and was emptying its contents on us and in this way it continued untill I was lulled to sleep

"By the ceaseless patter of the rain."

Navigation must be opened everywhere now except perhaps on the peaks of the Rocky mountains and a few other hills. This morning quite a different scene from what I had anticipated was presented. About two inches of snow lay upon the ground and now (9 A.M.) it is coming thicker and faster and more of it. Saw nothing of Col Thom to day, and hence am no nearer a settlement of my affairs or know no more of my fate now, in regard to engineering, or going home than three days before I was born. Col McPherson has given me a job which I worked at in the afternoon, but did not complete[.]

Thursday Jan 30
Had just completed Col McPherson's discriptive roll of a Co of Engineers he had enlisted, when much to my delight Col Thom came in, and after examining the maps approving them proceeded to settle my accounts. I could get pay at this time for my work in St Louis, in all amounting to near $32.00, a sum sufficient to pay my debts, but not all I wanted, for home & debts[.] The best event of the day if not of many days, was my success in getting a furlough[.] Dr [Joseph K.] Barn[e']s readily gave me a certificate of inability to serve in the field and on the strength of it Col Kelton gave me a furlough of ten days. Thus has God heard and answered my prayer, to soon be with my family & again taste the joys & delights of home.

Most of the day was passed in writing pay roll and commutation rolls, for the purpose of getting my pay for services in St Louis, but I could by no means get pay for the surveying at Rolla, as a consequence will go home with a limited amount to pay debts & supply my family[.] Visited the hospital and paid Miss Biddle's money borrowed of her, purchased a cap—wrote to Mr Harrison—and after an infinite [illegible word] of running around, went to the St Charles packed my satchel for an early start in [the] morning. Then retired to dreams of home & its idol.

9 | Furlough

January 31, 1862–February 8, 1862

Can it be that in so short a time I am bourn from foreign lands to my home & from sadness to happiness.

—Lyman G. Bennett, January 31, 1862

The happy recipient of a ten-day furlough, Lyman began his journey home on January 31, 1862. Upon his arrival, his family and friends engulfed him in a whirlwind of visits as they sought to engage as much as possible with him during his brief visit. For one of the few times since beginning his diary, Lyman, the faithful chronicler, missed recounting a day's activities.

Friday Jan 31
Was up at five, in order to be in time for the cars[.] At breakfast became acquainted with an army officer who was going to Springfield in Illinois[.] His gentlemanly appearance pleased me, and we walked & chatted together to the ferry, & sat near each other in the cars. Through his influence with the R.R. managers I was passed free to Chicago, thus adding another link to the chain of good fortion that has recently attended me. The sleighing [?] in St Louis was fine but as [illegible word] progressed northward into Illinois it almost disappeared untill nearly to Chicago. The wether was cold & I scarcely left the cars to notice the country through which we passed[.] I saw however that it was level & mostly prairie[.] We reached Springfield near noon. It is a pretty place, & with a fine country around. Bloomington also is a large and fine town. The other R.R. villages through which we passed were generaly small. Arrived at Chicago at 6 P.M. and in two hours more started for Oswego and arrived at home at 2 A.M. Saturday. Going to the door I wrapped until I supposed Melissa was awake & then asked admission[.] She heard me, but not recognizing my voice & supposing it to be a stranger, said not a word but was frightened out of her wits. Not hearing any noise

within I redoubled my knocking at the door which called out a frightened answer, and it was some little time before I could assure them it was Lyman & get in from the cold. I can not describe the joyous meeting[,] the warm kisses and the happiness of the hour[.] I retired soon to bed but not to sleep and the remaining hours of that night was passed in an uninterrupted and unceasing talking with Mellie, untill after the day was fully ushered in. Can it be that in so short a time I am bourn from foreign lands to my home & from sadness to happiness.

Saturday Feb 1 1862.

How pleasant to again be at home.—to feell that I am again under my own roof, with no one to molest or make afraid, and that what I see and enjoy is my own, and that none dare usurp this my fortress and my kingdom[.] In the army I am not myself, nor am I master of anything I wear or have, or of myself[.] At home I am free, am a man—am an equal with my fellows, and with those that love me there, what more of earthly happiness can I want. After a good breakfast (Melissa know[s] how that is done) I went down to father's. My arrival was entirely unexpected, and took them by surprise. Around the dinner table was the whole of the Bennett family, a spectacle which has not been seen for years, and this was a source of the greatest pleasure to mother. Called upon several old friends in the afternoon at Oswego, & in the evening received a visit from Guy [Bennett].

Sunday Feb 2nd.

Attended Mr. Rudd's church and heard an excellent sermon, met with a number of old friends & acquaintances and after prayer meeting & after noon services returned home benefited with what I had heard, & happy in the security of my folks at home[.]

Monday Feb 3d

While Melissa washed I choped wood untill tired out, then visited at father['s] and towards night went down to Oswego, and called upon many of my old friends[.] Returned in the evening expecting a visit from Drake but he did not come. Mr & Mrs Brooks came in and we had a pleasant little time. How many thousand questions I answered in regard to men and scenes in Missouri I can not begin to remember. It seems that newspapers can not satisfy the curiosity of those who are at home & the lucky man who gets a furlough for a short time is quizzed & questioned to death. How pleasant, how supremely happy to be at *home* and with friends and more. Never did home look better or wife and babe seem more Divine than now.

Tuesday. Feb 4 1862

Visited at Mr Sanderson's,[1] and had a very pleasant time, quite a number of friends were there and the day was passed in discussing our national affairs—and in enquires after those in Missouri. Good cheer and a general good time closed the day, but weary and almost used up Mellie & I returned home and was glad to find repose and a good night's sleep. I have slept but little since Thursday[.] The idea of *going home*, kept me awake nearly the whole of the last night in St Louis. The first night at home was entirely without sleep, and every evening since visits to or from friends have kept me awake untill midnight, and from present appearances, night as well as dawn [?] will be actively employed in much the same way[.] I should have 30 instead [of] 10 days furlough.

Wednesday Feb 5

Edgar [Lyon] came up early and all our folks turned out to get my winters wood. Guy drew and myself & the others chopped, and I have the satisfaction of knowing that my family is provided with nearly a year's wood, and will not suffer from the cold. My recent sickness and a lack of out door employment has made me tender as a woman, and one day's work at chopping wood has about used me up, back, arms, and my whole mortal corporosity feels somewhat achish [?]. It is a satisfaction to know that so much of the wants of my family are provided for. Mother passed the afternoon at my house, & in the evening a house full of youngsters & oldsters were there and gave us a sever [?] visiting[.]

Wednesday Feb 12 [appeared later in transcription but fits here]

Edgar came up this morning and my folks all turned out to chop my winter's wood. Guy drew and the others chopped, and although I was about a used up commodity yet I had the satisfaction of knowing that wood enough, for a year was nearly provided for my family. Although I have all along been satisfied that my family would not lack for food and the necessaries of life. Yet when I see with my own eyes, wood and other necessaries provided I can rest satisfied that all will be well with them. Mother [all of the above entry is crossed out]

Thursday Feb 6

Intended going to Big Rock, but a storm & cold wind prevented—Passed the day at home, writing and visiting for once with those with whom it is a pleasure to visit. Towards evening went to Oswego, saw Mr Cole,[2] Hollenback[3] & others[.]

Friday Feb 7 1862 [Bennett wrote no entry for this date.]

Saturday Feb 8 1862

The day for my departure dawned pleasant but cold, there being a slight fall of snow during the night. Packed my clothes, and after eating a hearty breakfast, went down to father's & harnessing prince & Sam, proceeded to Mr Gortan's,[4] whose sleigh I had borrowed. After a brief chat, went home & as soon as the folks were ready was on my way—first to father's—and then to Plainfield. Bade adieu to friends in Oswego, not as if for the last time for I have no idea but that I will return in safety—The sleighing was in excellent condition & we reached Plainfield about 1 P.M. without accident or suffering but little from the cold. Father & Mother stoped with Mr Bibbins[5] but Millie and I passed most of the afternoon at the Mr Roye's enjoying a good visit and a good dinner. Towards evening—in company With Phill & his sister we went to Mr Goodhues. They had been expecting us all day & were a little disappointed that we did not come before. A very pleasant evening we had[.] Minnie however worried with her teeth which are just cutting through.

10 | Return to St. Louis and Rolla

February 9, 1862–February 15, 1862

> *I know something of the troops sent against him [Sterling Price], and believe if a battle is fought it will be most bloody and desperate[.]*
>
> —Lyman G. Bennett, February 13, 1862

An exhausted Bennett left home "with the deepest sorrow & regret" on February 9, 1862, and arrived in St. Louis the next day. After reporting for duty and running various errands, Bennett boarded a train for Rolla two days later. The town had changed markedly since Lyman was stationed there in December of the previous year. Rolla had grown much quieter due to the departure of thousands of soldiers, a sign that active campaigning had started in southwest Missouri.

Major General Sterling Price took the pro-southern Missouri State Guard into winter quarters in Springfield, Missouri, in late December 1861.[1] Meanwhile, Major General Henry W. Halleck appointed Brigadier General Samuel R. Curtis to the command of the Military District of Southwest Missouri, a new administrative entity, on Christmas Day.[2] Curtis swiftly left St. Louis and arrived in Rolla the next day to take command.[3] Angrily, Brigadier General Franz Sigel resigned in protest as he insisted that he, rather than Curtis, should be the new district commander. Halleck soon calmed him down, and Sigel retracted his resignation. Curtis and Sigel cooperated throughout the forthcoming campaign.[4]

In mid-January 1862, Halleck ordered Curtis to begin a march toward Springfield.[5] Curtis targeted Lebanon, sixty-three miles from Springfield, as the initial goal for his Army of the Southwest.[6] Traversing the primitive Telegraph Road between Rolla and Springfield in wintertime, and venturing across other difficult terrain, proved to be a slow, challenging process.[7] By early February, though, the army had arrived in Lebanon where Curtis took the opportunity to reorganize his 12,100-man force. He divided the army into four divisions. Sigel, the army's second in command, was assigned to command the First and Second Divisions. Bennett's regiment, the Thirty-Sixth Illinois

Infantry, was assigned to Colonel Peter J. Osterhaus's First Division.[8] After a rest, the Army of the Southwest pulled out of Lebanon on February 10, 1862. Their march caused Price to evacuate his Missouri State Guard from Springfield two days later, allowing Curtis's men to enter the town with no opposition on February 13.[9] Soon, Bennett would be swept into the rapidly developing campaign.

Sunday Feb 9 1862

As the day arrived that was to separate me from my family, I could not assume that cheerfulness which all along has attended me. Minnie by her lively prattle & play beguiled the sad hours that intervened to church time. It was in fact a solem[n] sabath, which not even the consolations of the bible or words of cheer of the minister could make less sollemn. Mr Goodhue offered a seat in his sleigh to take me to Joliet, which gave father time to attend church.

The hour for separation came at last. I dare not trust myself to say good by. My feelings were such that a word would have unmaned [unmanned] me. A presure of the hand, a kiss and a tear, were the only indication that a heart was bleeding, that an hour of most painful agony had arrived. O! when shall I see my wife—My child again! Perhaps never. Whether I live to return, or fall in battle I think I have the assurance that all will be well, and that I shall meet my loved ones in Heaven. The ride to friend Goodhues was in silence but in tears. Never did such agonizing feelings penetrate my very soul. Towards evening we started for Joliet, but one of the horses giving out we went to a neighbors and borrowed another and without another mishap reached Joliet about dark. The train, from Chicago did not past untill 2.A.M. and the lonly hours as they slowly passed was full of thought and reflections of the past and the future. At the time our Regiment left, and farewells exchanged my feelings were different from what they are now. 1000 happy, and I may say noisy boys were around me, brim full of patriotism with a glorious future spread out before them. The hardships & privations of a soldier's life was not thought of and all felt as if they were to have a few months play day and then return covered with glory. But after becoming acquainted with the realities of war, and then for a week enjoying the festive cheer and happy smiles of home, it is with the deepest sorrow & regret that I leave. O gladly would I quit the army could I do so, and like the repentant prodigal return to my home, and never be lured from it by the din and glitter of war. I can not hope for this untill Uncle Sam or death discharges me[.] Farewell my Mellie. Let not my absence trouble you or anguish at our separation wring bitter tears from your heart, as flows from mine now. Look to God for friendship and protection[.] He will be to thee more than an earthly friend[.]

Monday Feb 10 1862

The train for St Louis was behind time and it was two oclock in the morning before I found myself on the way for St Louis. I paid $4.90 for my passage and then doubled myself up on the seat for sleep. The jolting and constant din of the cars was not favorable for sleeping, and I cannot say that I derived any benefit from the trial. Dreams of home, filled the passing hour and flitted through my nervous brain[.] On waking in the morning, an aching head and weary body, did any thing but prepare me for enjoying the ride, or the rapidly changing scenery through which we passed[.] As we approached St Louis, the sun came out warm[,] the snow disappeared, and winter seemed to exist only in name. What a contrast between the ice and snow of Chicago and the mud of St Louis, and how short a step, from lowering winter to smiling sun shine. Reached St Louis about 3 P.M. proceeded to head quarters and reported myself for duty and was discharged from engineering and ordered to my Regiment[.] Found Frank Briggs, and passed the night with him at his quarters. The din, bustle and muddy streets of St Louis look as natural as ever, and perhaps more so[.]

Tuesday Feb 11 1862

Awoke refreshed after a good night's rest. Which was scarcely ever needed more. Ten days of constant excitement and deprivation of sleep, and then a long tedious ride on the Rail Road, was enough to render rest and sleep acceptable. Found a number of boys from the 13 [Illinois] just from Rolla with prisoners, among which was Ketchum[10] from Bristol a former acquaintance[.] From him I learned many items of news, in regard to what had transpired during my absence.

Visited my friends, the Miss's Biddle at their home, and had a very pleasant time, returning I went to the hospital—then to a gallary and spent $1.00 for photographs, then to the pay master's office—thence to the Q.M. department—then the engineer office, then the St Charles, and other places which now I can not remember. Altogether it has been a busy day with me, and in addition to all this I have written a letter to Melissa & the Tribune[.][11]

Wednesday Feb 12

Was up early, and bidding good by to Briggs proceeded to the soldiers Resturant and ate a hearty breakfast. The Photographs for which I sat yesterday

One of the "spoiled photographs" of Lyman G. Bennett taken in February 1862.

were a portion of them spoiled[.] I had no time to wait for others but sent what were finished to Melissa, the artist promising to send the remander to Rolla. Found John Martin at the depot to give me a brother's farewell. A large crowd of soldiers were there for transportation to the various camps in the West, not a few of which were for Rolla[.] I found three of the Wet Glaze prisoners at the depot who had been discharged from sickness.

One could scarcely walk, and I trust his experience as a prisoner at Ft Wyman and at St Louis Arsenal will forever deter him from again taking up arms against his country. It was a lovely day more like april than Feb the softness of the air and of the surface of the earth so nearly corresponded as afforded no room for contrast. Mud seems to be the great staple of Missouri at this time. The train was not a large one, and yet some of the grades were so steep as to bring the engine to a dead stop, and at times when doing down hill the speed was break neck double quick to the discomfort [and] alarm of women and passengers generaly[.] Reached Rolla about dusk—the platform at the depot being thronged with soldiers and citizens for the news. The desire of the soldiers for news, can hardly be imagined by the sober stay to home people—Almost the first one I saw was James Roseman, a comrade at the fort. The 13" Regt had removed to their first camp, and the one ocupied by the 36" on the day of my arrival and I was not slow in finding my old friends among them. Friend Dickey[12] of Co I 36 Regt was at the hospital with sore eyes which caused him to suffer severely. Lieut [William] Walker of the same Co was at Mr Secores and I accepted his offer to lodge with him[.]

Thursday Feb 13

Rolla of February and Rolla of December are two entirely different towns. Then 18,000 soldiers were encamped in its environs and at all times of day and night. The tramp of horse and foot, the rattle of sabres and the word of command[,] the gleam of muskets[,] the soft sunshine—the rattle of cumbrous army wagons over the hard dry roads, with the constant activity and bustle of a city, would almost delude one into the idea that he was in Chicago or St Louis instead of "poor miserable Rolla" as the boys termed it. To day but one camp is visible that of the 13" Ill. which is destined to remain a fixture here in glorious inactivity. A patrol & a sentry here and there, a few officers around head quarters and a team of mules floundering through the mud are about the only signs of active life visible. Not a spade or a pick is being lifted at the half finished Fort which solitary and alone looms up on the distant hill. The arrival of a courier, from the advance or the arrival of a train from St Louis and an eagre rush for news are about the only event[s] which break in upon the dull almost chilling monotony of camp life here. About noon in company with Geo Walker I went to Ft Wyman, and took dinner with Friend Roseman scarcely a stroke of work has been done upon the fortifications for two months. Notwithstanding a vast amount of amunition in store in the block house of powder and arms, there is no guard stationed there. The guns can not be worked and in case of an attack, the fortifications would be entirely useless. It has never been doubted that secessionists are constantly lurking in and around the town, and it would be the easiest matter in the world to attach a fuze to a keg of powder and blow the whole thing into kingdom come. Why an efficient guard is not stationed there is a mystery to the uninitiated. I am told that but a few days ago a secessionist entered a house but five miles from here, robbed the inmates *of a gun and a trunk in which was quite a savings* of money. We have news that our army left Lebanon, for Springfield last monday,[13] and that 20 or more of Price's pickets have been captured and are now on their way here. And it is believed that to day or tomorrow, if Price will make a stand, the most bloody battle of the war will be fought. Price has shown himself a cunning and able commander and will never come to a fight unless the advantages are decidedly in his favor. I know something of the troops sent against him, and believe if a battle is fought it will be most bloody and desperate[.] While at head quarters in the evening, news of the completion of the telegraph to Lebanon was received thus bringing us almost within stone's throw of news from the army[.]

Friday Feb 14

I have written nearly all the time to day wether cold and a shelter within the tent decidedly agreeable. The boys in the 13 are decididly clever & were I to stay a month would undoubtedly be welcome[.] Looked for transportation to the Regiment but finding none that suited me concluded to remain untill the paymaster comes back to get my little Q M pay[.]

Saturday Feb 15

Friend Dickey's eyes are very bad and he has concluded to go home on a furlough and desired me to go to the Fort and get vouchers for his extra pay[.] Accordingly passed the day at the Fort except a few hours in the afternoon when Long & I went hunting saw a few quails & prairie chickens & shot a rabit—Drew a sketch of the Fort from the South looking towards Rolla—Mr [Isaac N.] Buck has received orders to resume the post of quartermaster for the 36 Regt and will proceed on monday and has asked me to accompany him Which I have concluded to do—Played checkers with James Roseman untill near midnight, coming out a little ahead.

11 | To Springfield

February 16, 1862–February 22, 1862

I noticed the blackened ruins of a large house near the crossing of the Piney, a sad comment comment [sic] upon the evils of war.

—Lyman G. Bennett, February 17, 1862

Upon his return to Rolla, Bennett decided to travel to Springfield so he might join the Army of the Southwest's encampment there. Within an hour's time, Bennett left with his "knapsack[,] gun & haversack" and hurried to travel with a small group of fellow soldiers. Over the course of six days, Bennett wrote excellent descriptions of the route's roughness, devastation wrought by soldiers, and the struggles of civilians still in the area. On February 22, 1862, Bennett arrived in Springfield, evacuated by the Missouri State Guard ten days before, and bearing "the dread effects of war."[1]

Sunday Feb 16

Took breakfast at the fort helping to finish the rabbit we shot yesterday, and then proceded to camp, and made out a set of vouchers for friend [Orrin] Dickey. Going to the depot I saw 18 prisoners guarded and conducted to the cars to be taken to St Louis. Thus squad after squad of secesh are taken and confined[.] The whole batch of Missouri secesh should be provided with comfortable & secure quarters then [?] a quietus can be put upon their maraudering and mischief. At the marshall's office I found the paymaster had proceeded to Springfield and would not be back for a long time. I also found a Lieut named Black going to Springfield as rapidly as rapidly as he could push through. Thinking this a good oportunity I resolved to go with him and hurried to camp to get ready. Geo Walker prepared dinner for me, and filled my haversack with bread and meat. It required an hour to get ready, and when I reached the Rolla house the Lieut had left a half hour before. My knapsack[,] gun & haversack made a large load for me but I hurried after them as fast as the road & my load would would admit, but it was seven miles before I overtook them, and was glad enough to get rid of my load. The two teams in the company were poor & weak & could scarcely draw their loads, let alone carrying the boys so I walked with the others, getting on the wagon only to cross the Little Piney a large & crooked stream. Once or twice the horse team was "stalled" as the Missourians call it and the mule team was put on ahead to help us out and by dint of lifting & prying we succeeded in getting out of our difficulties[.] Our whole party numbers 15 or 20. All on foot except once in awhile one that was sick or lame would ride a short distance—Thus we went on through mud holes, across creeks, over ridges—rough roads &c untill 10 P.M. The night was so dark that it was difficult following the track and as a consequence our progress was slow and difficult. One of our number and the owner of the teams had a quantity of liquor with him and was so drunk he could not help himself and once while riding over a rough road he was dumped into the mud, hurting a hand. One fellow wished it had been his neck, and to put a stop to an occurrance of the same thing emptied all the whiskey he could find. We made 20 miles and at 10 reached a log cabin and laid ourselves upon the floor to sleep. My back ached some from walking, but the hard floor made it ache more, and what little sleep I got did not seem to refresh me.

Monday Feb 17 1862
I took the lead as a sort of pilot for the squad in the night and was first to make the discovery of the cabin where we staid. Going in I asked if any corn could be got there for our jaded horses. A long hared puke[2] "reckoned not for they hadent got no corn and did not raise but a little sprinkle of a crop." When I told him that if there was an ear of corn about we would have it, he mellowed down a little and going to the door and thrusting his head out into the darkness, shouted "Ho pap!" Ho pap!" In a short time "Ho pap!" came from a neighboring cabin, and after a consultation with his interesting son the conclusion was arrived at that we could sleep on the floor & that a few ears of corn could be had.—————[hand drawn line]

 Started shortly after sun rise and in about an hour came to the Big Piney a broad rapid stream, and about three feet deep[.] The mule team crossed without any difficulty, but the old dilapidated horse "team "stalled" in the middle of the stream and for half an hour the wagon remained a fixture there like a small inhabited island. Finaly the mules were sent to the rescue, but it required a little management to attach them to the wagon without getting into the water.

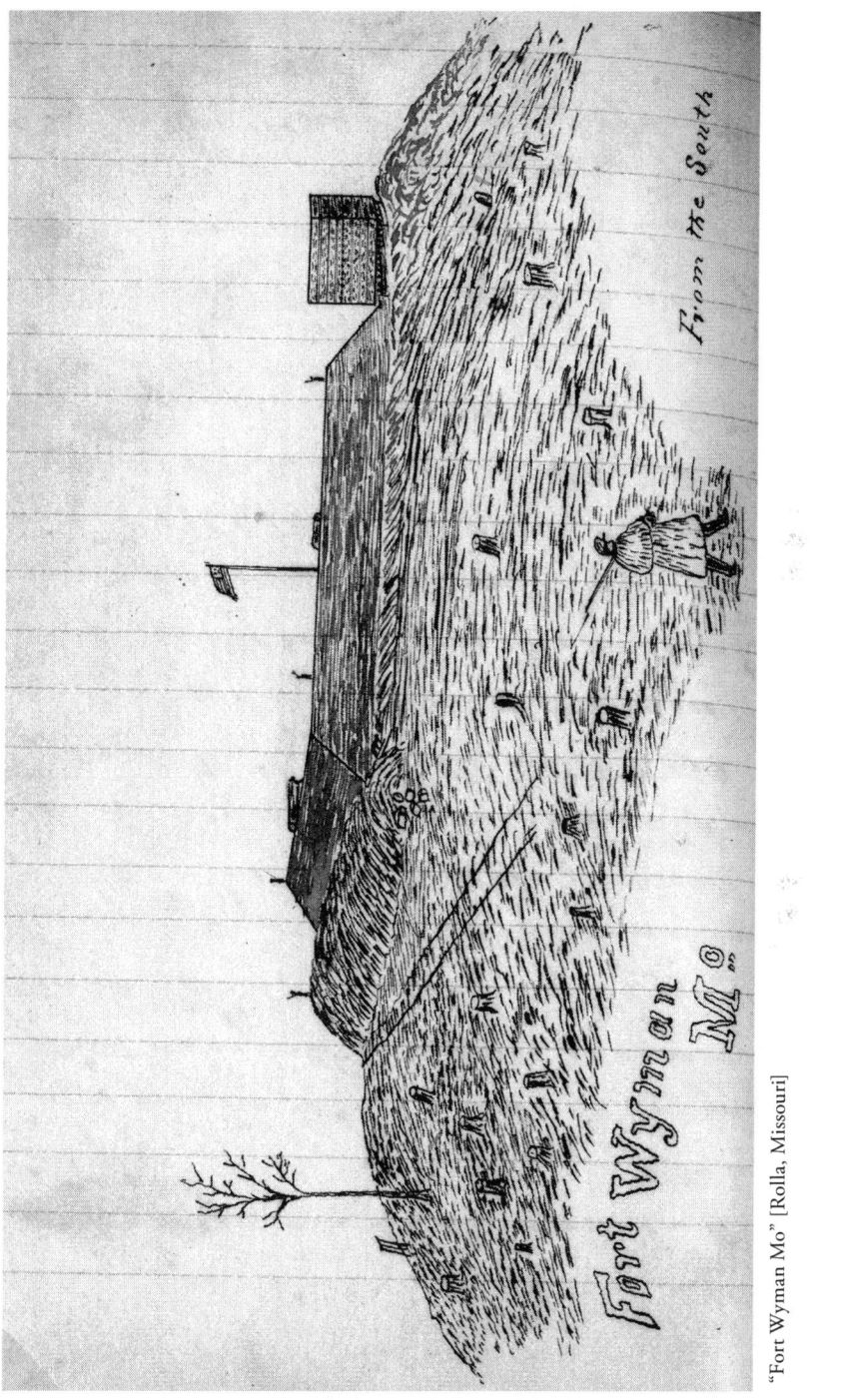

"Fort Wyman Mo" [Rolla, Missouri]

Shortly after getting over We came across a flock of wild turkeys, but not near enough to get a shot. These were the first I ever saw[.] I am informed there are large quantities in the country. About 9 A.M. we came to a large house, and a good one for Missouri where we took breakfast and rested ourselves & teams. Our road from thence to Waynsville was over a ridge and about as crooked as a rail fence. It was all timbered and for miles not a house appeared. I noticed the blackened ruins of a large house near the crossing of the Piney, a sad comment comment upon the evils of war. If this was the doings of rebels or not I can not tell. Waynsville is a miserable apology for a village the county seat of Pulaski and on a fine stream called the Reubedoux creek. This is now being bridged[.] About 25 of Wood's battalion[3] are there as a sort of garrison. The inhabitants have all left except 7 families a sorry looking place it is. From Waynsville on the country is not so rough, and looks more like living but that too is unsettled[.] We stopped six miles west of Waynsville at the California house, which is a fine mansion beside of Missouri houses in general. One characteristic of nearly all the log cabins on the road, they are without windows and the door must be opened [in] order to admit light. The cracks between the logs are generaly window enough for all practical purposes. Slept on the floor as last night[.]

Tuesday Feb 18
Were on our way at sun rise. The air was keen, and we were obliged to walk briskly to keep warm. The wagons were to[o] slow for me, and I hurried on, and reached the Gasconade and crossed it an hour before the other boys came up stopping at a large house untill their arrival. We stopped to feed at a log house [illegible] mile from the river, the owner of which was secesh to the core, and did not hesitate much to make it known. He was an old man or we would [have] left in such a manner as to give him to understand that secesh doctrin[e] was not good policy for him to advocate. The country from the Gasconade to Lebanon was a paradise beside that about Rolla, and more thickly settled. The fields were large, well cultivated & fenced and the buildings though universaly of logs were far better than the shake roofed, crazy concerns we passed the first part of our march. One of my comrad[e]s & myself left the wagons and almost double quicked it to Lebanon which place we reached before dark. I found 7 of the boys of Co E in town and proceeded to their quarters and was heartily welcomed by them. Dave [G.] Cromwell, was commander of the 7 and from appearances, wore his honors with becoming gravity. A hearty supper, a good *family* visit, and a good camp bed was provided, and I passed a pleasant night with the boys.

Wednesday Feb 19

Found a "smart sprinkle" of snow on the ground and the weather decidedly squally[.] Passed the day in reconoitering the town chatting with the boys and writing a duo decimo letter to Melissa. I have been so anxious to press forward, and so elated with the idea of meeting the boys that I have not thought of home, wife and baby more than half the time since leaving Rolla. The varied scenery where I have passed has served to divert my attention and mind from everything but what I saw. Three of the boys of Co E are detailed to run a steam grist mill,[4] which is kept running night and day manufacturing flour for our armey. There are large quantities of unthreshed wheat in the country and a number of machines are running by the boys and thrashed out and brought to mill & ground. The wheat of union men is paid for but that of secesh is not paid for. A can of oysters was raised among us and in the evening two of my traveling companions (Lieuts Black & McBride) invited in and supped with us. Early in the evening James Hatch[5] came into the tent lugging a box containing 100 pounds of tobacco, which he had "cramped" from our drunken sutler's wagon. His wagon and goods were left in the street unprotected and it is a wonder his whole lot of goods were not stole[.] About midnight another of the boys brought in a 24 pound candy box. I was astonished at such barefaced thieving by boys whom I had no idea would be guilty of such things[.] The tobacco is worth $50 and the candy $8 or $10. Had I expected [?] them it would have been perfectly right[.] J. [Isaac] N. Buck arrived at Lebanon to day.

Thursday Feb 20

Found a train of 7 wagons, and about 200 men about to start for Springfield and soon made an arrangement to have my knapsack carried, and I could run my chances on foot. Bidding good bye to the boys, I started with the train about 9 A.M. For an hour we had good walking, but the sun soon thawed the snow & frozen ground untill it became slipery & not easy traveling. A short distance from Lebanon we crossed a prairie 2 or 3 miles in width. It was uncultivated except now & then a small patch. Why it is that the best land is the least improved I can not tell. The country was generaly rolling but not rocky and ragged. We traveled 17 miles & camped in a field near a spring and close to the timber. A party of Phelps' men built a large fire in the bushes & wrapping up in their blankets lay in the open air. I lay down with them & for all I know slept as soundly & rested as well as if I had been in a tent. When one awoke he would replenish the fire & we passed the night very comfortable.

Friday Feb 21

Started earlier than yesterday & marched 21 or 22 miles. Met a train of more than 100 wagons, on their way to Rolla for supplies. Day was cloudy but the roads good. Passed a few small prairies[.] The houses mostly deserted. In many places the fences were burned and rails scattered about showing that the demon of destruction had for a while reveled there. Orchards destroyed and the charred ruins of buildings told of fire and sword. Camped on the banks of a small stream a sketch of which I give on the next page. Slept in a tent but no more comfortable than the night before in the open air[.]

Saturday Feb 22

I started shortly after daylight ahead of the teams. The morning was dark & foggy, dampning the ground & making it so slipery that it was difficult traveling. The train did not overtake me untill 1 oclock. I saw the place where Price's first pickets were encountered at the forks of two roads. The secesh rapidly retired to within five miles of Springfield, where being reinforced to about 3000 they made a stand. The ground was rolling prairie, now a cornfield, an orchard and a house. Sigle's advance of cavalry charged upon and dispersed them.[6] Tis said that Gen Curtis ordered several howitzers to open upon them which Price heard in Springfield and immediately commenced his retreat. The rebels had a large supply prepared and were about to distribute a large amount

[the next page consists of an unlabeled drawing of a camp scene]

of clothing & had no idea of leaving untill morning but the roar of artilery hurried his flight, and his supplies remained uneaten. Tis said that if Curtis had not opened fire with heavy guns Price would in a measure been surprised. At any rate his escape would not have been so easily made[.] A part of the prairie was low & muddy and difficult to travel. It is a wonder to me that the heavy guns were hauled through the mud. We reached Springfield about 3 P.M. The place was thronged with soldiers who made free with the property of secesh as if it was theirs[.] Most of the houses were empty & many were ocupied by the soldiers[.] I soon found some of the boys of the 36 and was at home again[.] Springfield is scattered over a large area and save around the public square but few of the buildings are compact[.] The Court house, a large church and a large brick hotel are used as hospitals for the sick of Price's army, who number about 400. A green or yellow stripe sewed on their sleeve is the only indication of uniform or marks of secession soldiers.[7] Otherwise they are the same miserable butternuts I have seen all over the country. Many buildings, many of them large and elegant have been burned.—some say by our soldiers and some say by the owners themselves who have fled with Price to the South. Whoever may be the guilty party, the act is heathenish, and touches of the barbarian. The large stone chimneys common to the country generaly remain standing. No place that I have seen tells so fearfuly of the dread effects of war as Springfield and its environs, and no picture however vividly drawn, has ever conveyed to my mind its sadning realities, so clearly as the desolation I have seen with my own eyes[.]

12 | From Springfield to Pea Ridge

February 23, 1862–March 5, 1862

> *We passed through Keitsville the scene of the fight a few nights ago: The houses, particularly one where most of the men were quart[er]ed was completely riddled with balls some even passing through three or four ceilings. It was suggested to use the house as a pepper box.*
>
> —Lyman G. Bennett, March 3, 1862

On February 14, 1862, the Army of the Southwest left Springfield to pursue Major General Sterling Price's Missouri State Guardsmen.[1] Later that day, Federal artillery bombarded Price's rearguard along Crane Creek, thirty miles southwest of Springfield, much to the displeasure of Brigadier General Samuel R. Curtis, who desired a stealthy advance in an attempt to trap the enemy.[2] Alarmed by the bombardment that he mistakenly thought indicated large enemy numbers, Price ordered a hasty retreat.[3] It became so rapid an exit that Price's men covered the nearly fifty-mile distance between Crane Creek and Little Sugar Creek in thirty-six hours.[4] A sharp skirmish between Price's rearguard and Federal cavalry occurred at Cross Timber Hollow, just one mile from the Arkansas border, on February 16.[5] The two sides continued to clash with another skirmish occurring the following day, this time further south at Little Sugar Creek.[6] Curtis assumed that the enemy had decided to hold Cross Hollow still further south on Telegraph Road and sent out troops to outflank the position.[7] Confederate Brigadier General Benjamin McCulloch, however, had persuaded Price to give up a position that he regarded as a trap.[8] Instead, Confederate troops trudged south until they ended their retreat "at Strickler's Station, seventeen miles south of Fayetteville."[9] Curtis then went on the defensive and stationed his 10,500-man army near the Little Sugar Creek Valley.[10] By early March, detachments operated grist mills in the area, and others conducted raids, one of which resulted in a three-day occupation of Fayetteville.[11] Forage, though, grew short, and on March 3, Curtis ordered a concentration of his army at Little Sugar Creek.

Bennett reached Springfield on February 22 and the next day trailed behind the advancing army, commenting on devastation along the route, observing the sites of skirmishes, and visiting the Wilson's Creek battlefield where he drew a map of burials and collected two flint stones as souvenirs. Like much of the army, Corporal Bennett traveled on Telegraph Road, also referred to as the Wire Road or the Military Road.[12] In 1858, John Butterfield had selected Telegraph Road as a part of his Butterfield Overland Mail route that stretched between St. Louis and San Francisco.[13] His selection stimulated business enterprises along the road, and in 1860 workers installed a telegraph line. In the summer of 1861, however, the line was removed south of Springfield after Federal troops occupied the town.[14] Although significant, Telegraph Road remained unimproved and difficult for armies to traverse.[15]

As the Army of the Southwest advanced, Bennett participated in an expedition from Cassville, Missouri, to Newtonia, about twenty-eight miles to the northeast, so that he might collect supply wagons loaded with flour and salt left behind by Price's men. Bennett proudly commanded twenty-five men and ten wagons during this journey. In his February 28 entry, Lyman mentioned writing to the *Chicago Tribune*; the newspaper published his letter on March 18 with the byline "From our Special Correspondent." Although unsigned by Bennett, he obviously wrote the *Tribune* article as some of its observations matched those in his diary entries. After the Newtonia expedition, Bennett caught up to the army at Little Sugar Creek on March 3 and then walked twelve miles further south to Cross Hollow the following day so that Colonel Grenville M. Dodge could sign his pay vouchers. On March 5, Colonel Dodge introduced Bennett to Curtis and recommended that he be detailed as a topographical engineer. Curtis only had one engineer officer on his staff, Lieutenant Arnold Hoeppner, and he had no detailed map of northwest Arkansas.[16] Bennett, however, wished to rejoin his regiment to "pass a couple of days," and Curtis agreed to detail him to his staff after that time. This turned out to be a fateful decision. Bennett made his way to the Thirty-Sixth Illinois Infantry that same day, one day before the regiment engaged in a skirmish near Bentonville. Bennett followed that up with combat at the Battle of Pea Ridge.

Sunday Feb 23

A portion of the train were ready for a start early in the morning to convey provision[s] to the army and as I had decided to move on with it, I had but little time to write to Melissa. We went about 3 miles on the Wilson Creek road and halted untill [it] was the middle of the afternoon to await the arrival of the remander of the train. We passed several houses that had been burned I am sorry to say by our own troops. Among them was a large brick mansion belonging I am told to a union man, who was driven away last summer by the rebels and is now at Rolla. Such proceedings is a burning shame, and officers allowing it should be cashiered and discharged from the service. While the teams were waiting, a small party of us went to the battle ground at Wilsons Creek. On the way we came across an old darkey who seemed overjoyed to see federal soldiers. His account of Price's retreat from Springfield was rather ludicrous. A party of secesh secreted themselves in the brush near his master's house for the purpose as they expressed it of giving the "dam[n]ed yankees hell." At length a party of our cavalry appeared in the distance and the leader shouted "Here they are boys trying to surround us—run for your lives" "and sich a skeedadling fur de brush, I'll be burn fired if I ebber seed in all my dog on days" were the old darkey's expression. Some lost their hats and never stopped to pick them up, but with long hair, coat tails & blankets streaming in the wind rushed pell mell to join the flying army. The old darkey told us of a secessionist living ¾ a mile away who had a large drove of horses & suggested that it was too bad for us to go on foot. The hint was sufficient for Jake Horne[17] and at once he was in for a horse & urged me to go along and steal one also. He had lost a horse by secesh & was bound to make it up in some way. Finding him bent on the project I went along, not to assist but to look on. We found the horses in a large corn field but after a chase of an hour, Jake was obliged to relinquish his project, not being able to catch one. We went in to Wilsons Creek and walked over the ground, where on the 10" day of last August one of the most terific and destructive battles ever known on the continent occurred and which will occupy a conspicuous page in American history. I found graves thickly scattered along the bank of the creek for a mile or more, and in some places I was told scores were buried in one common trench or grave[.] A sort of cave upon the hill was filled with the bodies of the dead and lightly covered with logs & dirt.[18] The trees in many places were scored with bullets and many entirely cut down. One oak tree about 18 inches in diameter was entirely perforated by a cannon ball, going completely through its center. The white bones of horses and their tough hides lay scattered all over the battle field. Old shoes, hats, and other dilapidated portions of wearing apparell was strewed all over the field. The small ravines entering the valley of the stream and the woody slopes of the hill sides were admirably adapted for concealing the approach of the enemy, while our forces were exposed on the tops of the hills where the trees were few and the land open. And notwithstanding these advantages as well as an overwhelming superiority of numbers, the enemy was driven about a mile, the ground covered with their dead & drenched with their blood, and were it not for the fall of ~~Sigel~~ Lyon, the enemy would have been driven into an ignominious retreat[.] Indeed a lieutenant, now a prisoner in our hands has told me that McCulluch advised a retreat, and a few more rounds from our cannon, and rigorous blows from the infantry would have sent them hurrying over the hills. The ground has been thoroughly searched for bullets and other mementoes of the fight, the balls have been hacked from the trees and I could find nothing to take from the field but two flesh colored flint stones which I will preserve to remember Wilsons Creek & remind me of the martyred Lyon. The train did not arrive untill sundown & it was late before the teams were cared for and our supper eaten. Dan Whitney[19] and myself crawled into a wagon but nearly froze and could sleep but little. At three oclock we were up, the teams fed and everything ready for a start at daylight[.]

Monday Feb 24

A long steep hill near the camping ground was the first obstacle to surmount and it was with much difficulty that many of the teams hauled their loads up the steep ascend. Our day's drive was up and down a succession of hills and vallies and through a rough, stony and barren country. I saw but few attractions to tempt intel[l]igent men to settle and cultivate the country. We passed by Dug Springs[20] where a battle was fought previous to that at Wilsons Creek. Fences were torn down, rails burned and very many houses deserted along the whole route, showing that here too the fierce tornado of war has raged. We camped near a large house occupied by secession women, the men being gone. A fine spring being near by affording an abundance of water for man and beast. During the day we passed the scene of a skirmish at Cranes Creek.[21] The advance of our army overtaking the rear of Price's retrating force. A few shells were plunged into thier frightened and disordered ranks, but beyond the wounding of a few, and hurrying their flight nothing was accomplished. Sigel, by a detour of several miles was getting in their front & cutting off their retreat, and would have had them in a trap had not Col Ellis[22] & others spanked them behind & hurried them off. My opinion is that Curtiss had better be sent home and Sigel assume the position rightly belonging to him & we will have fewer blunders to record.[23] Slept on the ground in the open air and a great deal more comfortable than last night.

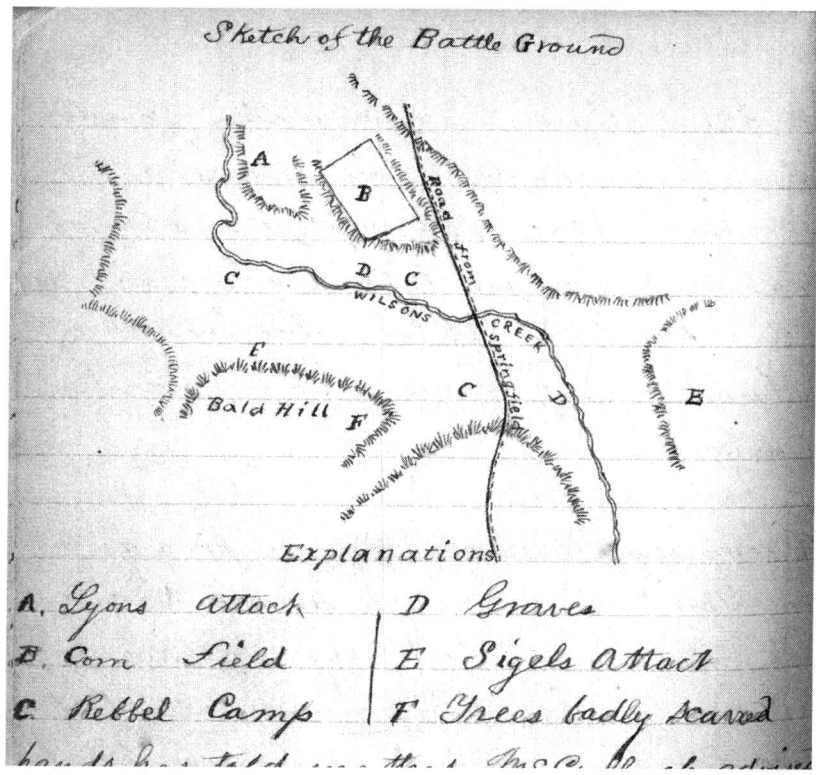

"Sketch of the Battle Ground"

Tuesday Feb 25

Our route most of the day was up the valley of Flat River or creek. The valley was broad, and the soil rich, and in many places good farms were in cultivation. Here too the flying secesh had been hurried by a few loud words from some mountain howitzers. We reached Cassville, the county seat of Barry County about two oclock. Lieut Col Holland[24] of Phelps' Mo Regt is in command and has about 100 men, a rather weak garrison in my opinion, to hold so important a post. The mill at Cassville had been set to work grinding for our army and was doing good business. I had the pleasure of meeting friend Harrison here who has charge of the Mills & commissary stores. Another mill at Gadfly 13 miles west has been occupied and set to work by the army. Col Holland told me of an expedition on foot to Newtonia 25 miles west of Cassville in Newton County where he had learned a large quantity of flour was in store for Price's army & that in his hurry to get away he had left it, and he wished me to stop and take charge of the expedition. He had that day ordered Lieut Moore[25] with a portion of the force at Gadfly to go to Newtonia or neighborhood, to ascertain the facts in the case. I concluded to go on the expedition and accordingly left the train, which moved on 8 miles to a little town called Keetsville and encamped. Here Capt Montgomery was stationed with his company of 80 or 90 men. I was much fatigued with the tiresome march from Rolla and retired early with Mr Harrison & was soon in a sound sleep. At midnight Col Holland waked us and whispered that the train had been captured at Keetsville and an attack threatened upon the town. We were up and formed in short meter and about 40 men mustered into line and marched to the out skirts of the town and formed in line of battle on both sides the road to give the enemy a warm reception in case they came upon us. We each examined our arms and with an almost breathless silence listened for any sound of an approaching enemy. Soon we caught the clatter of a wagon, which came more near, and finaly with horses upon a gallop drove into town, and then another & another untill the court house was completely surrounded with wagons, horses & mules. In about an hour the whole train but one wagon had arrived and been carelled [corralled] about the court house. Soon after came a squad of [Bacon] Montgomery's men, then Montgomery himself. Then another squad & still another mostly on foot, and some minus hats and shoes, and it was not untill after sunrise that the last came in. All was excitement and noise, and fears of an immediate attack prevented sleep, to the citizens and soldiers of Cassville that night. Each had his story of dareing & miraculous escape to tell & it was with difficulty I could get at the truth of the matter.

It appears that Captain Montgomery had no intimation or fears of an attack and consequently neglected to post proper guards around the town. had no pickets out whatever. The men had gone to sleep in the empty houses about town, when suddenly they were surrounded and fired upon by about 500 secesh, who had been lurking in the thickets among the hills to cut off the trains carrying provisions to the army. Before the men could dress and arm themselves, volley after volley was fired, crashing through the windows & into the buildings. Finaly their fire began to be returned and two or three of their number shot dead. They then ceized what horses they could and made put into the brush. Some of our men at the outset were badly scared and without hat coat or shoes, broke for the woods and in their attempts to escape two were killed and one wounded. Two of the enemy were found dead, and it is presumed that others were shot. Montgomery lost about 70 horses.[26] This surprise is keenly felt by the Captain and he swears a terrible revenge. The train was camped a mile from the town, and at the first alarm the teams were harnessed and pushed through to Cassville on double quick. One wagon was broke down on the road and was left, but recovered again in the morning. There was much anxiety felt at Cassville untill the saf[e]ty of the train was assured. This is not the only instance of cowardly bushwhacking in this vicinity. Last sunday one soldier and three home guards were returning from the Gadfly mills when they were fired upon from the brush and one instantly killed. The others were all wounded—one quite severely. The mother of the slaughtered home guard when she saw the body of her son, stiff and cold in death raised her tearful eyes to Heaven & thanked her God that he had died so glorious a death, in defense of the old flag of our country. She had before lost a husband, who was hanged for his devotion to his country and her only remaining son is a soldier in the army. How many firesides and hearts are made desolate by the cruel and fratricidal war begun without cause by the demagogues and plunderers of the South.

Wednesday Feb 26

Yesterday morning, on the strength of information brought him, that a large quantity of flour belonging to Price was in store at Newtonia, Col Holland sent a detachment of 60 men under Lieut Moore to ascertain the truth of the report and if so take possession of it for our army and hold it until teams could be sent to bring it away. A messenger came in the evening confirming the report and this morning 10 wagons and an escort of 25 men was placed under my command, with orders to push through to Newtonia as soon as possible, load the flour and return as soon as possible. It was thought by some to be impracticable to send so many men away, when the force at Cassville was so small and the danger of an attack so iminent. But in case of an attack the detachment at Newtonia was needed, and an order for their return could just as easily bring the flour as to leave it. Accordingly I set out with my wagons & escort about 10 A.M. Most of the way was among hills, thickly covered with brush and in a neighborhood notoriously disloyal[.] We drove briskly on, reaching Gadfly at 1 P.M. But few knew our destination and were surprised that a halt was not ordered at Gadfly. We kept a sharp lookout untill we reached the prairie about 7 miles from Newtonia and then pressed our way more leisurely reaching Newtonia before sun down, and loading the wagons for an early start. Newtonia is situated in the midst of a beautiful prairie and though small, is a neat and flourishing little village. A large grist mill belonging to Judge Ritchie[27] has been pressed and kept in operation by Price, and a large quantity of flour was in store for his army, but his departure from the country was so hurried that he had not time to attend to it. Threats had been made during the day that we could not hold the place during the night, so a good guard was kept on the lookout. A high stone wall surrounding the barn yard,[28] within which was our teams, wagons & camp was a good protection against an attack. A dozen other teams were pressed in the neighborhood, and in day light the whole, comprising a train of 30 wagons was loaded and under way to Cassville, which was reached in safety by most of the teams before night[.] A few of the teams and horses rather heavily loaded had remained at Gadfly and will come on tomorrow. I am much pleased with the success of this my first command and first expedition. We also made the discovery of a large quantity of salt and bacon, also in store for the rebels. We had not force enough to bring it all away, but presume another visit will be made to the neighborhood.

Thursday Feb 27

I have jumbled yesterday and today together and given the events of both in one. I will say however that upon returning to Cassville I found Major Wright's battallion[29] there and reinforcements from Springfield sufficient for holding the place. The train had also left under a strong escort from the main army[.]

Friday Feb 28

The result of our Newtonia expedition is the capture of 451 sacks of flour and 2100 pounds of salt, and the discovery of of nearly as much more.

Today that portion of Phelps' Regiment here was mustered for pay. A sutler passing down to the army, sold a large quantity of liquor to the soldiers and a general drunk was the consequence. Citizens were insulted and in some instances men were drag[g]ed from their houses, and threatened to be shot or hung, but Col Holland soon put a quietus upon this & guarded most of the houses. If the sutler was mob[b]ed and his goods destroyed I would say amen. It is amusing to see the Home guards. Most of them are long gaunt six footers mounted on the smallest kind of ponies, scarcely tall enough to keep their riders feet from the ground. Their long rifles when shouldered look like liberty poles[.] Tis not uncommon to see one with a short stub of a gun, that when carried at a support will scarcely reach to the top of the seedy & dilapidated stove pipe, worn in place of a hat. Some hats are brimless, and some almos[t] crownless. Again will be seen a 200 or 300 pounder seated on a little filley that reels under its burden of man flesh and then a boy or pigmy man will be found perched on the back of a large rawboned horse. Falstaff's army of ragmuffins is not to be compar[ed] with them. When scouring through the country with seedy butternut coats, long hair and blankets streaming in the wind tis enough to make a horse laugh.

Wrote to the Chicago Tribune and passed the day in writing.

Saturday March 1

A sutler has been stopping here for a few days, awaiting a train or an escort to go on. They have opened shop around their most common articles of [which] is whiskey. Mongomery's men feeling sore over their snubbing of Wednesday night, dround their chagrin in liquor and to day they are drunk and noisy[.] Three or four went to Dr Rodeboy's house and took the Dr into the brush for the purpose of shooting him. Col Holland getting wind of what was going on, promptly rescued him, and took him to the Courthouse and detailed a strong guard to protect him. A guard was also placed at the Dr's house. I found Mrs R crying, for she believed her husband dead, but when assured that all was well she dried her tears, and expressed herself satisfied with Col Holland's course. One of the sutlers, (the same whose tobacco was missing at Lebanon) was ordered outside the lines. It has been a most beautiful day, clear and warm almost like summer. I wonder if it be so pleasant at Oswego and if bright skies, bright hopes and joys makes my dear Melissa glad. Have written a long letter to her.

Sunday March 2nd

Our weather has taken a sudden cold. During the night the wind veered to the north, and a slight fall of rain & sleet made it quite cold. I passed the day in reading & writing and ventured but little from the fire. A train arrived from Springfield but will not be permitted to go on till morning and then I will go on with it.

Monday March 3d

The train did not get away from Cassville untill 8 A.M. and its progress was not rapid making only about 20 miles to Sugar Creek. Here Gen Davis' division[30] was encamped and the sight of a large body of men and tents and the sound of drum & fife revived the scenes about Rolla in December. We passed through Keitsville the scene of the fight a few nights ago: The houses, particularly one where most of the men were quart[er]ed was completely riddled with balls some even passing through three or four ceilings. It was suggested to use the house as a pepper box. Country broken & rocky. The night found me several miles in Arkansas or Dixie. I found quarter Master Carr at Sugar Creek. Slept out doors in the open air, and got my head and blankets wet from a slight fall of rain.

Tuesday March 4

Friend Edwards[31] & I walked to Cross Hollow 12 miles from Sugar Creek which place was reached at 12 M. Passed along the place where a skirmish with Price occurred a few days ago. The road was strewed with dead horses. In all about 20 or 30 mostly our own. Our loss was 14 men killed and 7 wound[ed].[32] Secesh not known. Among our wounded was Major Bowen[33] & Captain Switzler.[34] The secesh after a few rounds fled. Col Dodge signed my vouchers for extra pay & I staid with him during the afternoon & day. Col Dodge is a gentleman & a man.

Wednesday March 5

In the morning I took the Col's compass and located the camp of the Iowa 4" to agree with the meridian. The Col went to Genl Curtis's head quarters and gave me an introduction & a reccommend to be detailed as topographical engineer which was done with the condition however that I might go the Regt and pass a couple of days, and in accordance with this agreement I made preparations for going to the 36". The Gen advised me to wait a day or two because they would be at sugar creek and the country in that direction was more secure than about Bentonville. But scouting the idea of danger, Edwards & myself set out for the trip on foot and unaccompaneyed by others.

I think Cross hollows by nature is strongly fortified. A deep ravine intersects the Fayetteville road at right angles and several branches runs into this ravine from both sides, and there are other smaller ravines entering these branches so that the whole country is but a succession of deep, narrow, rocky and crooked ravines and ridges between them. A single cannon at the head of a ravine would do terrible execution among an advancing foe and an escape by the flank would be next to impossible from the precipitate nature of the sides. Besides the ground is covered with trees with but little under brush, so that guns would have fair sweep & a wide range. If Price had the force reported I cannot see why he did not make a stand here.

The road to Bentonville was crooked and blind through the woods with scarce a house on the whole route. We saw scarcely any one on the road except here & there a butternut at his own cabin. Osage springs where Sigel was two days before is a fine camping ground, with plenty of wood & several very fine springs. At Bentonville we tried at several houses to get something to eat, but every one said they had nothing for themselves, that between the two armies they were eaten out of everything they had. A small piece of cold corn bread was at length procured which in a small degree satisfied our craving hunger. Many of the best houses in Bentonville have been burned by our men (the Benton Huzzas)[.][35] I was told that a detachment was placed there and one of their number became some what intoxicated, and when they were ready to leave & were mounted, he returned declaring he would have another drink before he left. His companions went on for a short distance & halted, but he not coming up they went back & instituted a search & finally found his mutilated remains in a privey. He had been killed it is supposed by an ax. In retaliation for this outrage another was commit[ted] by burning the business part of the town.[36] The remains of valueable furnature such as a piano, was not pleasant to behold. Disguise it as you may, war is but a system of atrocities and barbarities. We found the camp of the 36 about 4 P.M. & was glad to get food and rest. Boys all glad to see me and eager for news[.]

13 | The Pea Ridge Campaign

March 6, 1862–April 4, 1862

I have now mingled in the dread scenes and seen the sad sights of the battle field and I must confess that I have no particular relish for its frequent repetition. Heaven grant the speedy restoration of peace, and a deliverance from such scenes again.

—Lyman G. Bennett, March 8, 1862

Corporal Bennett recounted his first combat experience in this set of diary entries, probably one of the longest contemporary accounts about the fighting at Pea Ridge by a soldier in the ranks. Through his narrative, the reader senses the shock of combat experienced by Bennett as he described in detail the sounds, confusion, hunger pangs, and graphic descriptions of the dead. In the nearly month-long stretch after the battle, Bennett joined Brigadier General Samuel R. Curtis's staff and engaged in surveying duties while further supplementing his standard military pay by drawing maps of the battlefield for various officers.

Brigadier General Franz Sigel had overall command of the First and Second Divisions of Curtis's Army of the Southwest. Colonel Peter J. Osterhaus commanded the First Division, and its Second Brigade, led by Colonel Nicholas Greusel, comprised the Twelfth Missouri Infantry and Bennett's own Thirty-Sixth Illinois.[1] The Thirty-Sixth Illinois's Company F missed the battle as it was sent with detachments from several other regiments to Maysville, Arkansas, on a mission that resulted in much fruitless marching.[2] Major General Earl Van Dorn sparked the Battle of Pea Ridge when he ordered his army to march around the Federal army; this required Curtis to turn his army completely around to meet this threat from the rear.

For the Thirty-Sixth Illinois, the action started on March 6 near Bentonville, Arkansas. There, Sigel commanded a six hundred-man detachment that supposedly protected his supply train several miles to the east.[3] Confederates under the command of Brigadier General

James M. McIntosh attacked this outlying group, but all escaped except for some sixty men of the Thirty-Sixth that guarded an ammunition wagon. Not long afterwards, reinforcements rescued about thirty-five of these soldiers.[4] This small rearguard action resulted in no serious injuries among Union soldiers.[5] Much more significant combat awaited them the next day.

In the late morning of March 7, Osterhaus led his division north through the village of Leetown. On a nearly parallel route, two divisions advanced further east along Telegraph Road toward Elkhorn Tavern.[6] After marching north of Leetown for a short distance, the Thirty-Sixth, along with other units, was directed to a position behind a rail fence that bordered the south side of Oberson's Field.[7] A belt of trees ran along the north side of the field that soon filled with Union cavalrymen and artillerymen driven away from their position by a much larger Confederate force. Companies B and G of the Thirty-Sixth cautiously advanced as skirmishers toward the tree line where in quick succession they killed Major General Benjamin McCulloch, in command of Confederate troops at Leetown, and then his second in command, Brigadier General James McIntosh.[8] The rest of the Thirty-Sixth advanced when the two companies engaged in combat against the Sixteenth Arkansas Infantry and then the Second Arkansas Mounted Rifles.[9] Although unaware of the deaths of his superiors, Colonel Louis Hébert was now the Confederate commander at Leetown and led his men in an assault through Morgan's Woods that bordered the east side of Oberson's Field. The Thirty-Sixth Illinois redeployed so that it faced east to meet this advance. Although they fired toward the woods, they played no significant role in repulsing Hébert's men.[10] The fighting was over at Leetown, and Union soldiers then moved to the northeast and camped near Ford's Farm.[11]

The next morning, Sigel's soldiers shifted eastward and then marched via Telegraph Road to connect with the other half of the Union army that had experienced heavy combat near Elkhorn Tavern the previous day.[12] Sigel's men connected to the left of the line with the Thirty-Sixth Illinois positioned near the far-left flank.[13] Confederate artillery fire killed a handful of men in Sigel's division that morning, including one soldier in Bennett's Company E.[14] Sigel had his finest day as a commander as he skillfully positioned artillery batteries near the high ground of Welfley's Knoll.[15] For two hours, soldiers experienced "the most intense sustained artillery barrage ever to take place on the North American continent up to that time." When artillery fire drove Confederate soldiers further back into cover, Sigel ordered his men to redeploy.[16] The Thirty-Sixth Illinois advanced "at the double-quick" as part of this turning movement, and instead of facing north, they now faced nearly east on Ford Road.[17] Skirmishers from the regiment fought briefly with the First Arkansas Mounted Rifles near "the base of the rocky promontory."[18] The target of intense artillery fire, this rocky hill was the scene of heavy Confederate casualties.[19] Up the steep, rocky hill the Thirty-Sixth marched, then down the slope north of Elkhorn Tavern, barely missing the retreating Confederate artillery batteries. As Union soldiers realized they had won, cheers broke out across the battlefield.[20]

Thursday March 6" 1862
At 2[?] P.M. an order came to prepare at midnight to march at two. None could conjecture the cause of so untimely a movement. I made up my mind that it was for the purpose of accostoming the boys to sudden moves, and night marches. Others thought "something was up" but *what* that something was none knew. The safety with which I came from Cross hollow through Bentonville convinced me that danger from an enemy was impossible. We were up on time, our breakfast eaten and Co E ready at the prescribed hour. But the regiment did not commence moving for an hour. The wind blew the snow came down, and the night was so extremely dark that it was judged prudent not to go till daylight so we returned to our old quarters, built up rousing fires, around which we sat and shivered, in the gusty wind & driving snow untill day began to peap upon us, then we were away for good. For three or four miles the marching to me was irksome enough—Now they would walk at a rapid rate, then, all would be huddled together and halt for five or ten minutes. To me this seemed extremely disorderly. We reached Bentonville shortly after sunrise & after a few minutes halt marched slowly through town & towards Sugar Creek. The amunition wagon broke down a short distance out and Co B left to guard it, while the remander of the Regt, except a few straglers marched on. We had gone, but little more than a mile, when the booming of cannon in our rear startled some of the boys, but it was generaly conceeded that Sigle who had remained at Bentonville was practicing with his guns[.] We soon found that he *was* practicing upon the secesh for we had marched scarcely four miles when some of the boys of Co B came up without hats & in a perspiration and announced that a battle was going on at Bentonville. Nothing could have been more

startling, but the arrival of one or two wounded men, and the continuous booming of cannon soon relieved our doubts. The rapid riding of couriers to overtake Col Greusel who was in advance[.] All these movements was ominous of battle to the boys, but never one flinched, but on the contrary when Col [Edward S.] Josly[n] ordered us forward more than one was indignant and inveighed loudly against the disgrace or retreating when the first sign of danger occurred. After marching about two miles Col Greusel came up and ordered us back. The order was received with cheers and, obeyed with alacrity, the boys urging each other forward to participate in the fight. We soon came in sight of the smoke of battle and formed in line of battle across the valley, throwing out skirmishers on each side. The fireing soon ceased and several regiments which had been in our rear, marched by, then came our artilery and Gen [Alexander S.] Asboth and Sigel who were received with loud cheers. The whole force then marched slowly & in perfect order down the valley. Soon the pattering of small arms in the rear announced the pursute & attack of the enemy. A few guns went back and a few rounds of shells effectualy silenced & dispursed them, so that our march to Davises division[21] on Sugar Creek was unmolested. From those engaged I learned that we had scarcely left Bentonville when the whole prairie to the South and bordering timber was swarming with secesh. They must have been close at hand & in the thickets bordering the road when we passed, but suffered us to pass in order to be more sure of the remander. Co B who was guarding the amunition wagon was suddenly surprised and surrounded by 500 secesh cavalry, and ordered to surrender. Seeing resistance impossible and the boys wholy unprepared they stacked their arms & surrendered[.] Just then a heavy firing commenced between them and the Mo 12th[22] which distracted their attention from the prisoners, and all but 24 of Co B made the escape and came in to camp.[23] Gen Sigle saw the enemy as soon as any others and poured shot & shell among them cutting his way through their ranks and joining those in advance, the enemy following after: but Sigel's cannon kept them at a respectable distance, and but little harm was done to us after the first attack. As near as I could ascertain our loss was about 25 killed 30 wounded and 26 taken prisoners. I can not find out the enemie['s] loss. We however captured 10 or 15 prisoners. We found Davises division had left the valley and gone to the bluffs bordering the North bank of the creek & had planted cannon & cut down trees to form a [illegible] brestwork against an attack from the South. We also camped on a neighboring hill and planted our batteries ready for an attack. We also found that the force at Cross Hollow hearing of the approach of Price had fled precipitately to Sugar Creek during the night destroying their stores and many of their wagons to prevent their falling into the enemies hands. Another detachment of 500 infantry and 200 cavalry were at the War Eagle Mills 42 miles distant under Col. Vandevere,[24] and were using the mills for our own army. Gen Curtiss hearing of Price's advance sent an express to Col Vandevere, who brought his handful of men to Sugar Creek in safety, marching 42 miles in 15 hours[.] Thus when evening closed in our whole force except those captured at Bentonville and a few foraging parties and Co F of the 36" Ills, who were on an expedition to the indian country were in position at Sugar Creek & ready for a battle.[25] During the evening there were many speculations among the boys whether there would be a battle. I firmly believed that if Price had ventured to attack us at Bentonville, he would not fail to give battle wherever he could find us. Most of the men[?] differed from me. During the night the moving [of] batteries, the rapid riding of cavalry convinced me that our officers were not ignorant of the movements of the enemy and were preparing accordingly. Tomorrow will reveal more than many will wish to know of the movements of the enemy. Where the attack will commence I can not tell. The boys talk laugh and are as merry as though nothing of danger was near.

Friday March 7"
As soon as I was up I began to look around for indications of battle. Soldiers cooked their breakfast, discussing in the mean time the probabilities of a fight. The sun arose lazily above the smoky horizon, dispensing light and warmth around. The cavalry horses munched at their scanty meal of corn and the grim engines of destruction, the cannon, stood where they were left the night before, silent as the grave and yet one look was enough to remind one of death and carnage. Directly couriers were flying from brigade to brigade. A horseman would come in, his steed reeking with foam. The officers, mounted their horses, and were seen in close consultation. I knew the time had come. Presently came the order *"Fall in"* shrill and loud, for you must know that both Col Greusel and Joslyn have clear ringing voices that can be heard [over] the loudest din of conflict. We marched a short distance in the rear of camp & in a Northwest direction, and formed in line of battle, overlooking a ravine and cleared off the brush and fallen timber so as not to hinder our movements. We stood here half an hour straining our eyes through the long forest aisles, and over the summit of the distant ridges to discover the enemy, but none appeared. We then marched Northeasterly to our right to the Springfield road, which was covered with a moving mass of wagons, horses, mules & men in almost inextricable confusion. A cavalry

regiment came from the North and filed off to the left, then came one or two Indiana regiments and then the 36 was ordered to fall in (except Co E.)[26] and followed after the others. And now to the north is heard the distant roar of cannon, announcing that the work of death had begun. A battery of artilery followed our regiment and our Co came after to support it. It seemed that we were marching directly away from where the cannonaiding was going on, but trusting to our officers we felt that we would be led to the right place. Entering a small clearing we discovered a yellow flag floating from each house and the surgeons standing around, ready for the sad task assigned them.[27] We entered the woods again and had gone but a little way, when the word was passed along the line "*look out for cavalry*," and sure enough, a squad of horses without riders came tearing through the brush on either side [of] the road, in the most headlong speed & confusion. Following after was another body with their riders, and it was difficult to tell which was the most frightened, the horses or the men. All, in the utmost confusion disappeared in the brush, one fellow cried out as he passed, "Turn back they will give you hell." We learned that a regiment of Cavalry (the Iowa 3d) crossed a field to support a cannon which was advancing where the enemy was supposed to be, and entering the woods on the north side was received with a close and destructive fire from all sides, and from a concealed foe. Saddles were emptied, horses killed and the artilery horses all shot down. A rapid retreat was the consequence and when we gained the field with our battery, the regiments preceding us were in the utmost disorder and on the point of flying[.] But the 36" Ill, marched to the front in as good order as if on parade. Their good behavior and fearless stand restored the confidence of the others and was just in time to prevent another Bull Run stampeed.[28] Battery infantry and all then slowly retired towards the other side of the field, in order for the enemy to show themselves. They thinking we were retreating came from their hiding places and the battery commenced playing upon them raining a perfect storm of shot & shell into their ranks, which caused them to disappear into the woods[.] A detachment of infantry & cavalry was sent in as skirmishers and succeeded in retaking the gun that was lost in the morning and hauling it out of the woods, but a fresh charge from the enemy compelled them to relinquish it. About this time a battery opened upon us throwing shot and shell at a fearful rate, killing and wounding a number of our men, compelling the 36 to give ground into the timber. The batteries remained in the open field and Co E was forced to remain there also to support it, and being the only infantry in sight drew the whole of the enemies fire. For a time the whizzing of balls, the bursting of shell all around us was appalling but we stood our ground never giving an inch. A shell plunged into the center of the Co. tearing off a limb from Ira Fuller and mutilating the lower portion of his body terribly.[29] His shrieks, and the firestorm of balls around us appalled the stoutest hearts. The poor boy was carried to the hospital and died in 2 hours. In the mean time our batteries were not idle and the enemy were obliged to slacken their fire. A regiment was sent into the woods to our right and advanced nearly to the position of the enemies batteries, and encountered an overwhelming body. For 15 or 20 Minutes an unceasing roll of musketry was heard in the woods, and the enemy was repeatedly compelled to give ground, but getting around to the right of the regiment, our boys were compelled to give ground. Our Regt marched to the front after the enemies battery was silenced to clear the woods and had arrived in gun shot range when eight regiments of the enemy was discovered behind the fence & in the woods ready to receive them[.] Knowing that the conflict would be bloody and uncertain Col Greusel ordered it to file to the left & the battery to open on them, which was done, and such a stampeed of secesh as ensued was amusing to [be]hold and our boys sent up cheer after cheer[.] The regiment to our ~~left~~ Right was hardly pressed & gave ground and large numbers of the enemy appeared in the field, capturing two guns, and forcing the 37" Ills & 25 to fly, but the 12 Missouri and the battery guarded by Co E hurled their shot into their ranks and soon put them to flight, and leaving the ground covered with their dead and wounded. A demonstration was also made upon our rear but the 36 & 12" Mo soon cleared them out, and none were left except in front and their position was unknown. Co B & D were sent out as skirmishers and when nearly to the woods was received with a storm of musketry which sent death and destruction among our boys. A squad of Co B, saw an officer ride up to the fence & shot him dead from his saddle and secured his gold watch but was too warmly pressed to get his pistols. The officer proved to be Ben McCulloch, and after his fall, but little fighting was done. Occasionally a regiment would appear upon the summit of a distant hill, but a few rounds from our guns would disperse them. A portion of our Co secured the two guns which had been taken a short time before. This closed the battle on the left in which Gen Sigel's division was alone engaged. On the right where the main portion of both armies were engaged, the booming of cannon & the rattle of small arms could be heard all day long. The Iowa 9" the 4" & 25" Mo under the immediate command of Gen Curtiss were raked with a cross fire from batteries on the hills, on the right & left, and were terribly cut up. Repeated charges of the enemy were repulsed with great loss on both sides, but finaly our men were obliged to give ground for more than a mile the enemy ocupying the ground vacated by us.

Towards night, the battle haveing ceased on the left, several regiments were marched to their support, ~~and~~ which with a heavy fire from the 2nd Ohio battery checked the enemy, and enabled us to hold our position. Our stores had however fallen into their hands, which supplied them with plenty to eat for the night. Some was destroyed & burned. The Cannonaid lasted untill 8 oclock at night, and the rapid flashing of the guns, as vivid as lightning, and the deafning roar reverberating among the hills was a grand spictacle[.] The 36 marched to a cornfield contiguous to the enemies position and remained untill 3 [?] oclock the next morning. The wearied men laying down & sleeping soundly on the cold damp ground and with no covering but the heavens. I have been told by prisoners that 3 rebel regiments were in the bushes not more than 20 rods distant. Thus closed this day of blood which was disasterous to our right and many of our noble soldiers lay upon the field cold and stiff in death. On the left where Sigel's division was engaged & in which was the 36, we were entirely successful and a glorious victory was achieved, and left us free <u>to assist the other division on our</u> right. [underlining may simply mark the division between this entry and the next one]

Saturday March 8"
About 3 A.M. we left our position in the corn field and after various meanderings through woods, around hills & ravines, we found the main body of our army, encamped on a small stream and prepareing something to eat for the coming day. The muddy rivulet was eagerly sought by our thirsty men, and never did water taste more good. Our fires were kindled and a few meal pancakes baked, but not half what our ravenous apetites craved. During the previous day I had not eaten but one small biscuit, and it is needless to add, that anything, however poor was relished now. The smoke of yesterday's conflict hung in drapery folds over field, wood and mountain, and the rising sun shone with a mellow radiance through the drifting sheen. While sitting around our fires, the sudden roar of a cannon burst upon our ears like the bursting of a thunderbolt, and so near as to seem just upon us. Then explosion after explosion followed in quick succession making the very earth tremble, the whizzing noise of shell[s] winging their way through the air just over our heads, and clipping the leafy twigs from the trees & underbrush reminded us that yesterday's fearful scene and tradegies of blood were to be enacted over again. One ball struck in the fire around which a score of men were sitting. Not long did the enemies guns monopolize the thunder of the hour, for our guns were brought into position and one continuous roll of

thunder, showed that no idle hands were in the work. Our line was quickly formed, and the field of sulphurous fire quickly reached. Sigel['s] batteries were added to those already in operation and our line formed on the left for their support. Other regiments arrived upon the field, and the lengthning column stretched away to the left, and in the intervals between fresh cannon, were placed untill our whole line was belching flame and smoke. The enemies guns were not idle or poorly managed for while we lay upon our faces, hugging the dusty earth a storm of iron was rained in our very midst. One shell came whizzing to the very spot where Col Greusel stood, and it was only by playing the dexterous dodger that he now lives to tell the tale of carnage and death which reigned around. The ball struck in the midst of Co E instantly killing John Ray.[30] He never moved a muscle after he was struck. In the mean time the enemies fire began to slacken and Co B was sent to reconoiter a small clump of timber which projected towards our line. It was found to be filled with secesh, and our guns hurled death and destruction among them. A rocky and almost inaccessible hill ¾ a mile in our front, was seen to be black with the enemy. Again our artilery searched them out and shells were distinctly seen to burst right in their midst, tearing up the earth and rocks, in a fearful manner. For a while they were not disposed to give up their elevated and advantageous position and spitefuly fired their rifles and shot guns into our skirmishers who advanced almost to the foot of their strong hold. Not long however could they stand the storm of shell which were hurled among them, and rapidly vacated the premises, pursued for a short distance by our skirmishers. Their guns had now ceased to play upon us, and ours were moved forward, together with the infantry towards their position. A sharp lookout was kept upon the hills for flying secesh and whenever a butternut squadron was discovered hurrying away, a few shells would give an impetus to their flight, and it was amusing to see them wildly scatter upon the receipt of a message from our guns. I am told by prisoners that about this time two of our shells killed & wounded sixty of their number. As we advanced our right was received by a scattering fire of musketry, and then the fireing ceased. The 36" marched to the rocky hill directly in front, expecting to receive a warm reception from some concealed foe. The base of the hill is reached and not a secesh appears or a shot exchanged. Its precipitous sides presents an impassable barrier but a detour to the left discloses a passage, and the summit is reached. Great God what a scene is presented. The mangled trunks of men are thickly scattered around. From each tree or sheltering rock the groans of the wounded arise. Muskets, saddles, horses, blankets, hats, and clothes hang on every bush, or in gory manner strew the ground. And now in the valley to the right ten thousand wild cheers proclaim the victory ours. Dead horses, dead men and dismounted guns, are strewed over the blood drenched field, and as some gun is taken or trophy secured, renewed cheering and shouts of gladness ring out upon the air. As we moved down the sides of the hill, burning camp fires, half eaten breakfasts, sacks of flour, sides of bacon, even [missing word?] scattered in profusion as well as confusion all around. Not one regret did I hear that we had interrupted their morning repast. Our hungry boys seized on everything eatable, and I was amused to see a short dumpy dutchman marching in the ranks with a large sack of flour on his back. The right and left wings meet upon the Springfield road, and then a shout which makes the wood resound is given with a will. All are feeling good. Sigel's ~~dark~~ eye has a more nervous but joyous twinkle than the day before. Methinks I can see a smile of satisfaction lurking among the mustach[e] which covers Asboth's stoic face. And Osterhous is more jolly than before, and his blue eyes fairly laughs with satisfaction. In the gladness which rules the hour wrecks of humanity thickly strewed around are not neglected and it is good to see the stalwart soldier giving water from his canteen to the wounded and thirsty southerner, who but now was panting for eac[h] other's life. They are enemies no more.

Not long have we for congratulation for the line is formed, and the column headed to the North in pursute of the flying foe. For miles the road is str[ewn] with rifles, muskets, clothes, blankets and every thing calculated to impede their flight. For 12 miles we march but no living foe appears nothing but their butternut paraphenalia to[ld] that they were ever there. We camped but a few miles from Keetsville in Missouri without supper or the means of getting one. I found the head of a hog that had been recently killed and obtained a few scraps of meat from its jaw and roasting them at a fire, was glad to eat without salt or bread. To sleep was easy, and came a welcome visitant to our eyes. I have now mingled in in the dread scenes and seen the sad sights of the battle field and I must confess that I have no particular relish for its frequent repetition. Heaven grant the speedy restoration of peace, and a deliverance from such scenes again.

Sunday March 9"

Up at daylight and again nothing to eat. Our march towards Keitsville was about the quickest [ki]nd of quick time, and but for the name and dis[middle letters illegible]nor dishonor of the thing I should have fell out of the [illegible] droped behind. It was a warm morning and the breath of the south wind did not cool the atmosphere and we all sweat profusely. Arrived at Keitsville a halt was ordered & during our stay I conversed with the prisoners, and among them found two Officers, both related to the rebel general. One, Col Thomas Price had been a prisoner before and if I have [?] not forgotten had taken the oath of allegiance so death will be his fate. The other had once been a candidate for Congress against Col [John S.] Phelps and found him a man of shrewdness and intelligence. They were all taken towards Springfield. A light rain occurred before noon after which [our] regiment marched back to the valley and halt[ed] during a severe rain storm. I spread my blanket over me and while the rain poured down slept soundly. A provision train arrived while there, and Co F who had been on an expedition towards the Indian nation[.] Hearing of Price's approach, they had after severe marches, managed to elude the enemy & reached Cassville in saf[e]ty & thence to the Regt. After the storm had subsided all marched on. I however loitered behind, kicking my way through the mud and across the swolen stream, and reached camp as soon as the others notwithstanding their early start. I could get but little to eat, but each morsel was worth so much gold. A half dollar for a mess of oysters was paid without grumb[ling]. What is money worth when hunger pinches the stomach.

Monday March 10"

I have written to & received a letter from Melissa[.] It came just when twas needed most . It is useless to describe the pleasure it gave me, how much of true love it breathed, or how endearing the ties which bind me to her. Tis useless now to regret that I am away from her smile & her presence for regrets will bring me no sooner to her. I have ascertained the amount of the loss sustained by the 36". It foots up 7 killed 40 wounded and 24 taken prisoners, this includes both the battles of Bentonville and Rose Hill or Pea ridge.[31] I walked over much of the ground where the battle was fought. Men were busy collecting the dead and burying them. It was a harrowing sight to see the disfigured and mutilated forms of the dead. I noticed one that a shell had taken away ⅓ of the skull and skooped it entirely of brain[.] Another, with head entirely blown to attoms other[s] with limbs entirely gone. A Mississippi Colonel[32] had lost half his head which hung in shreds to that which remained. Phelps regiment was almost destroyed and the 9" & 4" Iowa had their ranks terribly thinned. I can not get at the loss on either side, but each have suffered teribly[.] The trees where our shots were plunged are marked and shattered, some are blown entirely down, and attest the severity of the conflict[.] Cannon balls, fragments of shells, grape & cannister is strewn everywhere, and I picked up a few as mementoes of my first battle[.] On my return to camp I noticed horsemen, wagons and men hurrying in breathless to their respective camps & on enquiry I was told that the enemy was upon us. I was too fatigued to run & pursued my way leisurely to camp, learning however that the scare was caused by a party of secesh who were collecting & burying their dead[.] The various regiments were called to arms and put in readiness for any sudden emergency. Passed the remander of the day in writing & rest.

Tuesday March 11"
There is much wrangling among the officers of our Regiment. Col Greusel's imperious will & overbearing behavior to them have induced Col Joslyn & Major Barry to resign,[33] and I hear they will be with us no more. I regret this for aside from his drinking habits Col Josly[n] is a man & a gentleman, and his courage & bravery has been *tried on the field of blood before us.*

It has now been more than a week since I wrote the above and the events of each day are so confused and jumbled together that I will not try to keep each day separate, but briefly relate what has occurred as near as I can remember. I think it was on friday that I was definitely informed that I was needed to assist Lieut Hoepmer[34] in the engineering department. Gen. Curtis was about to move his camp on Sugar Creek, towards Bentonville and I hastily sealed up a few balls and relics of the battle to send home as mementoes, and then rode the Genls horse to camp. My detail was made out, and an acquaintance with Lieut H commenced, which thus far has been very friendly & I must say I like him much. A tent was assigned me and a horse and saddle promised. The first duty assigned me, was to survey the valley down the creek, accompanying a foraging expedition. A mile below camp the creek entirely disappears, the water sinking into the ground and no appearance of water is again visible for four or five miles, and then it comes out again, as large almost as a river. The valey is very rocky and wild, and the few who live here and there, look as if they were banished from civilized society for their evil deeds. We went about 15 miles before anything in the shape of forage was found. Here the bluff or mountains receeded from the creek forming a basin or ampitheatre of three or four miles and in this were some good farms and we found corn, oats and hay sufficient to load our wagons[.] Within this valley was some prairie land and altogether it looked more like live [*sic*] than almost any place I have seen in this wild region. About the center of the basin was an isolated peak which could be seen for several miles, and presented a fine appearance. It is called Bobbs Knob, rather an odd name surely. Below is a rough sketch of it from the west. The men comprising the expedition behaved more like a set of desparadoes than any thing else. Horses, chickens, geese, and pigs were taken and even the smoke houses robbed of the last vestige of meat. They wished me to take a horse but I have not yet so far debased myself as to become a horse thief. I however rode a stolen one back to camp. A severe thunder shower came on just about sundown, and a squad of boys who were guarding two wagons loading with corn at that place, concluded to stay all night. We got a good supper, and I was lucky enough to get a good bed. It rained nearly all night & in the morning every ravine was a roaring cataract. Sugar Creek was scarcely fordable for teams. A quantity of rebel arms (200) comprising rifles, muskets, and shot guns had been discovered about 16 miles from camp at the head of a hollow and in the midst of the woods an expedition was sent after them, and on their return the infantry were obliged to wade the swollen stream. I realy pittied the poor fellows, to see them come shivering from the water and then spatter through the mud scarcely less in depth than the creek. It was nearly night when camp was reached, and wet blankets added nothing to my comfort during the night[.] My next survey was up the creek to the Springfield road, after which the road to Bentonville was was next. The wether is rainy and disagreeable most of the time. Occasionly the sun is seen which adds a ray of cheerfulness to the dreary days and desolated region. I am now furnished with horse and saddle and can get around the country with some little comfort.

Tuesday March 18" 1862

To day the 13" Regt[35] arrived from Rolla and every body was glad to see them. Geo Walker brought a letter from Melissa, and the likeness of wife and little one. It does my heart good to see them once more. God bless the dear ones, and in mercy grant a quick and safe return to them again.

Made a platt of a portion of the battle field for Col Bussy[36] for which he gave me a dollar[.] Also commenced one for Col Dodge.[37]

Wednesday March 19"

Rumors of the advance of Price towards Springfield and around our position has determined the Genl to make a retrograde move to Keitsville and to day we marched and rode 18 or 20 miles. I am looking for a fight[.] My first foraging was done to day. As night came on and I had nothing for my horse to eat I went to a farmer's barn where others were taking oats, corn & c and took three or 4 bundles of oats & the same of corn fodder and went my way. The day has been lowry, muddy & disagreeable[.]

Thursday March 20"

The mistleing rain of yesterday turned to snow to day, but it thawed as fast as it fell, untill toward night when the ground was whitened & the desolation

"Bobs Knob. McDonald Co. Mo."

which reigns around covered up[.] Winter which I supposed had left this region has once again returned to give a parting farewell ere he speeds away to the ice bound poles. Col Heron [Herron] & Chandler[38] who were wounded and taken prisoners returned to day and were exchanged. They say Price is at Ft Smith feeling rather gloomy over their defeat. They confirm the death of McCulloch, McIntosh[,] Clarkson[39] & Slack.[40] Eighteen prisoners were brought in from about 25 miles distant[.] They sought to meddle with our foragers and were trap[p]ed before they were aware of it.

Friday March 21"

Awoke and found two inches snow & more coming[.] Wrote to Chicago Tribune, to Melissa and did up a little reading and a good deal of idling. Moved the Genls cooking stove in to our tent which made us warm and comfortable. *Received a letter from home*

Saturday March 22.

Commenced the survey of the telegraph road at Keitsville and ran six miles and platted three. Mud in great abundance and wagons, horses, mules & men in sufficient numbers to keep it well mixed. Lieut Heoppner goes to St Louis leaving the engineering entirely to my controll. He takes letters &c for Mellie.

Sunday March 23d

Did not know it was sunday untill I had prepared to survey, but gave up the project as soon as I found it was sunday. Washed, read and wrote untill towards evening and then went to the 13"[41] & Iowa 4"[.] George Walker returned and staid over night with me.

Monday March 24"

Col Dodge goes home in consequence of sickness. Gave him a platt of the portion of the battle field where he was engaged and bade him good by[.] Col Dodge has been a good and true friend to me, and I shall not soon forget his kindness. Surveyed the road to Sugar Creek and returned weary and half sick[.]

Tuesday March 25"

Platted all day long and completed a sketch of the road to Sugar Creek which Gen Curtis approved and it was sent to St Louis by the express.

Wednesday March 26"

Made a platt for Col Bussy which took me all day. In the evening went to the 36" & had a short and pleasant visit with the boys. Friend Wagner gave me a knife which I needed much. Friend Hiram [Wagner] is true & faithful. Several of the officers want platts which will give me business[.]

Thursday March 27" 1862
Rode to Cassville (13 miles) and had a good visit with [M. LaRue] Harrison & Col [Colley B.] Holland. Harrison's son has arrived & is fat and hearty as a pig. Nearly every house in Cassville [is] a hospital for the care of our wounded. Mr Harrison is beginning to understand Col Greusel's true character. He has promised him a position long ago and although abundant opportunities have occured he has entirely neglected him & turned a cold shoulder upon him. Meanness & a want of manhood will in time be found out and known. If I am to rely on others for favors let it be men of principal not who knows no one but *self*. Had a pleasant ride. Am troubled with the diarrea and am taking medacine. Wrote to Mellie. Expect to be paid in a day or [illegible word]. The wether warm and pleasant. Spring is realy upon us now. The leaves begin to appear, and all nature rejoices in God's pure sunlight.

Friday March 28"
Was quite unwell all day and took medacine, but nothing seemed to help me as much as a small swig of whiskey the first I remember to have taken. I do not think this a violation of my temperance pledges & principals, for as a medacine nothing is prohibited. Am better to night.

Saturday March 29"
Platted nearly all day on a plan of the battle field for friend [Eugene] Lake of the 2nd Ohio battery.[42] Will be obliged to work tomorrow to finish it in time[.]

Sunday March 30"
Finished Lake's platt and took it to him. He was about ready to start & gave me a few stamped envelop[es] and took a letter for Mellie. Sunday is no more regarded here than any other day, & I fear I am falling into the habits of the others in this respect. Warm dry & dusty[.]

Monday March 31"
It has been a rainy and disagreeable day. Commenced platt of battle field[.] Received assurance that our pay would be forthcoming tomorrow. Wrote a letter to Mellie in regard to it. Money is very much needed by me now, not for myself alone but the loved ones who are dependent on me.

Tuesday April 1" 1862

Towards night went to the paymaster's quarters and received four months rations of pay. I propose to send $50 home which will do me more good there than can possibly accrue here. I live only for those at home without whome life would be dreary enough. I *know* that *there* is a fountain of love and affection which no one can alienate, and which will never cease. May a kind father above provide for all the wants of my family during my absence. Our mess is now regularly arranged our rations drawn, cook installed and we are living *high*.

Wednesday Apl 2"

The mail of the 36" to go east was found broken open and the contents scattered in a neighboring gulley. Capt Stark's[43] negro was suspected but on being questioned about it denied all knowledge of the matter and intimated that George Raymond[44] had done it. George was searched but nothing suspicious, or implicating him found. After awhile the darkey was found to be the possessor of several dollars of U. S. script and other things combined to fasten the crime on him[.] He is now arrested and placed under guard. He undoubtedly is the guilty one. Several parcels of money has been lost, and some found on the darkey was identified. I hope the thieving rascal will get his just deserts.

Thursday April 3d 1862

Platted most of the day. In the afternoon rode three or four miles over the country which refreshed me much[.] Purchased some chickens for our mess of a dirty faced bare footed Missouri girl about 16 or 17 years old. Oh What a country for tangle haired girls and tumble down log houses. Should like to know how much longer we were to stay in our present camp. An early and rapid move away from home is to me the best guarantee of a speedy return home. I wish I was there now.

This evening the darkeys are giving a sort of theatrical entertainment[.] A fiddle is constantly being sawed by one and the others join in the dance and the way they hoe it down is a caution. I never saw better dancing, and all can got [*sic*] it about alike. They are a happy set of mortals.

Friday April 4"

Hereing that friend Scofield[45] of the Newark Co. was going home on furlough and would carry letters for the boys I wrote one to Mellie enclosing $20[.]oo and started for Cassville to see him. The day was warm and my horse sweat profusely.

Scofield agrees to take my letter, also my diary which is nearly full. If he goes via of Chicago he will call at my house and tell personaly what I have forgotten or omitted to write[.]

I have gathered two or three flowers and put them in this book, and though flowers may be in bloom when this reaches my home. Yet there is one who will regard them with some degree of intrest [*sic*] because they were plucked her by a hand, true as heaven and which, "God helping me" shall wrong not another. Mellie—as you read these hastily written pages you will see the marks of your own.

14 | Pea Ridge to the Mississippi River

April 5, 1862–August 17, 1862

To one used to the broad prairies and cultivated fields of the north this is a most forbidding region[.] Its only redeeming feature is the fine springs of clear cold water which gush from every hill side, and which sparkle in a thousand rivulets, meandering among the hills.

—Lyman G. Bennett, April 8, 1862

Bennett sent his diary home to his wife with flowers pressed between some of the pages. There is no surviving diary for the time period covered in this chapter, and possibly Bennett did not attempt to keep a personal account in this period. As described in the last chapter, Bennett conducted surveys near the Pea Ridge battlefield, and these survive as a set of maps, some incomplete, and a fragmentary map memoir titled "Route of the Army of the Southwest from March 1st to July 10, 1862," created for Major General Samuel R. Curtis. This oversized book is part of the holdings at the Kansas State Historical Society. Correspondent William Fayel of the *Missouri Democrat*, who accompanied the Army of the Southwest into Arkansas, left a vivid pen picture of Bennett at work in a May 28 article written in Searcy, Arkansas:

> Traverse the country where you will, along the route of our march, and an eccentric looking personage will be seen posted on the mountain peaks, in the valleys, along the margin of noisy frog ponds, taking aim across an instrument towards all points of the compass. This is the omnipresent Bennett, the Topographical Engineer on General Curtis's Staff. The other day, Bennett planted his surveyor[']s compass in the road fronting a dwelling house. Presently an intelligent young lady sent out a boy with her compliments, and requested the supposed organ grinder to 'play a tune on his instrument.'
>
> The last time I saw Bennett he passed me on the return from this place to Batesville, having lost his field notes of a survey made

of 45 miles of road, and being obliged to resurvey the whole route. It was raining pitchforks at the time.[1]

The campaign in which Bennett now participated started as a reaction to movements by Major General Earl Van Dorn's Army of the West. Following a relatively brief recovery period after its defeat at Pea Ridge, Van Dorn planned for his army to cross the Mississippi River at New Madrid in the southeast corner of Missouri. High water on the White River near Forsyth, Missouri, in late March prevented this effort, and the Army of the West instead marched southeastward toward Jacksonport, Arkansas.[2] Another portion of Van Dorn's army made their way from Van Buren to Des Arc, and the entire force crossed the Mississippi River and started to enter northern Mississippi at the end of April. Van Dorn not only moved his army across the Mississippi, but also stripped Arkansas of weapons, ammunition, food, machinery, and other supplies. From that point, the trans-Mississippi Confederacy could never again field an army that equaled the Army of the West in strength.[3]

Newly promoted to major general following his victory at Pea Ridge, Samuel R. Curtis moved his army eastward to protect southern Missouri from Van Dorn's advance.[4] The soldiers departed from their camps on April 5 and soon entered formidable terrain that Bennett described in his surviving map memoir. Primitive roads and streams and rivers full of spring run-off presented problems that impacted efforts to supply the army. Soon, the army foraged and "devoured everything in its path."[5] In late May, department commander Major General Henry W. Halleck learned that Van Dorn's army had entered Mississippi and soon ordered Curtis's army to the southwest to Jacksonport, Arkansas. There, they would advance on Memphis and help capture that city in June.[6] Reinforcements in the form of Brigadier General Frederick Steele's division from southeastern Missouri arrived at Batesville, Arkansas, in early May 1862. The move to Memphis had to be aborted when it was discovered that the land was too marshy between that city and Jacksonport. Halleck ordered Curtis to detach half of his infantry regiments and send them to Cape Girardeau, Missouri, where they were conveyed across the Mississippi. Among the regiments that left for Missouri on May 10 was Bennett's Thirty-Sixth Illinois Infantry, which he never rejoined.[7]

Curtis reorganized his army after the departure of these regiments and then prepared to advance on Little Rock.[8] However, he grew concerned about his supply line and his army's subsequent ability to protect its foraging parties from an increasingly active enemy.[9] In late May, Curtis abandoned the attempt to capture Little Rock and instead marched toward Clarendon, where supply vessels awaited them.[10] For two weeks in late June, the army advanced from Batesville to Clarendon without a supply base, marking "the first time in the Civil War a Federal army attempted such a daring maneuver."[11] Reaching Clarendon on July 9, the men discovered that the supply vessels had retreated, so the army marched on to Helena, forty-five miles away.[12] Slavery was effectively destroyed along with food supplies in the region as the soldiers advanced on Helena.[13] The march into the town ended a campaign that had covered five hundred miles over difficult terrain.[14] Heat, high humidity, and "subtropical diseases" made Helena a difficult and dangerous base for any soldier posted there during the war, and it was there that the Army of the Southwest was systematically dismantled as regiments were "incorporated into the Army of the Tennessee" that went on to campaign at Vicksburg, the Atlanta campaign, Sherman's March to the Sea, and the final campaign into the Carolinas.[15]

Excerpts from "Route of the Army of the Southwest from March 1st to July 10, 1862":

9 [miles]

Cassville, the county seat of Barry County, has but a few hundred inhabitants. Many of the buildings are old and dilapidated and since the war deserted. There are but few indications of prosperity and enterprise. Flat Creek valley is here from ½ to ¾ a mile wide.

April 7" 6 [miles] N. 79° [?] E.

Steam Grist mill, and Camp of Maj. Gen. Curtis East of the town[.] [Cassville]

6 [miles]

A portion of the country is rough and broken. After reaching the table lands it is level and covered with a thick growth of dwarfish Oak. Soil, a light yellow loam upon a subsoil of clay. Very rocky among the hills—No cleared fields and but one house inhabited.

Cold damp weather.

April 7" 3 miles, N.65° E

Road decends into a narrow and rocky ravine, and frequently obstructed by [illegible word] rated payments of rock from the adjacent bluffs. Ravine thickly timbered with oak elm sycamore and other timber. A mile below the head of the ravine a fine spring issues from beneath a ledge of limestone.

4 miles, N. 60° E

Field on right of road. Bluffs not as steep and as abrupt as at the enterance of the valley, and their summits generaly bare and free from timber or bushes. Water high from recent rains, and road muddy in places.

7 [miles]

Road very rough and rocky, leading over steep hills capped with sandstone underneath which lies a section of lime stone. Disentigrated fragments of a glistning flinty quartz rock cover the hill sides, and frequently extend into the valley. The vallys have a rich fertile soil, the cultivation of which is much neglected:—the few inhabitants appearing ignorant and indolent.

April 7" 3 [miles] S 20° [?] E

Road passes over irregular and broken hills, the steep ascent rendering it difficult for the transportation of baggage and supplies. These hills contain inexhaustible supplies of lime and sand stone. But little timber and of a dwarfish size.

April 8" 4 [miles] N. 57° E.

Pitched camp in the midst of as wild and sterile a region as any yet passed over on the banks of a small mountain rivulet of pure water[.] The surrounding hills is covered with a thick growth of pitch pine of a diminutive growth. Not a sign of vegetation could be found except the oak and pine trees that moaned in sadness over the surrounding desolation.

7 [miles]

For many miles the road is very crooked, leading over a high ridge and keeping on the highest summits[.] Deep ravines and gorges branch from either side and lead into distant valleys[.] Pine and jack oak of a stunted growth is abundant.

April 8" 7 [miles] N. 60° E

During the early part of the day, the march was on the crest of a high range of hills, the sides of which were generaly steep & rocky, overlooking deep gullies—to avoid which the road was nessesarily crooked: Broken payments of a kind of flint or quartz rock was thickly scattered over the surface. Some of the rocks were of a f___ous [?] character, indicating mineral, & being blackened by oxidation, harmonized well with the dark foliage of the low pines which crowned the ridge—From the higher points the eye plunged into a labyrinth of deep rocky valleys and high mountain peaks, which looming up in the distance, afforded many wild and romantic views.— Damp lowry weather.

April 8, 1862

A few log cabins, mostly deserted: constituted the town. I noticed the remains of a mill which had been partially destroyed by high water. It was a a a matter of wonder where land in sufficiant quantities for cultivation could be found, to raise means for the subsistence of the few inhabitants[.] Weather cold and damp.

To one used to the broad prairies and cultivated fields of the north this is a most forbidding region[.] Its only redeeming feature is the fine springs of clear cold water which gush from every hill side, and which sparkle in a thousand rivulets, meandering among the hills.

We passed Cape Fare [Fair] about noon. The waters of Flat Creek were deep and rapid and the crossing difficult, and many of the regiments were delayed in crossing.

April 8" 8.7 [miles] South course.

Galena, undoubtedly so named from the existence in the immediate neighborhood of large quantities of lead, but which has been but little worked in consequence of the difficulties & distance to market, and the people being mostly hunters, and wanting in enterprise for prosecuting mining operations.

We reached this place just before night and shortly after a cold rain storm set in, which continued during the night, swelling the already overflowing streams and rendering the roads almost impassible. It is an insignificant and dilapidated town, mostly deserted since the war commenced. Many of the inhabitants were true to the union, and furnished many soldiers for the federal army.

The march over the [missing word] has been cold and cheerless, the clouds drifting almost to the mountain tops, and threatening each hour to drench the earth with its overcharged burden of waters. Not a house was seen upon the rout[e] except at Cape Fare. The teams did not arrive till midnight, and we found quarters in vacant houses.

April 9" 3 [miles] S 70° E.

The storm of last night delayed the march untill late in the afternoon[.] The swollen waters of the James rushed madly over the rocky bottom, and considerable care was requisite to cross in saf[e]ty. A teamster was drowned while attempting to cross. Wagons were ranged one after the other, forming a bridge for the infantry to cross. We entered the valley of Raleys creek following up the stream and camped 3 miles from Galena[.]

Apl 10" 3 [miles] S 70° E.

Passing up the valley of Raleys creek which is formed by springs issuing from the hill side and heads six miles from Galena. We then left it for a ridge & so steep was the ascent that it was with difficulty the teams could climb its rocky sides[.]

April 10" 3 [miles] S 75° E.

This ridge extends but four miles, and then abruptly terminates in the valley of bear Creek which heads near this point[.] On the summit & hill sides was a thin growth of post oak[.] Many of the surrounding peaks was nearly bald and the views from the highest points was truely magnificent.

April 10" 3 [miles] S 75° E.

Bear creek is of considerable size, and after meandering six miles among the hills empties into Bull creek[.] We were obliged to ford it several times, and the teams were frequently stalled. A private of the Benton Hussars was shot by an artilery Lieutenant and killed. Weather cl[e]ar and pleasant.

April 10" 8 [miles] S 75° E.

In the valley of Bear creek were many good farms, but the surrounding peaks which hemmed in [the] valley looked barren and desolate, their rocky sides often perpendicular, towering up many hundred feet above the creek. I learned that game was abundant, consisting of bear, deer wild turkeys &c[.] Perhaps the abundance of bear gave rise to the name of the creek[.] This stream empties into Bull Creek, a wide rapid stream, almost a river of itself.

Gen Curtis here left the train & most of his suit with orders to halt for further orders & pushed on to Forsyth[.] We pitched our tents on the banks of Bull creek in a kind of ampatheatre formed by the surrounding mountains—Water and wood were plenty & of excellent quality. All the streams in this region are formed from springs and clear and cold. This would be a country favorable for the operations of gurillas & it would be most difficult to contend with them successfuly among these mountain fastnesses.

Weather fair and pleasant, with indications of rain.

April 11" 10 [miles] S. 65° E.

It was a cold, rainy and cheerless day and none but urgent necessity would prompt an army to move. A partial lull in the storm, after noon induced us to strike tents and move forward[.] The intervening ten miles to Forsyth was ~~not~~ over a high mountain. The roads were execrable almost in a liquid state and it was most difficult to get the teams along, and it was not untill late at night that Forsyth was reached. Some of the surrounding peaks were quite bare of trees, and by climbing to their summits some most extensive and magnificent views of the surrounding country could be obtained, and away down the rocky ravines at the foot of the mountain, occasional glimpses of White river could be obtained.

[Notes on "Route of the Army of the Southwest" ends here.]

> In addition to Bennett's map memoir, he submitted to the *Chicago Tribune* an account of activities around Helena in August 1862 including his participation in one reconnaissance. General Curtis's staff arrived in Chicago on August 31 so that the general could "attend the meeting of the Pacific railroad commissioners," and it is possible that Bennett accompanied the general there and visited his own family at this time.[16]

FROM GEN. CURTIS' ARMY

RECONNOISSANCE BELOW OLD TOWN—
EXPEDITION TO WHITE RIVER—FORCE
AND POSITION OF THE ENEMY.

[Correspondence of the Chicago Tribune.]

HEADQUARTERS ARMY OF THE SOUTHWEST,
CAMP AT HELENA, Ark., Aug., 17, 1862

Very little of importance is transpiring in this military district at the present time, our leaders, appearing to dread the intense heat of an August sun more than Hindman's[17] hord[e] of ragged rebels. Still we have camp incidents, and expeditions of minor importance to relieve the monotony of the hour.

A RECONNOISSANCE.

The operation of guerrillas, and the facillty with which they cross the most difficult portions of the country, and secrete themselves in the cane brakes, induced Gen. Curtis to institute an exploration of the forests and swamps between the Mississippi and White Rivers. Lieutenant Bennett, of the topographical engineers, with thirty of the 5th Illinois cavalry,[18] pushed a reconnoissance forty miles below our camps at Old Town, and discovered a trail leading to White River, by which messages were carried secretly to Little Rock, and guerilla parties made descents on our lines. Owing to swamps and dense cane brakes, it was very difficult getting through, but the object of the expedition was attained, and the boats for crossing the White River destroyed. Information of sufficient importance was gained as to determine the general to send an

EXPEDITION TO WHITE RIVER.

General Hovey[19] and command returned from Clarendon to-day, and I am permitted to examine his reports and gather the details of the expedition. The command left Helena on the 4th inst., and the second day out approached a thicket known as "Patterson's Deadening," where a regiment of Texan Rangers were encamped. Word of our approach reached them and they left their camp in a hurry, leaving their supper behind, which was devoured by our hungry soldiers. An early start in the morning interfered with the rangers' breakfast, as that was also left to our boys. Nothing more was seen of the rangers during the march, and the command reached Clarendon on the evening of the 7th. Some of the soldiers swam the river and were fired upon from the bushes which line the bank, but a few shells from this side cleared them out. A coal barge, which had been sunk some weeks ago by Col. Fitch,[20] was raised, and several companies crossed to the west bank, driving the rebel pickets from the river in too much of a hurry to care for the dead ones left behind; after which they marched in the direction of Des Arcs.

The enemy, thinking this to be a move in force, and not a demonstration, abandoned Des Arcs and all their positions on the White River, and rapidly concentrated their forces near Little Rock, in anticipation of an attack. Guerlila parties were called in, and we hope for a few days rest from their annoyances. The expedition was only intended for a reconnoissance, and accomplished its object in a creditable manner. Three of our men were killed and two wounded by guerillas during the stay at Clarendon, viz:

> John J. Winchell,[21] 13th Ind.
> John Rader,[22] 5th Ill. cavalry.
> Robert Laine,[23] 5th Ill. cavalry.

FORCE, POSITION, ETC., OF THE ENEMY.

Reliable information of the condition, numbers and situation of the enemy was obtained. The troops at and near Little Rock number about 20,000, besides the bands roving about the country, and are mostly Texans under Gen. Johnson. The Arkansas conscripts are under Gen. Rust, and the whole commanded by Hindman. All reports represent them as greatly demoralized, and much sickness and dissatisfaction prevails in their camps. A large part would surrender had they a chance. They have forty cannon, mostly of a heavy calibre. The soldiers are armed, except a few regiments, with shot guns, rifles, etc., and two-thirds are mounted. The country for many miles around Little Rock is completely stripped of corn and forage, and were we in possession, our supplies would have to be transported across the country, as the Arkansas River is too low for boats, and can easily be forded in many places.[24]

FORTIFYING.

Preparations are being made to fortify Helena. The heavy guns for that purpose, I understand are on the way. In the event of a failure of navigation on the White and Arkansas Rivers, this will be the only post for obtaining supplies.

L. G. B.

[Published on August 28, 1862]

15 | Recruiting in Dixie, Part One

November 26, 1862–July 5, 1863

> *The "cheager" is exclusively a southern institution. Every body had them—had em bad, and when not sleeping, we were engaged in scratching[.] The "cheager" is an infinitesimal insect that penetrates the skin with the same facility as a badger would burrow into a sand bank. The only remedy resorted to was scratching, and it was no uncommon spectacle to see one thousand men scratching at one time[.]*
>
> —Lyman G. Bennett

Following the campaign to the Mississippi River, Corporal Bennett was assigned new duty as a topographical engineer in the St. Louis headquarters of Major General Samuel R. Curtis. In November 1862, he asked Curtis "to be relieved and sent to my regiment" rather than go to southeast Missouri "without the pay which such services deserve." In pleading his case, Bennett explained that "it is with the most painful regret that I make this request, as I have received the kindest treatment from you who have treated me like a man regardless of my station as a private soldier."[1] Curtis apparently dealt with the matter as Bennett did not return to the Thirty-Sixth Illinois. Bored much of the time, Bennett wrote a revealing letter to Congressman Owen Lovejoy, a prominent abolitionist and friend of President Abraham Lincoln, in early 1863. Generally Bennett's writings contained little direct mention of politics, but in this letter, he questioned the depth of support for the recently issued Emancipation Proclamation by Arkansas's military governor, John S. Phelps. A prewar slave owner, a Unionist from Springfield, Missouri, a former state legislator and congressman, and a recruiter of a Federal regiment, Phelps met Bennett in Rolla in 1861. There, Bennett "had an interesting argument with him [Phelps] in regard to slavery" according to his Nov. 23, 1861, diary entry. In his letter to Lovejoy, Bennett promoted Isaac Murphy, a Unionist, as a better governor of Arkansas. Lovejoy forwarded Bennett's letter to Lincoln. The president's reaction to it is unknown, but Murphy became governor when Lincoln's Ten Percent Plan reconstructed Arkansas in the spring of 1864.[2]

Bennett left the dull St. Louis office behind when he reentered the field in April 1863 under a "recruiting commission" from Curtis. On December 8, 1863, Bennett was mustered in as a newly promoted first lieutenant and adjutant in the Fourth Arkansas Cavalry that he had helped recruit during the summer.[3] While performing this duty, Lieutenant Bennett learned more about the complicated politics and loyalties in Arkansas, particularly in the northern counties situated throughout the Ozarks region.

White Arkansians were divided in their attitudes toward secession, but some vigilance committees formed before the State's withdrawal from the Union to question northerners and threaten suspected abolitionists.[4] Unionists in Arkansas began to organize soon after the state's secession convention voted in favor of withdrawal on May 6, 1861.[5] Most Arkansas Unionists lived in the Ozark Mountain counties that stretched from the northwest part of the state eastward to Independence, Lawrence, and Randolph counties. Another cluster of Unionists resided further south in the Ouachita Mountains. Most of these loyalists were small, non-slaveholding farmers, many of them poorer than their counterparts in the much wealthier cotton-growing lowland counties where slave ownership was much higher.[6] Since most resided in mountainous areas, they were commonly referred to as "Mountain Feds," a label used by Bennett.[7] These peoples above all desired "to protect their homes and families" and feared, correctly as it turned out, that they would be forced into the Confederate military and thus be unable to safeguard their farms and relatives.[8]

The desire for protection led to the creation of the Arkansas Peace Society.[9] Not all Arkansas Unionists were members, but many were in north Arkansas where Bennett served on recruiting duty. The origins of the Peace Society are murky, but elaborate passwords, countersigns, signals, and so on were developed, and it was broken into small groups for added security.[10] Betrayed in November 1861, suspected members were rounded up and investigated in several northern counties. Those arrested were marched in chain gangs to Little Rock and given a choice of either standing trial for treason or enlisting in Confederate units. Interestingly, those who selected the trial option were found not guilty due to a lack of evidence that they had actually participated in "acts of rebellion."[11]

Some Unionists fled to Rolla in November and December 1861, an event that Bennett noted in his diary entries for that period, when about forty members of the Arkansas Peace Society enlisted in John S.

Phelps's regiment.[12] Not surprisingly, places occupied by Federal troops such as Rolla and Springfield became "safe havens for Unionist refugees."[13]

Confederate supporters in Arkansas grew increasingly convinced that men had somehow "been tricked into joining" Unionist groups, and across the South the number of suspected Unionists was inflated.[14] The capture of bridge burners in East Tennessee at the same time that the Peace Society was uncovered made authorities uneasy, and they increasingly regarded Unionists as dangerous.[15] The enactment of Confederate conscription in spring 1862 actually led to more men joining the Peace Society as they now perceived greater danger to their homes and families.[16]

A number of Unionists were forced into Confederate service. Company F of the Thirty-Second Arkansas Infantry was primarily raised in Searcy County, a Unionist stronghold. After fighting in the Battle of Prairie Grove on December 7, 1862, men deserted from the company and reduced its strength by one-third by the end of the month.[17] By war's end, only Tennessee had raised more loyal units than Arkansas. These Unionist Arkansans eventually became the backbone of the nascent Republican Party in the state.[18]

As mentioned in the editorial section of this book, Bennett wrote his self-titled "Recruiting in Dixie" narrative in the postwar period, although its vividness suggests the possibility that it was based on a wartime account. Divided into two chapters, this first part of "Recruiting in Dixie" describes Bennett's journey from St. Louis into the recruiting camps for Arkansas Unionists in Springfield. It includes some of Bennett's wittiest and most humorous comments from his body of work.

HEAD QUARTERS, DEPARTMENT OF THE MISSOURI,
ST. LOUIS, NOV 26" 1862

Maj Genl Curtis
Dear Sir

Capt [Arnold] Hoeppner had or was making arrangments for my going to Springfield on a survey, and I had set my heart upon it[.]

Capt Turnball talks of sending me with Genl Davidson.[19] In view of this I ask to be relieved and sent to my regiment. If I am to do the most severe and laborious duties of the engineer department without the pay which such services deserve I prefer to go to my regiment. I have always tried to do my duty and it is with the most painful regret that I make this request, as I have received none but the kindest treatment from you who have treated me like a man regardless of my station as a private soldier[.]

I had hoped to complete the springfield survey in time to pass a pleasant week with my wife and chil[d] who proposed to visit me about Christmas, but if I am to be sent a long distance away with the poor pay of the field I had rather go to my Regiment[.]
Very Respectfuly

Lyman G. Bennett

Bennett Compiled Service Record, Thirty-Sixth Illinois, National Archives

HEAD QUARTERS, DEPARTMENT OF THE MISSOURI,
St. Louis, Mo Feb 17" 1863

Hon Owen Lovejoy
Dear Sir

I write you on the strength of being one of your constituents and a slight personal acquaintance in regard to a matter to which I would most respectfuly ask you[r] favorable consideration. It is a fact known to every one acquainted with him, that the present military governor of Arkansas (Hon John S Phelps) is intensly pro-slavery, and it is said since the issueing of the emancipation proclimation he is far from being truely loyal. At any rate he denounces that measure in no very modest terms and says he will never enforce the execution of its provisions. It is also true that he has done but little for the interests of the union people of that state, or for the government. Of the three regiments and Battalions of Arkansas troops.[,] All the commanders and very many of the subordinate officers, have been citizens of other states and forced upon them against their wishes. They have been personal friends of Gov Phelps and identified with his pro slavery ideas, and who cares but little about the Union provided slavery is not touched. They have no interest in Arkansas or the good of the union people there.

Had the right policy been pursued, Arkansas would to day have been where Missouri now is, instead of which, this city and the state as well as Illinois and Kansas is full of refugees from the rebel reign of terror which prevails there[.] Give them clothes, food and arms and in one month or in a very short time, ten thousand Arkansans would be found under our banners. I have been through the State have talked with hundreds, and believe this to be a fact. North of Arkansas river the people are far more loyal than in any portion of Missouri outside of St Louis.

Gen Curtis and others are making an effort to have the present governor removed, and a better, more loyal and an honest man and citizen of Arkansas appointed. I refer to Judge Isac Murphy of Huntsville. Perhaps you have heard of him, but it will do no harm to give a sketch of his services. He was the only man with enough principle and courage to cast a vote against secession in the Arkansas convention. His very countenance marks him as an honest man. On the final vote on the Ordinance of secession, but five were opposed to it. The president of the Convention in a most earnest and passionate address, appealed to the five and especialy to Judge Murphy, to change their votes and make it unanimous. When he was through with his harangue four of the five changed their votes, amidst the cheers which arose from the excited multitude which thronged the hall. In expectation that the Judge would do the same, the convention and spectators for a time remained in silence, and then calls were made for "Murphy." He arose and stated in substance, "that he had always lived in and for the union[.] He loved his country, and could not in this dark hour of its history change his allegiance or his devotion to it. He would never so strangle his sense of duty as to vote for secession as long as God spared his life. If they wished to kill him for this, he was ready to die." He reaffirmed his vote and wish[ed] it so expressed upon the records[.] The scene of wild and tumultuous excitement which followed beggars description. "Hang him! Shoot him! Kill him! rang from [the] gallary and hall, and it required the severest efforts of the more moderate to prevent him from being mobbed.[20] His constituents welcomed him with open arms. He had not proved recreant to their votes and wishes, and were ready at all times to defend him against Texan Rangers and desparados who sought his life. But so often and so perseveringly were efforts made to assassinate him, that he at length yielded to the solicitations of family and friends and came into our lines while we were at Forsyth, and has done valueable service ever since, a part of the time without pay, and living among the common soldiers. I have no personal interest in recommending him to you, only as I wish to see valueable services and sterling integrity rewarded and the best interests of the country subserved. Since he left his home his family have been persecuted and insulted, his house plundered of everything and wife and daughters forced to flee to this city. If you have any influence—as I hope and believe you have, and can bring it to bear in favor of his appointment, I hope you will use it.

Judge Murphy is now and always has been in favor of Emancipation—he endorses the proclamation in its fullest sense—Is in favor of the vigorous prosecution of the war and desires most of all to have the refugees and loyal men of Arkansas armed, and believes after three months he can be able to hold the State without assistance from outside[.] Any thing you may do to bring about his appointment will be most gratefuly remembered. Very respectfuly
Lyman G Bennett

[Last page is an abstract written in a period hand.]
Letter from Lyman G Bennett St Louis, Mo
Asks the removal of Hon John S Phelps, Military Governor of Arkansas, on the ground that he is in active sympathy with Secessionists—is thoroughly pro-Slavery & refuses to enforce the President's Proclamation—

Recommends the appointment of Judge Murphy, who is a loyal man in every respect and favors the Emancipation policy of the Administration, &c &c.

Library of Congress, Abraham Lincoln Papers

Washington City Feb 21 1863

Hon Abraham Lincoln
President
Dr. [Dear] Sir

I respectfully invite your attention to the accompanying letter of Mr. Bennett. I know him to be a man of truth, and his comments on Gov. Phelps, are worthy of your consideration.

I am unacquainted with Judge Murphy[.]
Truly Yours

O[.] Lovejoy

Library of Congress, Abraham Lincoln Papers

In the rapid succession of events and changes consequent upon the civil war in our country, and during the most active and exciting period of the contest, it was my fortune to be detached from my regiment and assigned to duty in the Topographical Office at the Head quarters of Major General Curtis at St. Louis, then in command of the Department of Missouri. It is needless to say that after six months familiarity with the details of map drawing, planning fortifications, and in the performance of the varied duties pertaining to this branch of the engineer service, I found the labor intolerably irksome. The dull routine of office duties and the almost stales prison like confinement within the limits of St. Louis was insufficient to satisfy my ambition or the cravings of an active and nervous temperament.

To the drudging war worn soldier in the field, my position, apparently so full of varied charms, so free from hardship and danger, would have been as desirable as a first class position in Paradise. But the construction of paper fortifications and the tracing of army movements in India ink lacked the thrilling interest of actual campaigning, and had effectually taken the romance out of that old topographical office, rendering life within its dull precincts a very plain and stupid affair, which gradually yielded to an intense desire for the more active duties of campaigning and the excitements of the march and battle field.

Such was the spirit of my dreams when Col. James M. Johnson[21] arrived from Fayetteville, where in a brief period he had recruited a regiment of loyal Arkansians. He stated that Western Arkansas was practically loyal, that the Magazine and Boston Mountains as well as the hills and valleys of White river were thronging with loyal refugees from rebel proscription and hate: that their love for the Union was strong and their alacrity to enlist could not be surpassed. He gave me an introduction to Mr. Wm. M. Fishback,[22] a prominent citizen of Fort Smith who had been driven a fugitive from his home, hunted like a wild beast of the forest, and a reward offered for his head, simply because he was true to the Union as the fathers made it.

He had obtained authority from Genl Curtis to recruit another Arkansas regiment. He had lived long in the Southwest and knew the feelings of the people relative to the contest that was ravaging the whole South: of the malignant hatred of those with whom war meant license, rapine and murder: of the time serving policy of the few men in representative positions, and of the narrow selfish views of the opulent planters and their ambitious sons. He spoke with feeling interest of the insults, outrage, suffering and beggery of the loyal men of the border, generally the poorer class who had stood firm against the surging waves of rebellion; and now that it was no longer possible for them to freely range the mountains, to hunt the wild deer, or eat their corn bread and bacon in quiet within their own cabins, were ready to shoulder the musket and swear lasting vengence upon those who had cruelly robbed and persecuted them.

He was sanguine of speedy success in recruiting, and as an inducement for me to accompany and assist him offered a prominent position in his regiment as soon as it could be organized. My desire for active service, as well as for some of the honors which the momentous events of the times were bestowing with lavish hand, were sufficient inducements for me to accede to his proposition, and I was accordingly "booked" for Fayetteville.

General Curtis was at first reluctant to let me go, having planned work for me at Fort Scott[23] and other out of the way places of the universe—but his warm and generous nature would not interpose a barrier to the attainment of a commission, and with many kind wishes for my success, a recruiting commission was given, and an order of exemption from duty at Headquarters with permission to return at any time in case of failure in securing the much coveted commission[.]

The morning of April 13th 1863, the day fixed upon for our departure for the scene of future adventures, was ushered in with a drizzling rain, and, where but the day before great clouds of dust were caught up in the giant grasp of the winds and sent whirling through the busy streets, there now was a thin plaster of light sticky mud to which the denizens of their antiquated burg are by no means strangers to. Through the mud and rain our party proceeded to the Pacific depot, in a sort of soppy condition, in appearance not unlike so many roosters just emerged from an involuntary bath in a swill barrel.

The depot as usual swarmed with the peddling "lazaironi" of the city. Scores of slovenly dirty faced girls, squadrons of ugly old women, seedy men and thieving boys, thronged around annoying us with their ceaseless importunities to buy their apples, peanuts or jackknives[.] The weather and the surroundings were not auspicious omens in our favor: and as we rattled over the "South West Branch" at a little more than cart horse speed, we had abundant oportunity to speculate upon the dangers, trials and hardships of the future, and wonder what reward fortune had in store for us. The South West Branch cannot be forgotten by any who in war times had the misfortune to ride over it. Running through a rough and broken region requiring long steep grades to surmount the rugged hills, to a nervous man, the speed would be excruciatingly slow: But when the heights were surmounted and the down grades reached, a rate of speed was at times attained that lead to the suspicion that we would soon be landed into a country of sultry climate—where secession had its birth and will have [its final burial place] [part in brackets is from the Huntington copy].

Half the time the cars seemed to be running on one wheel, then on another, making fearful leaps from rail to rail, setting one's teeth to rattling like a set of castinets. But finaly after having survived the perils of "riding on a rail," we reached Rolla in saf[e]ty just at sundown.

Making our way as best we could through the crowd of soldiers and "male whackers" who crowded around the station we sought for a hotel and were fortionate in secureing quarters and something to satisfy the cravings of a ravenous apetite at the Harding House, the "only first class hotel in town" as the runner confidently assured us. Hotels, lodging houses, and I was told empty dry goods boxes, were filled with guests to overflowing. Beds were at a premium, and more generally filed upon in advance by from two to a half dozen preemption claimants[.]

The rough ride over the South West Branch had made us hungry as sharks, and we were by no means backward in rushing with the crowd to the supper table at the first tap of the gong. Our supper consisted of commissary beef, potatoes[,] coffee, sour bread and butter, the latter being a compound of fierce odors and variegated colors. However, hunger renders the quality of our rations a matter of secondary importance, and by seasoning the butter with a liberal sprinkling of curses, and a few muttered growls it went the way of all victuals.

Little more than a year has elapsed since my previous visit to Rolla and during that short period the town had grown almost beyond recognition[.] I visited the former camp ground of the 36th Illinois and found it transformed into a city of log huts and temporary structures, tenanted by a motely heard of gaunt yellow legged and thriftless refugees. Refugees, or "Refujesuses" as the soldiers irreverently termed them is an outgrowth of the rebellion, and always infects the posts and temporary positions occupied by the Union army, and nothing short of the artist's pencil is equal to the task of delineating in true colors a family of the peculiar institution known as "Refugees."

The larger portion of the "Army of Missouri" commanded by Maj Genl [Francis J.] Herron was encamped in the immediate vicinity, reposing upon the laurels so bravely won upon the hard [next page is a drawing of refugees] fought field of Prairie Grove.[24]

We left Rolla at 2 P.M. of the 14th, after securing a four horse ambulance and the necessary outfit for the journey, and drove twelve miles to the Little Piney creek which we reached at sundown. The grand event of the evening was the presence of a pretty woman, young and unmarried, the daughter of our hostess. I had never been a very devout worshiper at the shrine of Southwestern beauty[.] The usually cadaverous faces, unkempt carroty hair, slipshod appearance, and above all the snuff stained lips of Missouri belles, were not at all times sufficient inducements for] rendering homage to their fascinations as demoiselles: but here was an exception: and truth compels me to say, that when we caught sight of the amply provided supper table, gracefully presided over by this blue eyed houri [?], we were ready to bow in adoration not alone to her good looks, but to her good sense in appreciating the fact that the principal avenue to a hungry man's heart is through his stomach. Col. Fishback undertook the agreeable and delicate task of cultivating her acquaintance and was making rapid and satisfactory progress until arriving at that part of the drama where she became the affianced of a tall and athletic Missouri Lieutenant, when he suddenly recollected that it was bed time.

Col. Fishback is a medium sized man, with a well made and wirey frame, with laughing blue eyes and light hair. A lawyer by profession and a graduate of the University of Virginia. He has a keen sense of the ludicrous, and is constantly bubbling over with fun. With an inexhaustible fund of humorous stories and an inimitable way of telling them there will be but little moping, or few lugubrious faces in a crowd of which he is a component part.

Descended from an aristocratic Virginia family yet having no sympathy in common with the sham chivalry of the South: he is fearless and outspoken in his devotion to the Union. He is a fluent easy speaker, and when fully aroused, the whole poetry of his nature comes out with eloquence and power.

There were six of our party, namely Col. Johnson, Col. Fishback, two gentlemen, father and son named Worthington from Arkansas,[25] and Lieutenant George [B.] Raymond and myself from Illinois[.] Raymond has received the appointment [of] Lieutenant in Col Johnson's regiment the 1st Arkansas Infantry,[26] and was on his way to join his command and assume the duties of his new position. Early the next morning the journey was resumed. The day being bright and beautiful, and the scenery wild beyond description, sufficed to turn all tender thoughts of our hostess' beautiful daughter, bag and baggage out of our minds.

The road from Rolla to Springfield passes through a mountainous region of rocks and black Jack, now winding along stupendous ridges—now [illegible words] dizzy deltas [?] now down steep and rocky descents, crossing rapid mountain torrents and then winding along the narrow valleys of streams or dry ravines. It is a wild and sparsely settled region, the grand stamping ground of "bushwhackers" and rendezvous for marauders of the worst description: and wo[e] to the luckless traveller who falls into their hands, for there is no atrocity to[o] inhuman for them to commit.

Our armament consisted of six revolvers and a rifle loaded and ready for an emergency, and any number less than a brigade of the "butternut gentry" would have met with a warm reception. So thought we except when passing some particularly dangerous point where had been enacted dark and bloody tragedies: then the forest depths were closely scanned with strained visions to detect signs of lurking danger. I am free to confess that at times would come the mental thought, "What if some fellow should by mistake give me a dose of those indigestible blue pills? Who would care for my amiable widow? and take charge of my rising generation?" But the warm sunshine and balmy breezes of spring laden with perfume from blooming red-bud and wild plum soon dispelled all thoughts of danger and foreboding[.] Spring had carpeted the hill sides with grass, opened the leaves and clothed the forest with robes of beauty.

We drove thirty eight miles and found comfortable quarters near the crossing of the Gasconade. One of our horses was from the condemned stock at Rolla and "gave out" on the third day and being too far "gone up" to be of more service, we gladly exchanged him for a stray US mule we found upon the way. While passing over a rough portion of the road one wheel of our ambulance was so badly broken as to be beyond the reach of temporary repairs, within the scope of our ingenuity to make. We conscripted a fence rail which did duty instead of the wheel and the whole party except the driver was obliged to walk eight miles to Lebanon where the remainder of the day was used up in repairing damages, as well as in limbering stiffened joints and healing blistered feet. Notwithstanding our anxiety to push ahead, and fretful impatience at the delay, it seemed that the wheelwright and blacksmith took a fiendish delight in prolonging the job, and our stay at Lebanon as long as possible, they working slower and accomplishing less than it had been supposed possible for even army workmen to be capable of. Twere well that it was so, for had we been able to proceed we should have stopped at a point upon the road where two hundred guerrillas encamped the same night and not the most skilled use of our firearms or of our legs could have prevented our being "gobbled up." I could then see that the breaking of our ambulance and the consequent delay was but the mysterious interposition of a kind Providence in our behalf. We learned that many Missouri rebels under cover of the thick foliage of the forest, were making their way through the woods to their former homes: plundering the inhabitants, stealing their stock and committing all manner of depredation and outrage on their way through the country. The band we missed was of this description, and had proceeded unmolested from the Arkansas river. In due time we arrived at Springfield without other incidents than many hearty laughs and cheerful song or story called out by unimportant adventures by the way.

At Springfield our horses were exchanged for mules and additions made to the commissarant preparatory to an onward movement. While the mules were being shod, I strolled about the town. The marks of war were but too plainly visible. Business was stagnant, the people listless and idle. Dead horses and mules were mouldering in the streets, and the air was impure and sickening[.] Many buildings in the southern outskirts were perforated with shot where Marmaduke had left his autographs as mementos of the 8th of January.[27] The town is situated on nearly the highest land in Missouri, the surrounding country sloping away in beautiful undulations of prairie and woodland. Were it not for the blighting influence of war, this would be a delightful region.

The first act upon the stage was a mule comedy consisting of a grand charge through the streets in which mules, men, ambulance and harness were precipitated into a confused state of "mixity," doing some damage to the latter which caused an hour's delay. When again ready for a start, a cavalcade of "mule whackers" was called into requisition, some to guide, some to hold and others to maul the long eared beasts into a realizing sense of their duty—reminding me of a milking scene in the song of

"Go to Ould Ireland where you will know
"How many it takes to milk and old ewe
"Two at her head and two at her hams
"And two little paddies to keep off the lamps."

We soon found that in exchanging teams we had the worst of the bargain as it was only by a constant and most vigorous use of the lash that they could be forced into any movement faster than the slowest walk.

Bands of refugees were frequently met, mostly from Arkansas not only from rebel tyranny and outrage but from starvation. Their goods, chattels and effects were tumbled promiscuously into dilapidated squeaking Ox carts and wagons, drawn by jaded steers and horses more starved and lean than Pharaos kine. Springfield was the objective point towards which their weary steps were tending. It was the Mecca of their hopes! Each convoy was accompanied by a pack of lean wolfish yellow dogs and swarms of ragged dirty faced and white headed children. [The next page is a drawing called Fording] So much poverty wretchedness and rags is seldom seen outside this war cursed country.

When within a few miles of Cassville we came upon a bevy of barefooted native damsels who were proceeding to the latter place on foot. Coming to Flat creek which was at least two feet deep we halted and were arranging to ferry them over, notwithstanding the ambulance was already crowded, and we were in a quandary to determine just how the matter could be accomplished when the "ladies" themselves solved the difficulty by raising their skirts and wading over, without giving us a chance for an exhibition of gallantry in their behalf.

At Cassville were many friends and acquaintances of Col. Fishback, who all spoke encouragingly of his prospects in soon recruiting a regiment. In the interval since my last visit to this town many changes for the worse had occurred and it was now what might be termed a "god forsaken" town. Two thirds of the buildings had been burned and those remaining, were in a fearfully dilapidated condition. Windows were shivered to atoms[.] Clapboards were torn off or left to clatter in the wind, mules and horses were stabled in deserted houses their sober faces and long ears protruding from parlor windows. Piles of brick from ruined buildings, straw offal and filth abounded everywhere. A more wretched and cheerless place for soldiers to be quartered or for human beings to live can not well be imagined.

Cassville was wild with excitement relative to the battle at Fayetteville[28] and the audacity of guerrillas in the immediate neighborhood. Rumors of the fight had been detailed to us while upon the road, and were corroborated on our arrival, together with the by no means cheering intelligence to us that all communication with that place was cut off except with large bodies of well armed troops. The roads were picketed with bushwhackers, and it was reported that 200 were at Cross Hollows laying in wait for the train then on its way to Fayetteville: in consequence of which it was detained at Cassville for further orders. In view of the situation it was deemed prudent for us to remain until a sufficient escort for our safety could be spared to accompany us. Col Johnson however procured a horse and went through a party of express riders in the night. The telegraph wires were being constantly cut, communication uncertain and the most exaggerated reports relative to the condition of affairs at Fayetteville received ready credence. On the 26th a party of eight men while upon the road with dispatches were attacked from the brush and two of their number quite severely wounded. A few hours later a second party was fired upon from near the same point, and some of their horses killed[.] On making a detour through the woods they discovered the guerrilla Camp and Capt. Wimpie[29] of the 1st Arkansas Cavalry was sent to break it up. This was all that was realy reliable relative to the operations of the bushwhackers.

Just as we were in expectation of a forward movement, an order was transmitted from Department Headquarters for the evacuation of Fayetteville and the retirement of its garrison to Springfield. This sadly interfered with our designs and rendered the raising of the regiment a question of time as well as of doubt and uncertainty. While awaiting the arrival of the troops there was but little for me to do but visit the camps of the Missouri and Arkansas soldiers, in which I learned much of their primitive habits and modes of life. The following illustrates their ideas of "high life[.]"

In a back and out of the way settlement in Arkansas resided a family, largely composed of grown up girls and boys. By dint of hard labor they became what in that region was termed wealthy. But schools and the appurtenances of more enlightened society had not yet reached their secluded settlement: and, as a consequence children grew up uneducated and unrefined. Gourds supplied the place of china and silver and greased sheet skin the place of glass windows[.] One of the sons by being away from the paternal roof had rather better advantages than his brothers and sisters, could read and write, and was somewhat familiar with the ways of the world outside. Becoming acquainted with a lady of some refinement residing in a town twenty miles away, he courted and won her affections. As a matter of course the backwoods brothers and sisters were present at the wedding and much astonished and bewildered at the magnificence of the bridal costume and the sumptuous repast that was provided for the occasion. On their return, in reply to her mother's questions the backwoods daughter and sister "Reconed that she sot a heap of store by Jeem's wife" and that she was "none your no account folks.["] But she could not "quite understand them ar new fangled notions o' their'n," and the "grand names they had for victuals and things. ["] "Why they axed me if I would have *gravy* and when I allowed I would, they guv me as pure sop as was ever sot on pap's table."

From official dispatches, and from conversations with those who participated in the affair, I obtained the following details of the battle at Fayetteville on the 18th. The troops occupying the town were exclusively Arkansians and consisted of the 1st Arkansas Cavalry by Col. Harrison, and the 1st Arkansas Infantry under Lieut. Col. Searles, the latter an incomplete regiment and but partially armed with such rifles and shot guns as were found in the country: the whole force amounting to 1200 men, the whole under the command of Col. M. La Rue Harrison.[30] The rebel Genl. Cabell[31] proceeded by forced marches from Ozark with 2000 men and three pieces of artillery, passing over the Boston Mountains, capturing a Lieutenant and eight men of the 1st cavalry who, without leave were attending a party nine miles from town. At day light, the pickets stationed on the Frog Bayou road were surprised and captured, and the first intimation the garrison had of the presence of the enemy was the deafening shouts of the rebels as they charged up a deep ravine which ran through the eastern portion of the Town, not far from where the infantry were encamped. Col. Searls[32] and Major Ham[33] rallied their men and hastily forming them in line, by a few well directed rallies checked the further advance of the enemy until the cavalry were also under arms. The infantry not being uniformed were in danger of being mistaken for the enemy, accordingly Col. Harrison placed the cavalry in front, and center, forming a reserve of a portion of the infantry and stationing the remainder upon the right and left wings. The enemy planted their guns on the side of the mountain to the East and overlooking the town and camps and opened upon them with cannister and shell: but notwithstanding the sharp crack of rifles, the incessant blaze of musketry and the thunder of artillery, the loyal Arkansian troops not only stubbornly maintained their ground but recovered some of the positions occupied by the enemy in the early part of the conflict. At 9 o'clock A.M. Col. Monroe[34] with the best of the enemies troops charged our right wing, and for a short time a desperate conflict was maintained. A murderous cross fire was brought to bear upon them when they retreated in disorder to the mountains east of town[.] Lt. Robb[35] of the cavalry advanced with a detachment to close rifle range of their artillery, and under cover of rocks, poured so hot a fire upon those having their guns in charge, as to effectually silence their fire and compel their withdrawal from the field. A straggling fight was maintained until noon when the discomfited rebels retreated precipitately over the mountains and back to Ozark leaving most of their dead and wounded upon the field. Col. Harrison's horses were so worn out and reduced in numbers as to render pursuit impossible. Our loss was 4 killed 26 wounded and sixteen prisoners.[36] The enemy left upon the field 20 killed 50 wounded and 55 prisoners.[37] Citizens reported one Col and several men as having died on their retreat. There was no distinction between the raw infantry and the more disciplined cavalry in repelling the assaults of the enemy. All were alike deserving. This was the first battle of the war in which the loyal men of Arkansas were alone opposed to the organized treason of the State, and their brilliant achievement gave a decided reproof to the rebel slander, that the union men of the South would not fight: and testifying to the national troops from the North that Arkansians are not only willing but anxious to second their efforts in rescuing their State from the grasp of traitors.

It was generally believed that Col Harrison by exaggerating the dangers that surrounded him, and representing that he could not longer hold the peace without reinforcements, knowing full well that there were no troops in the Department that could be spared, had induced Genl Curtis to order the evacuation.[38] The true state of facts were near Fayetteville since the action of the 18th was in less danger of being attacked than at any period since its occupation. It is true that with the unfolding of the leaves, brushwhackers emerged from their retreats and became more bold and daring in their operations, but the main body of the enemy were disheartened at their failure, while the Union people were correspondingly encouraged, and there had commenced a heigera from the mountains to enlist in the national service[.] The retrograde movement produced a most damaging effect throughout the border. Enlistments were stopped, and the union inhabitants left without protection and subject to the tender mercies of rebel desperados.

On the afternoon of the 25th of April the evacuation began. At Mount Comfort, three miles away, was a large camp of refugees, who hearing of the retrograde movement, came thronging into town and in almost every conceivable kind of vehicle took their place in the moving line. The loyal citizens of the place hastily gathered together such of their effects as could be most easily transported and prepared for their sorrowful journey. Those who could not depart waited in fear and trepidation the occupation of the town by rebel marauders and vagabonds, an event that was speedily to follow. The horses of the cavalry had been worn out in the service and most of the troopers marched on foot. The infantry had neither transportation or baggage, and but few arms, many even without shoes, pressed forward forlorn and despondent. In the refugee train were more than two thousand souls. Family after family moved along, the father careworn and dejected—the mother anxious, yet patient—and the children with a curious mixture of wonder and excitement that served to buoy up, rather than depress. All were in the lowest stages of destitution. The rude cart pulled along by half famished oxen contained such tattered effects as had escaped the hands of plundering rebels. Their houses had been burned: their cattle stolen, their farms devastated, and themselves driven from the homes of a life time. Many had fathers, sons or husbands in the federal army and were now bearing northward the mute testimonials of their devotion and sacrifices. Such were some of the scenes as described to me by an eye witness[.]

In the selection of camp grounds on their arrival at Cassville a difference arose between the two Colonels, enough to show that there was not the most cordial feeling between them. Col Johnson was very free in the expression of the opinion that to Col. Harrison the retreat was directly to be attributed. What his motive for the movement I could not comprehend unless it was in of a Brigadier's stars which no doubt to him appeared as large as [a] full moon.

The Infantry here received their clothing and soon was metamorphesed from dirty and ragged tatterdemaians into Federal soldiers in regulation blue. Their transformation from "Butternuts" to clean and not unlikely looking soldiers was realy magical. Their former habiliments were burned in order to the more effectually exterminate the mass of animal life which tenanted their old rags. After which Col Fishback made a two hours speech, which was received with much enthusiasm. When his thoughts ran back along the stream of time, and he recalled the days when our flag waved in proud triumph everywhere, hated by tyrants and revered by free men. When he recalled the days when the men of the North and the men of the South stood shoulder to shoulder like brothers to repel a common foe—when the women of the revolution gave their all, their husbands sons and brothers to purchase the inheritance over which we, their sons are fighting. When he recalled all the glorious reminiscences of the past and refered in impassioned sorrow to the present distracted state of the country as one mourning the loss of a great possession, the feelings of his audience were wrought up to an extent that burst forth in tears: and, as he opened up the future[,] gory with rivulets of brothers blood: strewed with the lifeless bodies of the sons of a common mother and littered with the wrecks and fragments of former greatness: the bronzed cheeks of the soldiers bore evidence of how deeply they felt the sad condition to which the country was reduced.

We proceeded to Springfield with the soldiers and refugees, who when in the line of march formed a column fully three miles in extent of crying children, men swearing, women scolding, dogs barking, mules braying, cattle bellowing and sheep bleating, which formed a babel of unearthly sounds, unparalleled by any thing that has been upon the boards since[.]

"Music heavenly maid was young" and unequalled by anything that ever has been, or ever will be again unless all the children, cats gobblers mules and cattle in creation are assembled for a grand concert[.] Among the heterogeneous multitude were women, mounted and armed with revolvers and knives a la maid of Saragossa style, and their manner and language conveyed the impression that these appendages to their toilets would be used upon the least provocation, or when the exigencies of the occasion required.

In passing over the battle field of Wilson Creek,[39] I was accompanied by a federal soldier and a rebel deserter who both participated in the conflict and to whom the ground was familiar. They pointed out the positions of the contending forces—the places where distinguished officers fell, and where the lamented Lyon yielded up his life for his country[.] It was near the top of a hill, on the southerly side of the valley and those who visit the spot are rearing a monument to his memory that bids fair to rival in its proportions any structure in the country. Thousands have rudely engraved their names upon the rough stones abounding in the country and deposited them upon the spot until it is gradually assuming colossal proportions. I also added a stone to the structure marked with my name and to the regiment to which I belonged which will no doubt soon lie deeply buried beneath the accumulating rocks which are constantly being added.[40] Time will unravel the threads of romance woven around this spot and I will venture the prediction that when the piping times of peace returns: when armies are disbanded, and soldiers no longer tramp over this hallowed ground with reverent hearts filled with cherished memories of the slain hero: some Missouri farmer or clod hopper with ruthless hands will desecrate this honored pile by carting off the slabs and fragments of which it is composed for less sentimental but more practical uses than perpetuating the spot of earth that drank the best blood of the nation, and they who look for the names rudely engraved on these rough stones, will find them treasured up in some cellar, or foundation wall of barns or pig stys.

For ten long weeks the Arkansas troops were encamped in and about Springfield: during which time military matters remained in status quo. All were impatient of the restraints and ennui of camp life, as two months before they were of their fatigues of marching. Only about thirty or forty men were recruited for the second regiment, and consequently little for Col. Fishback and myself to do but to remain in a state of "passivity—" as our spiritual friends would term it, waiting idly in camp for something to turn up: our only onerous duty being the absorption of hard tack and fat bacon. Ye Gods! how I loathed that cursed pork! as its stench arose from a hundred frying pans, its greasy slices covering every platter: and, as it lay in piles of superlative nastiness about the quarters of every mess. At times I devoutly wished that some second Moses could by an imperative edict given from the thunders and flames that in Divine sublimity covered Mt. Sinai, strike forever from the list of things tolerated on the tables and in the stomachs of men, that filth loving incarnation of all uncleanliness, the *Hog*!

A little affair in which Col Fishback and I were personally involved disturbed the usual serenity of camp, and for a time created no little excitement. An artillery company of Arkansians[41] had been recruited at Fayetteville under the auspices of Col Harrison. A portion of the men were from Fort Smith, and had not been formally mustered into the service of the United States, and being neighbors and warm friends of Col Fishback they were naturally desirous of joining his regiment[.] We hesitated to receive them until instructions could be had from St. Louis. In the mean time twenty men left the battery and came in a body to our camp, while at least thirty more were ready to come if the first succeeded in being transferred to our command. Capt Stark[42] arrested one or two who had strayed from our camp and had them confined in the guard house at Springfield as deserters, but Col Fishback procured their release. Forty men were at length sent to our camp to arrest the remainder but our men seized their arms, surrounded the detachment who sought to make the arrest and took them prisoners, but finally released them and sent them back without their so called deserters. Col Harrison getting wind of these operations caused the arrest of Col Fishback but a telegraphic order from St Louis procured his release. The recruits were finally induced to return to the battery on condition that their cases should be enquired into and they be allowed to join our regiment if it could be legally done. In retaliation Col Harrison induced two of our recruits to join his regiment. A polite request for their return was unheeded, but an order from the Post commander brought the Col to terms and our men were soon back and in the performance of their accustomed duty. Such of the recruits as we could not thoroughly depend upon to remain with us were sent eight miles away to guard a saw mill on James river, to prevent their being further tampered with.

While matters were quiet in camp, bushwhacking was active, small squads of "butternuts" were ever on the alert to plunder and murder those whose business called them beyond the lines: and at length they became so bold as to commit their depredations within the limits of the Post.

The family of Col Johnson resided three miles from camp, and it was his custom to pass his nights with them. Attempts were made to waylay and murder him on the road to and from camp. These attempts were frustrated by friends accompanying him well armed. Occasionally soldiers who straggled from camp mysteriously disappeared, and no trace of them were to be found. A private soldier from Col Johnson's regiment was employed as detective, and by passing himself off as a secessionest, became acquainted with the numbers and plans of the gang. All of them had taken the oath of allegiance to the government, and some were employed as clerks and laborers in the quartermaster's and other departments. Their principal operations were stealing horses and collecting information that would be of benefit to the enemy and forwarding it to Dixie. Col Johnson, Major Ham and other prominent Union men and natives of the South were especially selected as victims to be sacrificed at the shrine of secession. A night was fixed upon for the gang to meet in the woods not far from the Col['s] residence, to concoct a scheme for his murder. Timely notice of their designs was received from the detective and shortly before the hour of meeting a Lieutenant and myself were in the immediate neighborhood with a select and well armed squad of men. The night was favorable for us, being intensely dark and damp, the leaden clouds hanging low and sprinkling their aqueous contents over the leaves and bushes, and if by mistake they were rustled or a twig broken, the sound mingling with the moan of the wind and patter of rain drops, excited no suspicion of our presence and effectually concealed our movements. Our spy played his part well. And when it was supposed the assassins had all collected a simultaneous rush was made and the whole party captured. One fellow, a leader attempted to escape but ere he had gone five yards was shot dead. The trap had been sprung a little to[o] soon as only six were caught. These were secured, and I think would have been shot on the spot had I not dissuaded the men from carrying out their bloody purpose. The prisoners were marched through the darkness and mud and confined in the guard house at Springfield. Our detective was allowed ten days exemption from duty and not thinking of the probable results of his agency in bringing the desperadoes to grief he passed much of his time in the country. One day while on his way to camp an attempt was made to shoot him from the brush. The assassin's gun missed fire and before he could get in another shot or retreat from his covert, the soldier shot him in his tracks. This soldier was subsequently arrested and imprisoned upon the charge of murdering a citizen and then robbing him of five hundred dollars. None of the Arkansians believed him guilty, but supposed that some person connected with the outlaws who were privy to the murder had gained the ears of the Post commander and Provost Marshall and instigated proceedings that resulted in his imprisonment, he was a fine athletic soldier, and at the time I left Springfield was still in confinement his young and beautiful wife often sharing his prison with him.

In June, Col Fishback at the unanimous request of all the Arkansians proceeded to St Louis and Washington to represent to the authorities the state of affairs, and to urge a speedy movement of troops into the State. This step was taken not alone in our behalf but in the interest of the union people who were hemed in by guerillas, and who were suffering all the horrors of famine and siege among the mountains: and it was with no little anxiety that we awaited the result of his mission. Days and weeks passed without favorable news from him. The hot weather confined us closly to our tents or within the shade of apple trees. I rather enjoy respectable hot weather, but the intense heat that prevailed during the month of June, was slightly more than was laid down in the Bills. I will not undertake to say how many degrees the thermometer indicated as we had none in camp, but will venture the rough guess that it was somewhere between 500 and 1000 [degrees] in the shade. Now and then a Missourian or a "Rackensacker" was induced to enlist at about the rate of ⅓ a man per day. The hot weather and the *"cheager"* rendered it an onerous duty to attend to the wants of even this small number[.]

The *"cheager"* is exclusively a southern institution. Every body had them—had em bad, and when not sleeping, we were engaged in scratching[.] The *"cheager"* is an infinitesimal insect that penetrates the skin with the same facility as a badger would burrow into a sand bank. The only remedy resorted to was scratching, and it was no uncommon spectacle to see one thousand men scratching at one time[.] In short, scratching was reduced to a science, and not infrequently the order was heard—"Attention company! Guide right! Prepare to scratch in nine times,—Scratch!" Inspection of arms was responded to by inspection of finger nails. A man shouldered arms with one hand and scratched with the other, occasionally dropping his piece and scratching with both. Whatever else might be charged to the Arkansas soldiers it could not be said that they failed in coming well up to the scratch.

At the outset, there were persons not wanting in courage to venture alone and on their own responsibility through the cordon of rebel pickets that were stationed all along the border guarding every avenue of access and egress to the mountains. Such were furnished with recruiting papers and entered upon their perilous enterprise. Some were killed in the attempt, while others eluded the guerrilla rifle by penetrating the trackless forest, swimming rivers and climbing mountains on their way, guided only by the sun and stars: arriving at their destination they very soon succeeded in collecting a band of recruits.

Among those commissioned as recruiting officers was one Smith Brown a private of the 8th Mo Cav,[43] an ungainly Missourian, as lank and about as tall as a New England bean pole. In a few days he collected from the brush covered Ozark hills a number of recruits, among them one Shade Walden,[44] a denizen of the White river hills. Walden reported that a "right smart sprinkle" of boys were "laying out" in the woods and caverns who required but little coaxing to come out and join the Federals. Brown concluded to visit the locality, and, at his own request Walden went in advance to prepare the way.

With four of his recruits Brown proceeded from Cassville to White river reaching their destination about noon. A smoking dinner awaited them, and the boys were not slow in sitting down to it. They had scarcely commenced eating, when the house was suddenly surrounded by two score or more of yelling fiends belonging to Col Hunter's[45] rebel regiment, and commanded by Major Hammick. Walden rushed from the house and joined his rebel friends, and Mrs. Walden was following when Brown closed the door and threatened to shoot her if she attempted to leave. His comrades wanted to surrender, but were finally persuaded to stand their ground. The fight soon commenced and for two hours the hills resounded with the crack of rifles. The rebels finally withdrew, leaving three dead men and their commander mortally wounded. Not a man among the recruits was shot or injured. Major Hammick before his death, stated that Hunter was near at hand and would soon be there with his regiment, and advised Brown to leave the neighborhood. The advice was heeded and they left Walden's and hid in a cornfield on Roaring river until night, when, under its protecting shades, they made their way back to Cassville.

Another [recruiter] was Walter Brashears[46] a resident of Pope County, a tall stalwart and gray haired mountaineer of sixty years. He had three sons then in the Federal service, and the fourth, a stripling of seventeen was one of our first recruits. Brashears had in a short period raised a company of men who had been driven from their homes hunted like wild beasts from mountain to mountain and from cave to cave. They were like deer that have been incessantly pursued by hounds and hunters, timid and distrustful and wanting in courage to attempt their escape to our lines without aid[.] Mr. Brashears returned to Springfield alone, hoping that a scouting expedition might be sent down to aid his men in making their escape[.]

16 | Recruiting in Dixie, Part Two

July 6, 1863–November 13, 1864

> *After much deliberation, it was determined to proceed with only a small party of picked men, whose courage had been tested, whose ready ingenuity would suggest means of escape from any unlooked for peril and whose prudence at the same time would be a restraint in running blindly into danger.*
>
> —Lyman G. Bennett

On July 6, 1863, Lieutenant Bennett accompanied a small group of soldiers from Kansas and Arkansas on a perilous six-week mission into enemy-held Arkansas. The goal: to recruit Unionists for newly forming regiments. Leaving Springfield, the party made a dangerous crossing of the rain-swollen White River after convincing Confederate sympathizers to convey them across it. The men struck deep into the Ozarks, and while they were in the area, Bennett commented on the rugged landscape, the weather, Unionists, Confederate supporters, gunfights, and deeds of bravery and cowardice.

By December 1863, the Fourth Arkansas Cavalry was partly organized and stationed at Little Rock, Arkansas.[1] Several months later, on April 7, 1864, the War Department promoted Bennett from first lieutenant to major of the regiment.[2] Unfortunately, no surviving personal account exists of Bennett's experiences from the end of his recruiting expedition into Arkansas to his resignation from the Fourth Arkansas on August 22, 1864. Regrettably, his incomplete service record makes it impossible to ascertain his activities. The Fourth Arkansas did, however, scout regularly in the Little Rock area,[3] and detachments served at other locations, including Dardanelle and Helena as well as Montgomery and Prairie Counties.[4] It was perhaps sometime in this time period that Bennett sustained the loss of a fragment of his left radius caused "by a glancing gun-shot" that drove a "wristband button" into the bone.[5] In March 1864, a group of men from Company L requested that cavalry be sent to Montgomery County, west of Little Rock, "for the relief of our families and friends," and at least one detachment was sent there in June.[6] Ultimately, though, Bennett resigned his commission for the reasons stated in his letter of resignation.

Others of our recruiting officers had crossed the Arkansas river and were now in the Magazine Mountains and Walnut Hills of Montgomery County with their recruits anxiously awaiting an opportunity to escape to our lines, or hoping that a movement might be made in their behalf. As day succeeded day and still no indications of Northern Arkansas being occupied, a project for a movement of our own, independent of other troops was discussed[.] Our numbers were not sufficient to march openly with a fair prospect of successfully encountering the swarms of guerrillas that ranged the country between Springfield and the Boston Mountains. I likewise distrusted the coolness, courage and shrewdness of many of my men in an enterprise fraught with so much danger: and requiring all the fortitude and skill of the most intelligent and experienced among us. After much deliberation, it was determined to proceed with only a small party of picked men, whose courage had been tested, whose ready ingenuity would suggest means of escape from any unlooked for peril and whose prudence at the same time would be a restraint in running blindly into danger.

There was not a man among our recruits who was not eager for the enterprise. There was no difficulty in securing the requisite number, but rather in the selection of those on whom I could implicitly rely. Lieutenant Waugh[7] of the 2nd Kansas, and acting Judge Advocate, warmly seconded the movement, and rendered valueable assistance in procuring horses, arms, amunition and an outfit for the expedition. Eighteen men were selected: four from the 2nd Kansas,[8] the remainder Arkansians. All except myself exchanged their uniforms for citizens clothes—the ubiquitous butternut: so that we resembled in all respects rebel soldiers or brushwhackers except in our armament, which was the best that the ordinance department at Springfield could furnish. Imagine eighteen men mounted on bony steeds of all imaginable descriptions each clothed in a way exactly unlike any other, a few considerably more ragged and like scarecrows than their fellows, and one has a tolerable idea of the men composing the expedition, and of our appearance when the outfit was complete.

It was not until the afternoon of the 6th of July that our preparations were complete, our young Kansas friends occupying more time "than the law allowed" in writing to their Jerusias and Mary Anns which caused much delay in getting the whole party together. Parson Priddy[9] was on foot and cumbered with more luggage than the others combined: his pack projecting above his head, reminding one of an organ grinder with music box and monkey slung on his back. It is needless to add that the contents of that pack were strewn along the road for fifty miles as the toils and difficulties of the trip increased.

A thunder storm—one of the most terrible I had ever witnessed occurred before we were fairly under way and in a few minutes every man of us was wetter than if he had been ducked in a horse pond. The thunder rolled over the prairie as if each little eminence had become endued with the voice of a volcano. One may well imagine that traveling under such circumstances was anything but romantic and we were driven for shelter into a farm house but five miles from Springfield where we put ourselves to soak and our horses were turned in a field to graze. During the night it rained as if an ocean had been upset, and the sky seemed for hours one incessant blaze of ghastly flame. Before morning the storm abated and the next day's sun was as fair and smiling as the face of a woman.

Walter Brashears[10] was unanimously selected as captain and guide, his experience and familiarity with the country better fitting him for the position than any other of the party. We halted at Ozark for dinner and to remove a shoe from my horse he having become quite lame from what was supposed to be gravel, but which subsequently proved to be a bruse. During the day a lean raw boned horse was picked up and Parson Priddy mounted, his baggage in the mean time having been considerably reduced.

We followed down the valley of Bull creek, a rapid brawling mountain stream whose banks were full from recent rains, making it quite difficult to ford. One of the men named King being mounted on a diminutive filly was twice washed from its back, loosing his blankets and narrowly escaping from being drowned. This valley during the previous year had been the theatre of active bushwhacking and the inhabitants had been entirely exterminated by being either killed or driven away. The few remaining cabins being untenanted, the fields uncultivated and grown up to weeds. Deer and wild turkeys being about all the living objects we saw aside from our own party. Our voices and the clank of horse shoes upon the rocky road were the only sonnets that echoed through the forest and reverberated from the high hills which bounded and hemed in the valley, which was apparently as solitary and undisturbed as when it first emerged from the hand of creative power.

White river was higher than had been known for years, its banks being submerged and the valley overflowed, which obliged us to take to the hills and proceed as best we could through woods and thickets, now making a detour around projecting ledges and masses of fallen rock, stumbling over logs and timber until we reached a cabin four miles above Forsyth, built upon the hillside high above the madly flowing water. The family were in the lowest stages of destitution, and correspondingly ignorant, and secesh of course. We were informed that the only boat in the neighborhood was secreted upon the opposite side of the river and in possession of one Bird,[11] a noted bushwhacker and desperado.

We proceeded to the river bank and shouted for the canoe but could elicit no response. Then three or four of our party possessing deep bass voices gave a tremendous shout, and this not producing the desired answer the chorus was taken up by all, and in three seconds the whole crowd was shouting with the force of a thousand mule power, enough to shake the rotten confederacy from center to circumference. This had the desired effect and a long gaunt specimen of the genius [genus] butternut emerged from the brush and timidly approaching the bank in a drawling tone responded

"Ho thar what do youans want?"

"We want to cross the river, bring over your boat."

"Who be youans?"

"Southern men from Missouri."

"W,h,i,c,h!

"Missouri bushwhackers"—in a louder tone "going to Dixie to join Shelby[.]"[12]

Thereupon our interlocutor appeared to be in consultation with other parties who were concealed from our view. Presently three natives, more ragged and fiercer looking than the first and armed with knives and shot guns joined him and again demanded who we were. We repeated the story that we were Missouri bushwhackers from Saline County—that the "Feds" had made it to[o] hot for us to remain longer, and that we were now on our way to join Shelby. The river was at least forty rods wide and its swift roaring current rendered it necessary for the above conversation to be carried on in a high key. Even then our explanation was not satisfactory, for after a consultation among themselves we were informed that the boat would not be sent over. One of the Kansians named Kelley[13] proposed to swim the river and endeavor to arrange terms with our butternut friends. There was too much risk and danger in encountering the deep swift current for his plan to be successfully carried out. Indeed it is doubtful if the most expert swimmer could have successfully accomplished the feat. Accordingly we commenced negotiations anew, and appealed to their sympathies as true southern men "for God's sake" to help us across the river. That we would starve while waiting for the water to subside, or that some federal scout might "gobble us up," and that it was absolutely necessary for us to cross *now*. Then we were again subjected to a course of rigid questioning regarding our numbers, names, our leader, where we were from, where we were going, and finally

"Whar d'ye git them ar fedral clothes?"

Taken in 1864 when Lieutenant Bennett served in the Fourth Arkansas Cavalry (rather than the First Arkansas Cavalry as written on the photograph). Standing five feet, eight inches tall, Bennett had black eyes, brown hair, and a dark complexion according to regimental records found in his pension application.

"Cramped them:"

"And yer big critters, and revolvers and sich?"

"cromped [cramped?] them"

This answer was more satisfactory and an irresistible argument in our favor, and they finaly reluctantly consented to send a boy with the canoe on condition that our captain should return with him and give an account of ourselves, and if "youans are all right" they would assist in getting us over[.] To this we assented, and the boy pushed out into the stream, landing some distance below. Kelley volunteered to go envoy and negotiate for our passage. The guerrillas in the meantime were joined by others, and by the time Kelley reached the opposite shore, they numbered a dozen fierce and desperate looking devels as could well could be collected outside the bottomless pit. Kelley nothing daunted by their numbers greeted them as warmly as though they were dear friends and he a brother just returned after a ten years absence. Some of our party trembled for his saf[e]ty and brought their long ranged "Minnies" to bear upon the most conspicuous of the opposite party, with our fingers upon the trigger ready to fire upon the first hostile demonstration. The interview was successful, for after a short consultation Bird himself returned with the boat and after a careful inspection of our outfit, entered readily upon the work of ferrying us across the river. But two men and the equipment for one horse could be carried in the frail canoe, requiring eighteen trips to transport us all to the opposite bank. The horses were driven into the river in a body and by loud yells and belabouring some of them with cudgels they were directed to the opposite shore, which was gained by some without difficulty others were carried down stream nearly a mile, and three were drowned.

We were now fairly in the country of our enemies surrounded by bushwhackers and shareing the hospitality of cut throats. All retreat was cut off by a foaming and impassable river. The night was not pleasant: dense clouds curtained the forests and swept the mountain tops. The damp mists and sombre surroundings were sugestive of serious thoughts, and there was little moping and but little sleep for us that night. A feeling of insecurity pervaded the whole party[.] It would have been the easiest matter in the world for a lurking foe to surround and slaughter the last one of us. Kelley and myself were the only one[s] that appeared cheerful though costing a severe effort to do so. We passed the greater portion of the night in retailing fictitious stories of our achievments in bushwhacking in order to deepen the imposition we were practicing upon the credulity of our new found friends, and to rally the spirits of our own men. The rustle of leaves or the hum of nocturnal insects caused a tighter grasping of our pieces, and a closer scanning of the dark forest debths: and I am free to confess that I at least drew a long breath of relief when the morning again dawned brightly upon us.

We did not delay our departure from that dangerous locality and when the sun arose we were toiling up the steep sides of the White river hills and winding over ridges on our way southward, stopping at nine o clock to allow the horses to graze upon the tall blue joint grass growing in the jack oak barrons. An hours rest enabled us to more rapidly widen the distance between us and that worst scourge of the border, Bird and his band of outlaws.

"Halt thar!" was the rough and unexpected salutation that greeted us as we were suddenly confronted by a rebel picket stationed upon the way ten miles from the place of crossing the river. He demanded who we were, from whence we came, and our business. The ubiquitous Kelley again acted as chief spokesman, and repeating the same old story of our flight from Missouri, succeeded perfectly in "wooling" the sentinel, and placing us upon the most cordial terms with each other. He represented that he was one of a party of bushwhackers that were watching the road to intercept Union men who were passing from the Boston Mountains and that they had already summarily disposed of eight men. We were advised to avoid the mountains where Vanderpool[14] and Brashears had collected large numbers of "Mountain Feds," and, fortified within caves and behind rocky ledges had bade defiance to the forces sent against them. He stated that the lower country had been scoured by them for horses, arms, and provisions: and, added he, "if yer git into the clutches of that old d—l Brashears you are gone up sartin sure, as he makes mince meat of all southern men he can lay hold on." It was a little amusing to watch the expression of Capt Brashears face as he listened to these exaggerated statements of himself and his band who were yet in the mountains more than a hundred miles away, and it required an effort to keep our risibles down. We halted an hour, pumping information from the obliging and communicative rebel picket during which time sixteen of his ragged gang made their appearance. We had no means of knowing how many others were concealed in the dense mountain thickets, and as they detailed their bloody deeds of murdering union men, and outraging defensless women, the temptation to shoot them, almost mastered our prudence and we thought it a pity that such miscreants should be left to desecrate the earth.

After full directions regarding the roads, the location of gerilla camps, and the names of prominent secesh upon whom to call for food or aid, we proceeded on our way, thanking them for their information, and kindly wishes and advice, taking care however to avoid some of the localities indicated by him, not wishing to become anyways *burdensome* or for a closer acquaintance with our *"friends"* than was absolutely necessary for our several healths.

Avoiding the roads as much as possible, we made our way through the woods directing our course with a pocket compass, and camping at night in some secluded spot in the depths of the forest. Our progress was necessarily slow, and in climbing mountains and crossing rocky valleys and streams our horses were worn down and lamed. Instead of getting better, my horse had grown worse every day, and it was with the utmost difficulty he could be induced to move at all. The limited quantity of provisions with which we set out from Springfield becoming exhausted, we lived principally upon whortleberries and coffee, with now and then a wild turkey thrown in as a relish. Reaching a settlement on a branch of Long creek and not many miles from Carrollton[,] Kelley and another scoured the country for provisions and succeeded in procuring a sack of corn meal, from a citizen who returned to our camp with them. From the first the old man distrusted the truth of our story of our being Southern men, and finally with a good natured twinkle of his gray eyes he exclaimed

"Boys, old as I be, yer can't fool me." ["]Youans are no more secesh than I be, and God knows if there is anything I hate it is a dog and secesh[.]"

We were surprised at so open an avowal of loyalty in a region so notoriously disloyal: and when told that in Missouri such as he were usually treated to a halter and a limb, he was in no way disconcerted, but remarked that men did not usually hang their friends, and that he had no fears of so sudden and unromantic termination of his mortal career from our hands, and added that for years he had not dared breathe his true sentiments to his own family. He then informed us that if we were realy Southern men, by continuing our present course for four miles we would perhaps fall in with *friends*, as Captain Love[15] and about eighty rebels were encamped in the next valley: but if we wished to avoid them, we must make a detour to the left over high rocky ridges to the headwaters of Crooked creek, and from thence by the way of Gaither mountain to Jasper. It was deemed prudent not to avow our true character, but the old man was more than ever confirmed in his suspicions when we changed our course toward Crooked creek, and with moistened eyes he wished us Godspeed and success in our enterprise whatever it might be.

The rough character of the way before us was enough to cause the stoutest hearts and firmest nerves to hesitate. But there appeared no other way to safely extricate ourselves from the dangerous foes that environed us on every side, so we entered the timber, the green tops of the trees locking their branching arms over our heads: and then commenced a march, which I doubt if an equal number of men has ~~scarcely~~ ever accomplished and successfully overcome as many difficulties as we had to encounter. Our course lead over a succession of high chert ridges as rough as though dismembered by earthquakes[.] Now we climbed almost perpendicular heights, then plunged down narrow ravines, littered with boulders crossing small mountain streams, and then again toiling up ascents steeper if possible than the former, the rough stratas of rock cropping in ledges so high as to require a long detour to get around them, and then down again over trunks of fallen timber, and through dense thickets[.] Not a cheerful sound escaped the lips of the men, only savage oaths as sometimes horse and rider were rolled into the gullies worn by the rains, or dashed against the jagged points of chert rocks. At noon we had made but seven of the most laborious miles ever marched by man, arriving at length at one Lewis, on Crooked creek, an ardent secessionest who readily gave credence to our statements, and set before us the best his house afforded.

While enjoying his hospitalities and luxurating upon corn bread and honey, we were startled by the appearance of forty or fifty armed men who riding up demanded who we were and our business. Some of our Arkansians recognized Capt Love the guerrilla leader, the last man we wished to meet, to avoid whom we were now aching from many a bruise from the morning's march. Not wishing to cultivate a closer acquaintance with the man they most dreaded, some of the men instantly broke for the timber, leaping fences, logs and stumps like deer. Luckily they were not observed by Love, and by the representations of Kelley aided by "mine host," the Capt readily believed our statements and soon we were on the most friendly terms. So completely was he imposed upon, that we were not subjected to the usual amount of questioning. We were warned to avoid Jasper and the neighboring country as Vanderpool and his "southern Yanks" were lurking in the mountains and it was extremely dangerous passing over them to Clarksville. He offered a guide to pilot us through which was just what we did not want. As he was on an expedition to Yellville I suggested that he might need all his men and that Mr. Lewis could serve us just as well. The Capt was in a hurry and soon left, much to our relief, and after hunting up the "skedaddlers" we too were off, and truely grateful that we had not been detected. It was a narrow escape for us, as Capt Brashears and a number of the men were well known to Love. Brashears by means of a red handkerchief tied around his face to nurse an imaginary toothache, and by partially ducking his head in a clump of weeds, was not recognized. Our escape was regarded almost providential, as an unguarded expression or the recognition of a familiar face would have probably led to our extermination.

We found it very unhealthy for Union men in this neighborhood and accordingly "lit out" from Lewises as soon as Love and his blood hounds were well out of sight, pushing through the woods directly for Gaither Mountain, avoiding the roads and camps of the marauding bands that infested the country. Our horses were much worn and jaded and when late in the afternoon we reached the mountain, it was with the utmost difficulty they could be driven up the steep ascent. Soon Capt Brashear[s]'s horse gave out entirely and the others were but a few degrees removed from the same condition, and wholly unable to surmount the difficulties before us without rest. From our elevated position on the side of the mountain we had an extended view of the country and observed a commotion among the inhabitants in the valley below. Women and children were seen hurrying from house to house and men mounting their horses and riding rapidly away, a number collecting just before dark at a house a mile or more distant. It was evident that our presence in the neighborhood had alarmed the natives and there was no telling how soon a force would be collected and pounce upon us. We made another effort to proceed, but the horses being unable to move, we sought a black jack thicket where concealed from observation we camped for the night[.] Eight of the more timid ones, not likeing the situation and lacking the nerve to face and fight an enemy in case of attack pushed on to the top of the mountain, traveling until midnight, and we saw no more of them for some days, until after our arrival at Jasper. The weary men who remained, sheltered themselves behind logs or rocks and spreading their blankets upon the wet ground in the tall grass and weeds, in a few minutes were sound asleep, and executing a nasal chorus of the loudest, if not of the most enchanting character. I volunteered to stand guard to give the alarm in case of a visit from the neighboring valley. But vain were all my efforts to keep awake, whether walking or standing it was all the same. A ton of lead seemed resting on each eyelid and down they would come in spite of the most strenuous efforts, so wrapping my blanket around me, and secreted behind a log I too was soon sleeping as sleeps the innocent[.]

At daylight we were up and moving and in two hours had attained the top of the mountain, where we halted for all parties to graze, the horses on grass and the men on whortle berries. A parapet of perpendicular rocks surrounded the mountain top, through which there were but two or three openings where it was possible to descend. Unfortunately we struck the wrong one and entered a valley the waters of which flowed north in an opposite direction than the one we wished to go. Capt Brashear[s]'s knowledge of the country was at fault and we could not determine our precise latitude and longitude, accordingly I proceeded down the valley four miles on foot before finding a house and people who could give the desired information. To our surprise we were within a short distance of Carollton and pursuing a course that would take us to Missouri instead of Arkansas[.] Returning to the mountain we at last struck the right trail which we followed until night, camping as usual in the woods. A yearling calf whose guileless and unsuspecting nature unfortunately brought him in contact with an ounce bullet, furnished us with choice steaks upon which we regaled heartily. We laid by the following day, partly to refresh the horses, and to guard the carcass of the slaughtered calf, and prevent the wol[v]es from gorging themselves with beef that would inevitable [have] been left behind had we been in a hurry. Some of the Kansians jayhawked three or four horses from a heard they found straying in the woods, by which means the whole party were remounted, and we again pressed forward following a bridle path down the mountain into the valley of Buffalo creek. The inhabitants were mostly union men and within the protecting care of Vanderpool and his dreaded mountaineers, and for the first time since leaving Springfield we felt safe in avowing our true character, and purpose.

We had yet to cross an intervening range of the Boston Mountains before reaching Jasper. The condition of our horses caused grave doubts of our being able to achieve the ascent, but a firm resolve to see the end of an adventure now so nearly accomplished induced us to make the attempt. It may well be supposed that by this time we had thoroughly mastered the science of hill climbing, so dismounting we urged the horses up the rocky spurs of the mountain by winding and unfrequented bridle paths amidst the solitude of the primeval wilderness which covered these rocky heights. At noon of the 16th after weary hours of fatiguing travel we reached Jaspar and partaking of a sumptuous repast of corn bread, bacon and honey, we proceeded to the cabin of Capt Vanderpool to concert with him for future operations.

Capt Vanderpool was not at home, nor could we induce his wife to impart any information respecting him. We were strangers, and how should she know that our object in wishing to see her husband was anything but friendly. Fortunately we found Parson Priddy, one of the number who deserted us on Gaither mountain, who informed us that Vanderpool would be home in the evening[.]

It required but little time to explore the town. A dozen of the most primitive log cabins, which no one can adequately describe, or form a conception of, who is unacquainted with the prevailing style of architecture in the more remote regions of the south:, and about equally divided as to possessory rights between pigs, poultry, gaunt yellow dogs and still less prepossessing human beings: constitutes the town of Jasper, the capitol of Newton County and the rendezvous for persecuted loyalists throughout northern Arkansas. A log court House but little larger than the summer kitchen of a northern farmer was by far the most aristocratic edifice [drawing of Jasper on the next page] in the place, there being in addition to the interstices between the logs, one small glass window to illuminate the understanding of the dispenser of public Justice. I was told that no other building in Newton County was deformed by such a appendage as a glass window. In the classical dialect of the country "Whars the use of winders when a right smart light can git thrugh the chinkins[.]"

The town is quite in keeping with its wild surroundings, being situated in a narrow rocky valley or gorge through which Hudsons Fork, a considerable stream, goes brawling over rocks and cascades to White river, and nearly surrounded by overhanging escarpments of rocks and towering mountain heights which effectually intercept the rays of the sun except at midday when they descend nearly vertically into the narrow valley.

Captain Vanderpool returned in the evening, accompanied by a half dozen of his ragged but athletic followers. The Captain is a tall and powerfully built man, with mild and pleasing blue eyes and none of the savage brutality about him which rebel bushwhackers so freely attribute to him. He was a regularly commissioned officer in the Federal service as Captain of Co. E 1st Arkansas Infantry and already had more than sufficient followers to complete his company to the maximum, and was quite willing to turn over his superfluous members to the second regiment, and would aid in recruiting others for the same purpose.

In order to embrace as wide a scope of country as possible, it was arranged for Capt Brashears with nearly one half of our party to proceed to Pope County, forty or fifty miles away, while Mr Spivy,[16] King[17] and others were to cross the Arkansas river and operate in Hot Spring, Montgomery and other counties south of the river. Myself and the balance were to remain with Vanderpool. After partaking a hearty supper of fresh pork corn bread and milk, the whole male portion of the party took up their lodgings in a neighboring thicket, as a precaution against prowling bands of bushwhackers who might at any moment swoop down the valley and gobble any who were so unfortunate as to be at home, and inasmuch as the aforesaid "whackers" were in the habit of haltering all who fell in their homes and ornamenting trees with their bodies, and the idea of being made into buzzard feed was not a particularly pleasing one: as a consequence the adult male inhabitants and refugees in the country had not for months experienced the luxury of a bed within the walls, and under the roofs of their own cabins.

Among those who lodged with us that night in the woods were two who had recently been confined in the rebel prison pen at Little Rock for being accomplices of "Wild Bill"[18] and sentenced to be shot, but the night before the expected shooting, they had excavated a hole under the stoc[k]ade and crawled out and making their way to the mountains had reached Jasper a few hours before our arrival. These were our first recruits and by eight o,clock the next day, twelve more were enrolled and sworn in to the 2nd Arkansas,[19] which as Capt Brashears remarked was "mighty encouraging[.]"

Our party now separated and proceeded to the various stations assigned them. Leaving my lame horse at Jasper I proceeded on foot over Mount Judah to Big Creek where most of Vanderpool's followers were encamped. Such a motely collection of shoeless, hatless and desperate set of rag muffins I never before had seen. The dilapidated fragments of rags called clothing which at unfrequent intervals shielded their persons from the elements, looked as if they had passed through a threshing machine or saw [?] mill, and then pitched onto them with a hayfork. Falstaff's historic brigade however scanty their wardrobe were extravagantly attired when compared with these rough though loyal hearted mountaineers.

An expedition to Johnson County and the Arkansas river valley was set on foot, for the purpose of securing recruits and giving the "Jonnies" to understand that raiding was not altogether a one sided affair, and that blows could be given as well as received. In short I can call it by no milder term than a freebooting expedition for the purpose of "Jayhawking" horses and such other plunder as, should come in our way: particularly such articles as would replenish the exhausted larders of the mountaineers. My object was men for the 2nd Regiment, not plunder.

Upon setting out the expedition numbered about one hundred men, but as we proceeded up the Valley of Big Creek and over the mountains we were joined by others until in numbers we were formidable. A portion were unmounted and unarmed and I must confess that I felt some misgivings as to the courage and efficiency of so undisciplined a rabble, and but little confidence in a successful result.

No one, not even a wooden automaton could restrain its laughter at sight of an original company of stalwart, skin clad, or next to no clad "Mountain Feds." As a natural curiosity Barnum[20] has nothing that will surpass it. Imagine one hundred and fifty men on horses of all imaginable descriptions, each one clothed in a way exactly unlike any other, and considerably more ragged and like a scare crow than any of his fellows: each carrying closely hugged an old shot gun or squirrel rifle as long as a liberty pole, and with long unkempt locks and bushy beards streaming in the wind, and you have a faint idea of the prepossessing as well as formidable aspect of our cavalcade. The system of tactics and orders by which their unwieldy evolutions were performed was equally original and ludicrous. Imagine Casey's or Hardee's[21] look of blank astonishment at hearing yelled out an order like the following

"Now fellers git into a long straight string and then light out end wise."

After a long day's ride over mountains and through defiles, we entered the valley of Piney creek, the inhabitants of which were represented to be intensely bitter towards the "Mountain Feds," consequently everything that could be subservient to our wants was regarded as lawful plunder and at once confiscated. Horses, saddles, old [illegible word], bacon, meal and salt were scarce articles in that valley after we had passed. We also captured four rebel soldiers who were at home on furlough.

The next day we crossed another range of mountains and descended into the valley of Limestone creek, the inhabitants of which had equal cause to remember the tender mercies of the "Mountain Feds" with those of Piney creek. The thieving propensities of the men completely disgusted me and even at that long distance from home I became heartily ashamed of myself for being caught in such company.

In the afternoon information was obtained that Capt. Crisman,[22] a noted desperado and guerrilla, was at home. The

[The next two pages are missing from Bennett's account.]

three of our men were dead and three others more or less severely wounded. When Crisman fell a general rush was made and his body completely riddled with shot. He was a powerfully built man, with physical and mental endowments that should have fitted him for other and better employment than that of bushwhacker, and who deserved a less ignominious and tragic end. I could but admire his cool intrepidity and courage. Yet if one half the tales of his brutality was true he was stamped all over with ruffian and cold blooded murderer: and in his death the country was rid of a terrible scourge, and we did his wife a favor in releasing her from a disgusting profligate and brute.

The ghastly bodies of the dead the groans of the wounded—the grass and fence besmeared with blood and the frantic shrieks of Mrs Crisman formed a scene of horror never to be forgotten. All stood in mute dismay at the bloody picture[.] The house escaped the usual ransacking: only the Captain's arms and a favorite horse a most beautiful animal, was taken. The horse was subsequently given to me by one of my recruits and often bore me in saf[e]ly among the mountain fastnesses and from many an impending danger. Parson Priddy must not be forgotten. In the melee he was thrown from his horse and nearly killed. Our dead were wrapped in blankets or sheets a shallow grave was dug in a neighboring field, and though aching from his bruises Parson Priddey said a short prayer over their remains which were then consigned to the earth to await the final requisition of the Judgment.

Our march was continued until ten oclock at night in order to surprise and capture one or two prominent characters who were reported at home. But the birds had flown and we only received the benison of a trio of she vipers who wished our speedy exit into a region the climate of which is more tropical than pleasant. They were politely requested to "dry up," whether they heeded the simple request I have no means of knowing as I was more interested in securing food and rest for myself and horse in a neighboring corn field where we finaly halted and went into camp.

The succeeding day we passed over another mountain ridge as usual and down the valley of Mulberry creek: reaching the house of Major Farmer[23] about noon. As he had held the position of Major in the rebel army and participated in some of the most important events of the war, it was supposed he was a rebel still and a legitimate object for plunder. He had a fine garden and a number of beehives, and we being hungry did as hungry men are apt to do under such circumstances and naturally "went for them." Mrs Farmer[24] soon made herself acquainted with the character and politics of our "outfit" and directed one of her daughters to proceed to the woods and "call pap." In a few minutes the Major made his appearance with two of his sons and to our surprise greeted us as friends and directed the men to help themselves, as anything he had was free to loyal and union loving men. We learned that the Major becoming disgusted with the rebel service, had thrown up his commission immediately after the battle of Pea Ridge and retired to his home in the Valley of the Mullberry: where he had hoped to remain undisturbed, a quiet spectator of the events that were transpiring around him.[25] But no man in the South could long remain neutral. Threats and petty annoyances were resorted to, to provoke him into acts or expressions of disloyalty to the South, as a pretext for his arrest. These failing to disturb his equanimity, more desperate measures were resorted to. For an unguarded expression, one of his sons was hung up like a dog, to a tree and died as martyrs die for devotion to principal[.] The father and two sons only avoided a similar fate by fleeing to the mountains where for months they had remained though hunted like wild beasts from one place of refuge to another. At times his house was watched for days, and often his place of concealment had been discovered by his wife and daughters being followed at night when carrying provisions for his sustanance, and narrowly escaped falling into his enemies hands. Other men too sorely persecuted to remain at home, and deserters from the southern army had joined him and at the present time he was the nominal leader, or rather adviser of a large number who like him were stealthily fleeing from mountain to mountain and from cave to cave to elude the relentless conscript officer, or escape the halter or bullet of the assassin. The major's simple story enlisted the sympathy of the whole command and not another bee hive was despoiled of its contents, or a potato or onion taken from the garden only as they were pressed upon us by the kindly hospitality of his family. Addressing himself to me, he asked what he should do? To remain longer at home, or even in the country was impossible. Would the authorities forgive him for the part he had taken, and his year's personal service in the cause of the rebellion?

I told him of the hopelessness of their cause. That Vicksburg and Port Hudson[26] were ours—that Rosencrans was thundering at the gates of Chattanooga[27] and cutting the so called confederacy in twain and that the last lingering hope of rebels in Arkansas had been extinguished on the field of Prairie Grove. The loyal men throughout the south were awakening from their lethargic sleep and rallying in their power and might in support of the common cause. Neutrality was impossible and if his heart and sympathies were with us there was no other recourse for him than to join his fortunes with ours and fight it out to the end.

"I will do it!" said he, "and in ten days will join you with a hundred men." Then calling the men around him, he addressed them with such words of earnest eloquence as I little expected to hear in that far off region. He was a man of culture: had been an active participant in the politics of the state, and withal a fluent speaker: and warmed up by the importance of his theme his words were listened to with thrilling interest. At the conclusion of his little speech the valley rang with shouts of applause. Our prisoners, some eighteen in number caught the enthusiasm and were eager to enlist, and were accordingly enrolled for the 2d Regiment, and to their credit I will add that throughout the remaining long and dreary months of the war, with but two exceptions they were true to the oath that day voluntarily assumed.

Towards evening we left our warm hearted and new found friends and proceeded eight miles down the romantic valley of the Mullberry, which was flanked by the Boston Mountains on the North and the Mullberry Mountains on the South. Their summits softened by hazy mists loomed upon the horizon many miles away. The valley was dotted with farm houses, and at intervals were well cultivated fields: but by far the greater number of the farms was covered with a dense growth of rank unsightly weeds[.] From prisoners which were constantly being brought in, we learned that an organized force of the enemy was at no great distance down the valley, and to guard against surprise vidett[e]s and flanking parties were thrown out, and as much precautionary formality assumed, as though we were the best disciplined army in the service, instead of a roving mob of scarecrow mountain rangers. From them we also learned full particulars of a hard fought, and to the enemy, terribly destructive battle at Helena on the 4th of July in which the whole rebel force in Arkansas under [Theophilus H.] Holm[e]s, Price[28] and Hindman,[29] after gaining the bluffs overlooking the town, and capturing some of the outworks, were repulsed with heavy loss when within the range of the guns of Fort Curtis.[30] I felt no little pride in this achievement, as but little more than a year had elapsed since I had selected the site and planned the works of this Fort. The surrender of Vicksburg was also confirmed, vague rumors of which had already reached the mountains.

By some unexplained process our stock of horses was rapidly augmented and the ranks of long haired dejected gray back prisoners was likewise increased. Considerable booty in the way of corn meal bacon salt and iron was also collected, for every wagon, plow, or other implement composed wholey or in part of iron was considered lawful contraband of war and was immediately confiscated, the wood worked [?] knocked into pieces [?] and the iron speedily transferred to the pack saddles, for the purpose of being transformed into horse shoes. The people were secesh to the core. Even "delicate ladies" evinced their refinement and their politics as well by the most abusive and often profane and filthy language, bestowed in no stinted allowance upon us. I saw one she viper deliberately walk up and spit in the face of one of the men. The nasty—well I won't dirty my paper with epithets applicable to such a woman.

The condition of the roads, the darkness and exhausted condition of the command compelled a halt. We camped in a wheat field which had but recently been harvested and surprised a half dozen soldiers belonging to Hindman's Command who were at home on furlough to attend to their harvest. They were sleeping among the grain stacks and were much chagrined at being taken in by the detested "Mountain Feds[.]" Upon consultation it was judged expedient to proceed no further with the expedition but return to Jasper by a different route than the one by which we had come. Accordingly the next morning we climed the precipitous mountain by a rough unfrequented and almost undistinguishable bridle path[.] While toiling up the steep ascents the irrepressible gayety of the men broke jubilantly forth in wild shouts and songs which made the forest aisles echo and the welkin ring. To cap the climax a tall specimen of the genus "Arkansaw," a hatless, shoeless, skin clad mountaineer, with several yards more or less of legs dangling down the sides of a diminutive pony, suddenly broke out with "Gaily the

Trubadour touched his guitar.["] "When he was hastening home from the war: Singing from Palestine hither I come. Lady love! Lady love! Welcome me home."[31]

How in the name of the murdered muses the "gay Trubadour" found his way among these intermmable hills was too much for my benighted intelligence to fathom. I could do nothing but laugh[.] It was enough to provoke the stern gods of heathen mythology to laughter.

The fortunate results of the raid wrought a wondrous effect upon the spirits of the men, with panniers plethoric with plunder and provisions for the relief of suffering Betsys and starving babies felt like conquering gladiators marching home in triumph.

While threading our way among the hills, by paths which appeared to me would never lead us out of the wilderness, we were suddenly enveloped in one of those terrific storms which often sweep these mountains with fury. The rain poured down in torrents the forked lightning flashed, and the thunder rolled continuously. It was as if the entire aerial artillery and all voidless forms and forces of the world of spirits and space had been brought together for a grand field day. The scene was fearfully sublime, vastly magnificent in scale and wildly tumultuous in its uproar. Great trees fell crashing in the forest. Rocks loosened by the waters went thundering down the mountain sides, wakening echoes which vied with the roar of the elements. The heavens were black as fabled Erubus[32] except when riven by the lurid lightning. No shelter was at hand: and when at length the storm had spent its fury and the heavy masses of black cloud moved slowly across the heavens unveiling the sun and the blue sky which smiled serenely upon us a more drizzled, woebegone, storm washed and thoroughly soaked set of bipeds never were fished from mill pond or sea.

On reaching Hudsons Fork we went into camp, and while a portion of the men were engaged in the distribution of the products of the raid among the needy ones, the ballance were busy in wielding the scraps of iron collected on the expedition, into horse shoes: and at the expiration of a week all were in readiness for anything that might turn up in the shape of plunder or "scrimage." In the meantime, Lt Fesperman[33] arrived from Cassville with orders for Vanderpool to return immediately to his regiment, which command the Capt was disposed to obey, without awaiting the return of Capt Brashears, or of Major Farmer who were daily expected to join us. We were even upon the march and had proceeded a mile or two when a messenger arrived in hot haste from Big Creek with the intelligence that Capt Cecil[34] at the head of a gang of rebel marauders was devastating that valley, burning the houses of the union inhabitants and murdering many who fell into his hands. Our informant was a comely lass of sixteen who rode a barebacked steed, and with hair streaming in the wind looked like the picture of Mazeppa[35] or some of the wild goddesses, as she dashed along the valley at breakneck speed. She had crossed the mountains a distance of fifteen miles in a little more than two hours. One hundred of the most trusty and best mounted of the command were detailed, and directed to cross the range and strike Big Creek valley well up towards its head in order to intercept the enemy, who, it was conjectured, would raid the whole valley and escape to the mountains which bordered White river. The ballance of the men were directed to go into camp and await our return. On reaching Big creek we learned that Cecil had penetrated the valley but a short distance and then returned the way he had come, and that he was four hours in advance of us. At midnight we had rode forty miles since two oclock P.M. Horses and men were exhausted. It was difficult following their trail in the darkness[.] In the morning the pursuit was renewed, and after two miles we found the smouldering embers of their camp fires. They had been apprised of our approach and fled before day light. Citizens informed us of the situation and it was at once apparent that we could not overtake them. We were now in a hostile neighborhood, and from a pursuit of a flying enemy, the expedition was changed to a raid. Again the familiar sight of led horses laden with sacks of meal and salt, with now and then a side of odorous bacon, fat and oozy, anointing the animals' sides with its oleaginous drippings, greeted our eyes.

During the day I was attacked with severe pains in my side, which increased until every joint and muscle was aching and at 4 P.M. I could proceed no further. The severe ordeal of a forty mile gallop by night over mountains and paths scarcely practicable for a goat, together with scanty and unwholesome food and the fatigues of the last few weeks were too much for ordinary human endurance. Almost the only food to be had in this famine struck region was "corn pones" and fresh pork, unseasoned with salt or butter. However hunger is not particular, and, after a twenty four or forty eight hour fast with plenty of hard riding thrown in for a relish, even fat and fierce odored meat roasted at a camp fire upon the sharpened end of a stick, seasoned with a liberal sprinkling of honest growls can be made to go the way of all victuals. I have lived for days upon fat pork alone, and when at intervals a "corn pone" was attainable it was regarded in the light of a rare luxury, and bolted down with all the gusto and voracity of a halfstarved cannibal dining upon a nice new infant. Heaven preserve these mountain dames who from their scanty stores were ever ready to divide with the stranger in their midst.

The command was halted and went into camp at Adair's on Flat Rock creek[.] Not withstanding the secession sentiments of Mr Adair[36] he was kind and hospitable to me, and during three days of acute suffering from the Pleurisy he and his excellent lady bestowed every attention, and apparently manifested as much interest in my welfare, as could have been bestowed upon an only son. Nor was Vanderpool and the men of the command wanting in sympathy and devotion and though fifty miles from Jasper and in the midst of enemies, never a suggestion was made to leave me behind as long as there were hopes of recovery.

Vanderpool's camp was near a spring a half mile from the house. At day light of the second day I was awakened by the tramp of horses over the stony road and the voices of men in and around the house. Looking through a crevice near my bed I was surprised to see a body of secesh filing into the yard. Soon the voices of Mr and Mrs Adair was heard pleading with Captains Cecil and Love, for my life. No mother could have pleaded the cause of her child with more ernestness and as effectually than was her intercession in my behalf, until a promise was exacted that I should not be molested until sufficiently recovered to be removed from the neighborhood. Cecil came to my room and was catechising me in regard to my condition, the number and location of our men, and our designs when a rapid firing in the direction of our camp called him unceremoniously away without taking my rifle and revolver which were plainly in view. In a few minutes the whirr-r-r of shot was singing about the house, and then a skedaddle of rebs, with Vanderpool and the mountaineers pouring shot into their rear succeeded. They soon left the road crossed a cornfield, and made for the mountain bordering the valley on the north. So congenial a task as the pursuit of a retreating foe could not be resisted by the most timid of our recruits. "There they go!" "I reckon we've got em now. Hurrah!" and yelling like fiends they charged after them like an avalanche. A dozen saddles were emptied and a dozen steeds bounded riderless in every direction: an example that was followed by the remainder in eluding [?] their riders, who incited them thereto by a vigorous application of spurs and gun barrels. In a short time the brigands were scattered in all directions[.] I think I am safe in saying that that portion of secession represented by Captains Love and Cecil's butternuts, which in the early morning came pouring over the hills so highly exhilarated at the prospect of "chawing up" the detested "Mountain Feds," experienced a sudden revulsion of feelings after so telling a demonstration of our affection for them, and their disappearance was at a rate of speed which far outdid their efforts in coming. As long as there was a tail or main or fluttering butternut rag in sight to shoot at, Capt Vanderpool kept up a lively rifle concert.

In this gallant charge Lieutenant Fesperman was killed and one of his men severely wounded. The brave Lieutenant, bareheaded and half crazy with excitement alone in and in advance of the command swooped down upon the main body of the enemy, firing his revolvers with the rapidity of a clock and met his fate amidst the piled up bodies of the foe[.]

The painful duty of his burial was performed in the afternoon, not a man but was deeply stirred both with pride and sorrow: pride at his gallantry and daring, and sorrow that when brave men were so greatly needed, he should so soon seal his devotion to his country with his blood. Parson Priddy performed the last sad rites, and at the grave poured forth an earnest prayer that moved the whole assembly to tears.

In this affair the Parson's horse was shot under him and killed. "I declare for it" said he in giving the details, "It was one of God's miracles that I was not shot instead of the horse." Hardshell as he is, all the haps and mishaps which befall him he ascribes to the direct interposition of the hand of Providence and the "wonderful miracles" daily performed in his behalf would if written out and canonized would add another book of Jonah to our present version of the Bible[.] One graceless wag, just from "Doubting Castle" who does not "go much" on the supernatural "allows that the Parson has a monopoly of the time and attention of Deity who has his hands full in getting him out of difficulty—and that the rest of the human race must suffer thereby.

The enemy left their dead and wounded where they fell. Those who went over the ground reported ten dead bodies and thirteen of their wounded were brought to Adairs, thus taxing a new the kind hospitality of these people who had so assiduously cared for me.

Another day was passed upon Flat Rock creek. Rumors of the gathering of secesh clans to drive us from the valley were constantly reaching us, and exciting the fears of the men who began to complain at being detained here so long. It was finally arranged that Mrs Adair should conduct me through the forest to the house of Mr Sholes a union man living near the summit of the range, while the command should proceed homeward. Blankets and pillows were strapped to the back of a horse old and lazy upon which I was carefully placed and then a farewell grasp from the rough hands of the men and then we seperated as many supposed forever. Mrs Adair led the horse six miles up the mountain, and felt as if a great load was removed from her heart when she saw me again safely in bed. But that ride was one of intense suffering. A hand saw in the hands of [a] giant tugging at my ribs and severing the quivering flesh could scarcely have inflicted more pain. Mr Sholes lanced both arms with a razor and after the extraction of a quantity of black thickened blood I felt relieved and sank into a quiet sleep.

On awakening I was free from pain, and better in every respect. My only anxiety now was the certainty of being captured should I remain here. Accordingly the following day I was removed to the valley below, and from thence by easy stages to Vanderpool's, a mile below Jasper. None of the command had returned, and as straggling secessionists might be lurking around, for greater security I was conducted to a "rock house," so called, a sort of cave upon the side of Mount Judah. Here I remained several days suffering from abcesses which were as hard to endure as the pleurisy. Friendly hands brought food and water, and the weather being warm, I was as comfortable in my lovely cave as though reposing upon a bed at home. I was visited by some of my own men and finally by Vanderpool, who after leaving Flat Rock, had ranged a wide extent of country for supplies and recruits in which he had been moderately successful.

One morning before daylight a daughter of Mr Harrison[37] came from Jasper and informed us that the town was occupied by a large force of the enemy. We seized our rifles and cautiously made our way through the woods to reconnoiter. A quarter of a mile below town the road crossed the creek and on the opposite side we discovered a strong picket of the enemy. One of their number rode down to the creek to water his horse. Instantly our rifles covered him and we both fired together. The poor fellow dropped from his horse and we saw his body drifting down the stream. We did not remain long to watch the result but broke for the hills. The twer r r of a bullet close to my head which the next instant was followed by twer.r.r twer r r twer r r r as if the whole air had become suddenly alive with invisible snakes with peculiar hissing propensities, accelerated the double quick which we were just then executing, until a safer distance and a substantial oak tree had been placed between our said hissing reptiles. We were satisfied to remain in ignorance of the result of our shots and to loose the beauty of the scene, provided we thereby avoided the necessity of meeting some of the leaden missels that were driving through the air as if vicious in their tendencies.

South of and near the lower end of Town was a high bluff or promontory overlooking the place the face of which was a perpendicular rock three hundred feet high. Information of the irruption of the enemy, in the valley was spread rapidly over the country and men began to arrive to our help. Fourteen of us proceeded by a circuitous rout[e] through dense masses of vines and underbrush to the summit of this bluff from which every movement of the enemy was revealed. We judged their numbers to be 250. Immediately below us was a log house and outbuildings, around which half of the enemies force was gathered preparing breakfast.

Sheltering ourselves behind trees and rocks, and taking deliberate aim we dropped fourteen ounces more or less of lead among them. Down went one man shot through the brain. Another went tearing around with a hole through his arm: a third cut some wild antics as a bullet plowed through his cheeks and nose. One fellow did some shouting over a hole through his leg, and another made wry faces over a similar orafice through his shoulder. The matter was too serious for "Secessia" to overlook and within the next minute a hundred bulletts came spattering the trees behind which we were ensconced, clipped the leaves from the shrubery which crowned the summit of the hill, and tore up the ground around us. Well, I expected to be hit in the legs, arms, head, chest, back,—somewhere at every discharge and almost wished myself away from there, say about 200 miles: But not a man was hit. Each hugged his tree for dear life, for if but a finger, a nose, or hat rim projected from behind them, said nose, finger or hat would not have been around very long afterwards to do any more projecting. We remained in this position more than an hour, occasionally getting in a shot which left its imprint in some miscreant's skin.

Our men began to collect in formidable numbers, and shot was rained upon them from nooks in the rocks, from behind trees and from every point which afforded a secure covert to shoot from picking off their vidett[e]s and galling them in such a manner that in a short time, they left Jasper and retreated precipitately up the valley, closely followed by an adversary nearly as numerous as they. Numbers of their dead and wounded were left at Jasper while scarcely a man on our side had receved a scratch. Two of Vanderpool's men were captured at the time the town was first occupied, and we found their dead bodies in a smoke house riddled with bullets. Three others who were on their way to join us, unfortunately fell into Cecil's hands. One, John Baint [Bount?], knowing that a halter awaited him at the first convenient limb, resolved to escape or perish in the attempt. Accordingly at a point where the road passed near the creek, he sprang down the high bank, plunged into the water and secreted himself under a mass of drift wood.[38]

Two brothers, Henry and Luther Pearce were not so fortunate, for at the point where the enemy left the valley, we found the body of henry suspended from a limb, dead. Luther soon after joined us and stated that his brother was mounted upon a horse, taken under the fatal limb: and when the rope was securely adjusted, the horse was driven away and his body left writhing in the air. He was also compelled to mount a horse and placed under the same limb, but while the halter was being fastened he sprang into the brush, was fired upon and slightly wounded but made his escape, from what Saxe would call "A dangerous noose"

"In the end of a cord that was dangling loose[.]"

Similar tragedies were enacted whenever any of our men fell into their hands and though in a few instances I have known men to be shot down after their capture, yet be it said to the credit of the Union men of the mountains, the halter was never resorted to, not withstanding the many provocations and atrocities committed by the rebel miscreants.

A time was fixed for our departure from the country and Vanderpool's volunteers and my own recruits were directed to rendezvous at Radcliffs Mill on Buffalo creek by the 15th of August. This valley had been raided but little, and supplies were more abundant than in any other portion of the country[.] We were reinforced by the arrival of Kelley with sixty men from Pope County[.] Their adventures were of the most perilous as well as dangerous description[.] The scene of their operations were not very far from Dover the county seat where at that time a rebel regiment was stationed. Detachments were constantly patrolling the country conscripting, shooting, hanging and torturing all who fell into their hands. Our old friend Brashears had induced many of his neighbors and others who were "laying out" to escape the relentless conscription, to enlist. The evening before their contemplated departure for Federal lines, the Captain went to his home, to bid his family good bye. The next morning at sun rise while on his way to camp he was waylaid and shot, his body horribly mutilated by the fiends who perpetrated the cowardly deed, and his friends prohibited from burying his remains which were left to rot where he had fallen.

An attack was made upon Kelley's camp and many of his men scattered: but he had succeeded in bringing the greater number from the dangerous neighborhood. An ex confederate Captain Speer[39] also joined us with twenty men. Our men were dilatory in getting to the rendezvous, and it was not until the 18th that we were ready to leave the country.

On the afternoon of the 17th while the men were scattered promiscuously about under the shade of trees, some telling yarns or playing "seven up," and others sleeping, the alarm was given that a force of "rebs" was not far off and preparing to attack us. Cards disappeared as if by magic, every man was awake with flint locks tightly clutched. A hundred slung themselves into their saddles and swarmed like a heard of buffalos after Vanderpool who went out to reconoiter. Kelley and I selected a favorable position on the side of the hill, protected by a fence and by trees, and drew up our men in line of battle. Vanderpool soon returned and announced the enemy less than a mile away, and approaching us in good order. Dismounting the most of his men, he placed them on our right, partly sheltered by a deserted log house. Presently a cloud of dust arose above the trees indicating a pretty numerous force. Then coming out of the timber they steadily approached us, but without making a single disposition to attack us. They were plainly in sight when one of Kelley's men contrary to orders fired a shot at long range, doing no damage however, as the ball stopped of its own accord after going about one half the distance. The column halted suddenly and a single horseman swinging a dirty rag came towards us and had passed half the intervening distance when some one shouted

"I'll be doggone if that aint old Farmer."

Instantly there was a shout and a rush[.] It required no order to "break ranks" to dissipate our admirably formed line of battle[.] Our cheers were received by an answering yell from the approaching party. It was Major Farmer indeed: True to his word he had come to join us with 100 men at his heels. I doubt if the valley of Buffalo creek ever before or ever will again echo such boisterous shouts of joy. Not a man among them whether armed with flint lock musket, revolver bowie knife or club, but believed himself a match for any seven secesh who had a mind to pitch in: and that as soon as the Federal lines were reached secession must be speedily annihilated[.] I had thought just the same once, but such extravagant notions had been long since dissipated[.]

Physically the men excited my admiration, but their clothing and murals [morals?] were in a state of utter demoralization, and to such an extent as would have excited the profound contempt of the raggedest crowd that ever haunted the Five Points, or the seediest beggar that ever hunted the gutter for bones. Few had any shoes, and as for hats, words fail to do the subject justice. Hunt up all the old hats that ever plugged the windows of poverties dirtiest kennels, select four hundred of the poorest and most dilapidated and they might convey a fair idea of the head gear of the assemblage at Buffalo creek. Ragged, dirty and ill fed as these men were, they would fight to the last as was afterwards proven on many a hard fought field.

On the 18th we moved in an immense column towards Fayetteville. The country over which we marched was rough and savage in the extreme. Now we were going up hill to the North, down hill to the East, through woods to the South, across big ravines to the west, over every possible kind of country but a level one, and to every remaining point of the compass besides some which no known compass ever indicated.

About dark we reached a corn field, a patch of potatoes, a peach orchard and a spring, and halted. The horses were tied in the woods close at hand: that cornfield was harvested before ripening, and then coiling ourselves in fence corners, behind logs, in the road anywhere, we slept. In the neighborhood of Huntsville the column was frequently fired upon by prowling "brushwhackers" but with no serious results save the killing of a horse or two. A few were captured and taken to Fayetteville and one notorious desperado was shot.

We reached Fayetteville the 21st and found the town, or what remained of it in possession of sixty Federal soldiers who had arrived there the day before on recruiting service on behalf of the 2nd Arkansas Cavalry.[40] It required considerable diplomacy on our part to convince the Lieutenant in command that we were friends instead of foes. Indeed the appearance of our tatterdemalians was enough to excite suspicion in any one not thoroughly acquainted with the natives of the country. However we could not complain at the reception accorded us, and were proceeding to the public square to go into camp, when a sharp firing was heard in the southwestern outskirts of the town. Word was brought that Buck Brown a noted bush ranger, was attacking the pickets, who were soon driven in closely followed by as outlandish and savage set of yelling miscreants as could well be raked up outside the realms of Satan. Vanderpool, Kelley, Farmer and myself attempted to form our men behind the ruins of the Court House and adjoining buildings to repell the assault, but a panic seized the men and not a dozen would stand by us, or heed our commands, and we found ourselves alone confronting the secesh, who in thronging numbers was charging through the streets. We gave them a few shots and then Vanderpool went after his men swearing some terrible oaths. The detachment of Federals suspecting the whole movement to be a ruse to take them in, fled precipitately up the telegraph road to a hill a half mile distant and formed in line of battle. The handful of mountaineers who remained and who had gallantly stood by us followed after. Thus was Col Brown in possession of the place after scattering a force superior in numbers to his own by four or five to one. From the position we occupied we could see them ranging the streets and appropriating the contents of a sutler's outfit[.]

Kelley volunteered to head a charge upon the enemy and found about twenty five willing to follow. Right gallantly he led the way and went tearing in like mad among them. For a few moments the crack of pistols, carbines and muskets was incessant. Saddles were emptied and in the end the secesh rabble was sent whirling away. Almost the last shot they fired struck Kelley near the knee, and following up his leg passed out at his back. His fall put an end to the chase, as his men were far more anxious for their daring leader than to follow Brown in his retreat. Those of us who from the hill witnessed the dashing charge, and the ease with which a few determined men had recaptured the place, were chagrined that there had not been a more determined stand at the outset. Kelley and the squad who participated with him in the charge, being almost the only ones who displayed a courage and intrepidity worthy of the name of soldiers.

Late in the afternoon Vanderpool and the badly demoralized recruits were soon on the summit of East Mountain two miles away. On being informed of the situation of affairs in town and that all was serene, the whole outfit dashed down the rocky descent[.] The stones and gravel flew, and so did they, and the hills fairly shook with the tread of three or four hundred horsemen, charging into the place. Fifty of the recruits never stopped running from the battle of the morning until they reached Cassville fifty five miles away. Upon the whole the affair at Fayetteville was not very creditable to us.[41]

Col. Harrison commanding at Cassville hearing of our arrival through the fugitives, ordered Captains Worthington[42] and [Joseph S.] Robb with two companies of the 1st cavalry to proceed to Fayetteville to render such assistance as was needed for our saf[e]ty—as well as to escort a large number of civilians who wished to leave the country for some locality where pillage and outrage were less frequent and human life not held so cheaply. On their arrival Fayetteville presented a decidedly military appearance.

Vanderpool's Regiment, the 1st Arkansas Infantry was with General Blunt[43] in the Cherokee Country, operating against Forts Gibson and Smith, and thither he proceeded with about one hundred of the men, leaving two hundred and fifty undisciplined mountaineers in my charge. Through the efficient cooperation of Major Farmer and others who had taken an active part in recruiting them, we not only held our position but materially aided those who were preparing to leave the country. Women and children were about all who remained and they were reduced to a state of the utmost destitution. Constant exhibitions of rebel diabolianism compelled them notwithstanding their love of home and kindred to look northward for relief. Some mounted on horse back, some on foot, and others in a rickety carryall joined the marching column, and as family after family filed along, mute and uncomplaining, the sight was distressing indeed. Their farms had been desolated, their houses burned, their cattle stolen, and now were they exiles from the homes of a lifetime. We marched ten miles and camped in an orchard burdened with ripe and delicious fruit.

Parallel with, and about a mile west of our line of march was the Elm Spring road, and here it was reported were several squads of bushwhackers. Capt Robb, Major Farmer and myself were directed with about one hundred men to proceed by the way of Elm Spring, cautiously feel our way, and "go for" aught inimical we should see. While Captain Worthington with his own Company, the refugee train, and the most of the recruits proceeded up the telegraph road. Our detachment had reached the road leading to Elm Spring and had proceeded two or three miles when a heavy fireing was heard to our right in the immediate neighborhood of Fitzgerald's mountain. Volley succeeded volley in quick succession and we knew that Worthington was hotly engaged. Our detachment was immediately headed to the right, through woods, and across fields, at a gallop for the scene of action fully a mile and a half distant[.] As we neared the scene of contest, the explosions of musketry and carbines grew louder and more continuous, then the fire would slacken for a moment, and perhaps a single gun would lead off then a dozen in a succession so quick that each succeeding sound lapped on the preceeding one, and then the lapping on would [be] indistinguishable, and the whole would be merged again in one terrific volume. There is a magical excitement about a battlefield which fascinates, and at the same time sends a tingle through the nerves like a shock of electricity. Words cannot adequately express the peculiar sensations which thrill the heart. A battle must be participated in to appreciate the extent of one half of its terribly awful character. As we approached the field, squads of men could be seen galloping hither and thither: horses were without riders, men were getting away to the timber, some hatt[l]ess, and other limping painfully along. Reaching the road we dug the spurs into our steeds and clattered over the dusty road like a whirlwind: seeing which the enemy hastily withdrew, but not before we were able to get in a few shot[s] and empty several saddles. The refugees and many of the recruits were scattered to the four winds. Capt Worthington was severely wounded, twenty of the men were dead, and many more or less severely wounded. Although in possession of the field the result was a heavy loss to us.[44]

 Captain Worthington had been ambushed and taken completely by surprise. The enemy protected by trees, and rocks, suffered but little. Many of our men fled at the first fire. Women and children frantic with terror mingling their screams with the roar of battle, added to the horrors of the scene, and appalled the hearts of the men. Some fought with manly courage: but the timely arrival of our detachment only prevented Worthington from being annihilated. We arranged with the citizens to bury our dead, and hastily collecting the wounded and stowing them away among the wagons, we hurried away from a scene of such fearful disaster.

 We avoided Cross Hollows by a long detour to the left, and thus as we subsequently learned, avoided another battle with the combined commands of Col Brooks and Buck Brown.

 Approaching Sugar Creek, we were on historic ground. By gone memories came crowding upon my mind as I recalled my lonely walk over this battle scarred ground two years before, then strewn with the carcasses of dead and decaying horses. Forest trees were rent and broken, showing the fresh scars of the recent battle.

Then we descended into the valley and crossed the creek. The defensive works thrown up by Jeff C[.] Davis the evening before the field day at Pea ridge were still in a good state of preservation[.] Here we stirred up a nest of bushwhackers and chased them a mile or more into the woods. Soon we reached the former site of the Elkhorn Tavern[.] It was here that the struggle with Price and Van Dorn on the 8th and 9th of March 1862 raged the fiercest, and which resulted in the glorious victory of Pea Ridge. One year had partially healed the shot shattered trees, but the rocks and hills over which we had bourn the flag in triumph was the same. All else had changed[.] Not a house or fence was left around the extensive fields. All was silent as the graves of the buried braves by which we were passing. I lingered for awhile on the same spot occupied by our regiments and batteries: and Major Farmer, in turn pointed out the various positions of the enemy on that, to them fatal day. As the memories of that struggle came welling up with thoughts of dead comrades whose ashes [next page is a drawing of damaged trees] were scattered here an indiscribable feeling perhaps of awe came over me and I had no heart to remain[.] We descended into the narrow valley of North Sugar Creek known as cross timbers Hollow and about sundown crossed the state line into Missouri. It was after dark before we reached a cornfield where forage could be secured for both horses and men. Our supper was roasted corn on the ear and that was all.

Early the next morning we were away. Each field bluff, spring or other land mark was familiar[.] They seemed like old friends or former comrades. I had surveyed and mapped them all, and was as much at home as among my native fields. Keitsville, once a flourishing village was in ashes. It had been noted since the oragin of the rebellion as a rendezvous for marauders of the worst description[.] Jo Peevie,[45] living a short distance west of the village was the leader of a gang, and no atrocity was too inhuman for him to commit[.] For this the Village had paid dearly. Not a house was left standing. The only remaining vestiges left were the tall stone chimneys which the incinderaries' torch had failed to consume, and all that Keitsville can bequeath to posterity, save the enduring name of Thomas Keit its founder—is a hard name.

We reached Cassville at noon and were in America again where the old flag floated, and men could lay down beneath its protecting folds and rest in saf[e]ty. How we proceeded to Springfield were clothed in regulation blue and sworn into the service of the country is of little consequence to those who trace these pages. But the brave achievements of these men, under Phelps, Stevenson and others are among the historic Records of our country.

With the mustering in of the men ended for the present "Recruiting in Dixie."

Hd, Qrs. 4th. Ark. Cav.
August 22nd 1864

Lt. Col Green
A.A.G. Dept. of Ark.

I have the honor herewith to tender my resignation as Major 4th Ark. Cav. for the following reasons[.]

1st_____An unjust prejudice against northern officers, has been disseminated by a few discontented and jealous persons, producing discontent and dissensions in the regiment, and impairing my usefulness as an officer until I cannot longer remain in my present position with honor and credit to myself.

2nd_____The discontent existing in the regiment and the causless prejudice against northern officers is destructive to good order and military discipline and renders my position particularly unpleasant-and impossible to perform the duties of my position without injury to the regiment and to the service.

I certify on honor that I am not indebted to the United States on any account whatever: and that I am not responsible for any government property except what I am prepared to turn over to the proper officer on the acceptance of my resignation and that I was last paid by Major Herrick to include the 30" day of June 1864[.]

Very Respectfully Your Obt Svt
L. G. Bennett
Major 4". Ark. Cav.

Bennett Compiled Service Record, Fourth Arkansas Cavalry, Fold3.com (website), accessed March 15, 2021.

Ft. Smith Ark.
Nov 13" 1864

Lt Col Green
A. A. Gnl
Dept. of Ark

I have the honor to request a copy of my resignation as Major in the 4th. Ark. Cav. I believe it was accepted August 26" 1864[.] The copy given me I lost during the march of Genl Curtis at or near Cross Hollows Ark. and, have nothing now to show date of Resignation in order to get the ballance of pay due me[.]

Please send copy to Oswego Kendall Co Illinois[.]

> Very Respectfuly
> Your Obt Svt
> Lyman G. Bennett

Bennett Compiled Service Record, Fourth Arkansas Cavalry, Fold3.com (website), accessed March 15, 2021.

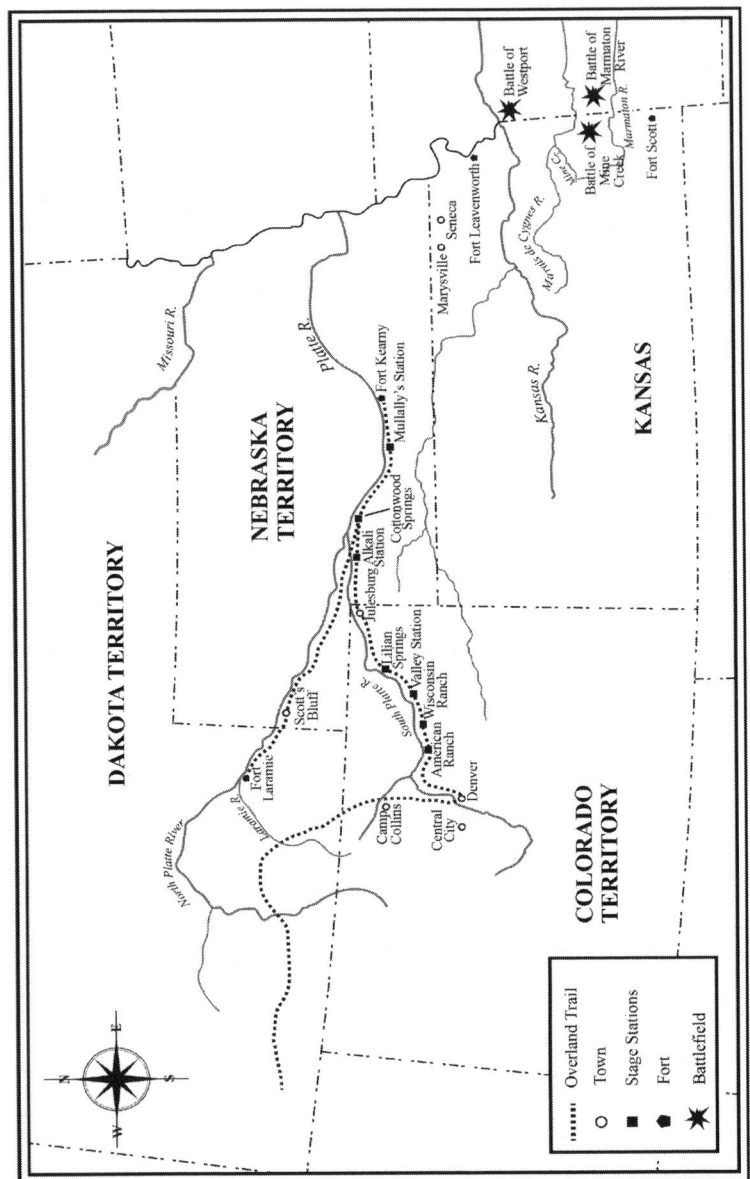

The West, ca. 1864.

17 | Mapping in Kansas and Missouri

January 1, 1865–February 14, 1865

> *After a plain camp dinner, our horses were brought out and we went to the battle ground on a gallop, the snow driving into our faces and in large flakes which made one feel like being pelted with snow balls.*
>
> —Lyman G. Bennett, January 8, 1865

Following the resignation of his commission as major of the Fourth Arkansas Cavalry, Lyman Bennett's historical record went briefly silent. Fortunately for posterity, Bennett picked up his diarist's pen again on New Year's Day of 1865. He was now a civilian employee of the Engineer Corps thanks to "the kindness of Major General [Samuel R.] Curtis." Lyman's wife, Melissa, and daughter, Minnie, had also moved from Illinois at some point and now boarded with Clara Bulkley, a relative, near Fort Leavenworth, Kansas.

During the first three weeks of January, Bennett visited two battlefields, Mine Creek and the Marmaton River, for the purpose of mapping them on behalf of General Curtis. In September 1864, Confederate Major General Sterling Price launched a cavalry expedition into Missouri with the hope of capturing St. Louis. Soon driven away, Price's army lurched westward across Missouri before losing a series of battles around Kansas City. The battles at Westport and Mine Creek proved to be particularly important Federal victories, and by late October, Price's army had been defeated.[1]

Curtis, now the commander of the Department of Kansas, submitted his lengthy report about the campaign in January 1865.[2] Four maps (*Westport and Big Blue, Battle-Ground of Westport, Osage or Mine Creek, Battle-Ground of Charlot*) prepared by Bennett accompanied his report, and all were published as part of Plate LXVI (66) of the *Atlas to Accompany the Official Records of the Union and Confederate Armies* (1891–1895). Two of those maps (*Westport and Big Blue, Battle-Ground of Westport*) listed Bennett as the creator, and he described the other two in his diary. Private Jacob Miller, mentioned in Bennett's diary, drew the map *Campaign Against Sterling Price, 1864,* also on Plate LXVI.³ Kyle S. Sinisi, whose book *The Last Hurrah: Sterling Price's Missouri Expedition of 1864* is considered the best modern overview of the campaign, faulted Bennett's maps of the Westport area. Sinisi believed that First Lieutenant George Robinson, Curtis's chief engineer, created maps that inaccurately placed some key fords and roads, errors Bennett repeated in his maps of the same locale.⁴

Besides his battlefield map project, Lyman surveyed government land near Fort Scott, Kansas, and prepared for a trip to Fort Kearney in the Nebraska Territory. Before leaving, Bennett helped a grieving father. Wilson Judson, a resident of Kendall County, Illinois, and an officer in the Fifty-Fourth United States Colored Troops, died of typhoid fever in October 1864 at Fort Gibson in the Indian Territory. Bennett retrieved Judson's horse at Fort Smith, Arkansas, and arranged for Judson's body to be returned to Illinois. On November 28, 1864, Bennett directed a telegram to Major General Curtis and asked him to order Captain Insley at Fort Scott to "send an ambulance with the remains of Lieut Judson" to Fort Leavenworth. Lyman had apparently taken other steps to see that the body was forwarded to Illinois and hoped that it had arrived to alleviate the concerns of Judson's father. Part of the "good death" ethos of the time, according to historian Drew Gilpin Faust, was to die at home rather than far away among strangers. Many families of Civil War soldiers made strenuous efforts to ascertain the true fate of their loved one and to recover their body, if possible.⁵ By recovering Lieutenant Judson's body and having it sent home, Bennett performed an important and thoughtful service, one that he probably regarded as a duty.

In his first entry of 1865, Bennett was "fearful these pages will lack much of the interest that would characterize the journal of the humblist soldier participating in the campaigns against the rebellion." He worried needlessly as these entries are characteristic of Bennett's vivid style and contain careful descriptions of the land, harsh winter weather, towns, and crimes.

Paoli Kansas

January 1st 1865

Though not connected directly with the army, and by no means expecting to participate in the great and important events of the coming year, yet

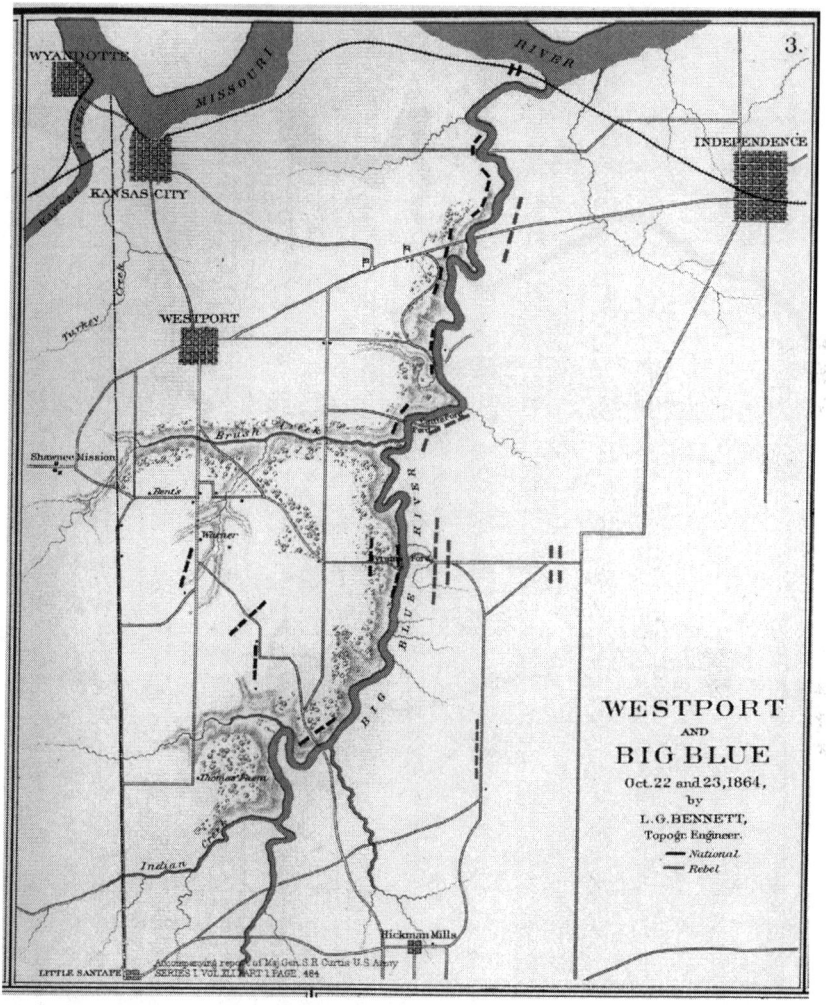

Westport and Big Blue. Prepared by Lyman G. Bennett. From Davis, George B., Leslie J. Perry, and Joseph W. Kirkley, *Atlas to Accompany the Official Records of the Union and Confederate Armies* (Washington, DC: Government Printing Office, 1891–1895).

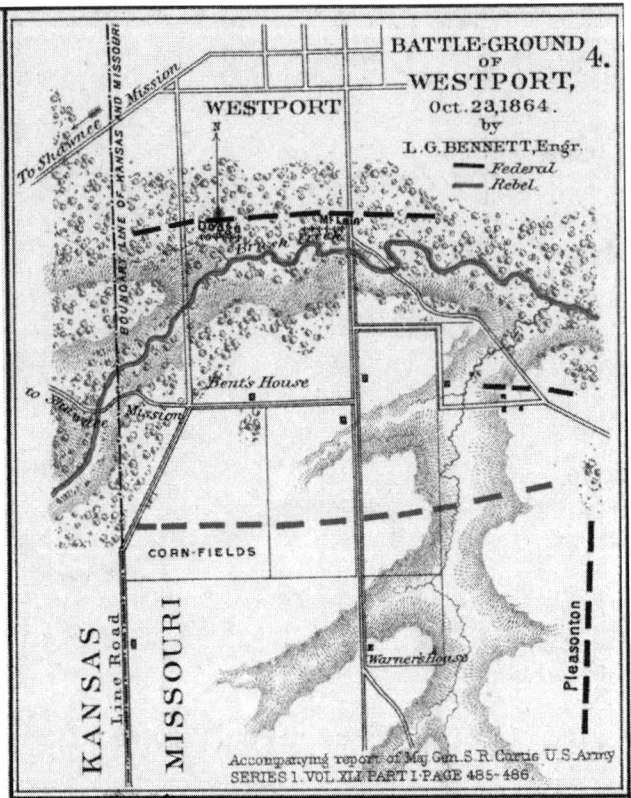

Battle-Ground of Westport. Prepared by Lyman G. Bennett. From Davis, George B., Leslie J. Perry, and Joseph W. Kirkley, *Atlas to Accompany the Official Records of the Union and Confederate Armies* (Washington, DC: Government Printing Office, 1891–1895).

there may be those that are interested in the personal history of the humble writer of these pages. For such and for my own personal gratification I shall endeavor to keep a faithful record of the transactions and observations of each day, and trust that in so doing I will not leave behind me a record of shame[,] misfortune, and crime. I have participated some what in the exciting events of the past three years of war, and now that I am a free man again and employed in the more congenial, and less hazardous pursuits of life, I am fearful these pages will lack much of the interest that would characterize the journal of the humblist soldier participating in the campaigns against the rebellion. The coming year promises to open up new fields of adventure in the uninhabited and but partialy explored territories of the West. Through the kindness of Major General Curtis commanding the Department of Kansas I have been employed in the Engineer Corps[6] at $150[.]00 pr month, and am now on my way to Fort Scott in obedience to instructions from Lt. George T. Robinson,[7] Chief Engineer, to survey and select lands for the erection of fortifications and quarter for a military Post, and make a topographical sketch of the battle fields in the recent campaign of Genl Curtis against the rebel general Sterling Price. In pursuance of these instructions I have surveyed the battle grounds of Westport and the Big Blue.[8]

Yesterday I arrived here from Shawnee Mission after a somewhat cold ride across the prairies[.] Our conveyance was one of Uncle Samuel's Ambulances drawn by four mules, and as we had a light load, we were enabled to make tolerable progress[.] The party consists of Private Jacob Miller[9] 3d Wis Cav, artist and draughtman, (Hans Nixforstay) a dutch teamster and myself. During our stay at Shawnee Mission Mr Miller sketched the place which make[s] a fine view. The brick buildings at the Mission are large and commodious, the architecture plane as was common in the days when they were built. I learned from the proprietor, that they are the oldest buildings in Kansas.

It has been a cold raw day. The winter wind came sweeping down from the North, over the bare bleak prairies, and I was glad to keep close within doors, at the Union Hotel, a rough, dirty shell, in which travelers and other poor unfortionates are obliged to content themselves. Hotels at best in Kansas are nuisances, but this is suprelatively so. The manner in which the table is furnished, and the victuals cooked is disgusting to anything but a half starved human. The works at this place are going on slowly. Lumber is supplied in very limited quantities and the mechanics for the most of the time have nothing to do. After dinner I inspected the ground selected for quarters, but the cold prevented a thorough examination as I intended[.] General [James G.] Blunt invited me to ride with him at evening, and with his fast ponies we soon made the circuit of the town, and its environs. General Blunt is in command of the Southern District of Kansas with his head quarters at Paoli[.] Col Jennison is also in town under arrest and and the quarrel between him and the General is rather too fierce to be pleasant. If half the atrocities that are told of Col Jennison are true, he is too much of a barbarian to live.[10]

Paoli is a place of some importance, has considerable local business, and is about half way between Leavenworth and Ft Scott. It is but six miles from Ossawattamie and much more of an enterprising and flourishing town. Ossawattamie, I am told is retrograding. There is an abundance of timber and fine prairie land, and Bull Creek waters the adjacent country. I noticed a grist mill, a saw mill[,] several groceries and stores, two hotels[,] a printing office[,] several blacksmith shops &c &c. A small fort with one gun is on a[n] eminence north of the town and commands the whole country. It is a small affair, but may be of service against brushwhackers[.] The soldiers are encamped in bottom, one from a mile distant[.]

Monday January 2nd 1864 [1865]

The day has been cold and cloudy. I borrowed Mr Wilkinson's horse, and was busy all day riding over the country, making the necessary topographical survey. All the roads, streams[,] fields[,] hills, houses and other objects were traced for two miles around. The fatigue incident to the survey, caused even the hard bed to appear comfortable, and tough beef was eaten without a murmur.

The country about Paoli possesses the same general features of other portions of Kansas[.] The prairies are high and rolling, in some places hilly[.] There is a range of knobs south west of the town, that stretch away towards Ossawattamie. I rode to the summit of one of the highest and the view was delightful. To the north and East Bull creek threaded its way, winding among the hills, and hid from view by the belt of timber which skirt its banks. Every where else the eye wearied in stretching across the sombre prairies away in the distance where sky and prairie met. A few houses dotted the country here and there, the long distances intervening makeing them appear like attoms on the wide expanse[.] A party at Ossawattamie empted the town of the gay and fast ones of both sexes, our land lord among them. During his absence the negro help about the house acted as if broke loose from some lunatic assylem. A "snow flake" and "snow ball," had been married during the day, and a cow bell, tin horn and tin pan band was extemporized and for an hour the negroes and loafers about town made night hideous with their unearthly music.

Tuesday January 3d 1865

The sun shone bright all day, dispelling some of the cold of the past few days. I surveyed a tract of about fifty acres of land for government purpose and fixed the corners. An appalling and fatal tragedy occurred in the afternoon. Very many of the soldiers stationed here, have been drinking to excess since Christmas, until drunken soldiers could be found both night and day. The result has been, any amount of quarreling, fighting[,] bruised faces and sore heads. This afternoon a drunken corporal belonging to the 11th Kansas[11] commenced abusing a negro. The negro wanted no difficulty and tried to avoid the soldier and get out of his way. At last the soldier cornered the negro in a stable and drawing his revolver put it to the negro's head and threatened to shoot him. The negro also had a revolver and in self defense drew it and shot the soldier down. He lived but a few hours, and before he died exonerated the negro from blaim. An investigation of the case called out the above facts and the negro was set at liberty[.] Very many cannot get over the idea that negros were made only to be kicked about. Were I the blackest, and most degraded of negros I think to save my own life I should shoot any one who was threatning me[.]

Wednesday, January 4th 1865

We resumed our journey after breakfast. The keen frosty air found its way in the open ambulance and we were obliged at times to get out and run on foot to warm ourselves. When riding stories were told to forget that it was cold, Miller notwithstanding his dutch sleepy look, has in store a number of good ones, that can create a laugh any time[.] The roads were in good order except a few miles across the bottom on the Big Sugar Creek. This is the finest belt of timber I have seen in Kansas. The sugar maple, oak, walnut, elm and cottonwood are large. Hickory is likewise abundant and their nuts covered the ground in many places. The points passed through were Twin Springs, Paris and Moneka all small and of no importance.

We reached Mound City at 3 P.M. too late to go out to the battle field, which was five miles away. Major Lang the commandant volunteered to go with me in the morning. At the hotel I found two prisoners under guard for Murder, Clark a citizen and Lt Murphy[12] of the 15" Kansas. It appears that on New Year's day every body at and about Mound City were drunk[.] A poor hard working blacksmith, after his day's work was done, purchased candies and toys as gifts for his children and started for his home four miles away. About dark he was seen to pass through Moneka. The next morning one of his children came to town in quest of him, but no one knew anything of him. After a search, his body was found about half a mile from Moneka, on the prairie, and perforated by five bullet holes. It was then remembered that shortly after the murdered man passed through town, Clark and Murphy were seen riding in the same direction both very drunk. Several shots were heard, but as this was so common an occurrence no one took any notice of it until the body was found. Suspicion immediately fell on Clark and Murphy and they were arrested, and placed under a guard of citizens[.] There can be no doubt of the guilt of the parties, and the people are somewhat disposed not to await a trial but hang them to the nearest limb. Clark is a mean looking fellow, and one would think him capable of any crime.

Liquor was also the cause of another death at Mound City the same night. Nearly all were drunk and a promisonous [promiscuous] fireing of guns continued both day and night during which a soldier was killed it is supposed by accident. It is not known who did the act[.]

Thursday January 5" 1865

At 8 A.M. our party, consisting of my own outfit, Major Lang, a captain and the Adjutant of the post started for the battle field at Mine Creek, which was five miles east of Mound City. We passed several fine farms the buildings, fences and improvements indicating thrift and enterprise in the owners. We struck the broad trail of the enemy and passing down it was soon where the strife of Oct 25 was most severe. A long row of dead horses laying on the prairie indicated where their line of battle had been formed. The surface of the prairie had been completely trampled up by horses and men, and several complete roads were formed where their artillery and trains had passed.

Soon we came to the body of a dead rebel laying beside the trail. The body was frozen and the features were preserved as fresh as though he had but just died. Wolv[e]s or hogs had eaten some of the flesh from the thighs and body[.] In passing over the field I came across the dead bodies of four men. One a staff officer of General Marmaduke.[13] I could not learn his name and rank. He was a fine looking man with black hair and mustache. I have seen scores of their dead on this campaign and with few exceptions they are sandy complected and with red or light hair. Nearly all by their appearance indicated a low ignorant class. But they are now dead and though they were enemies yet I do not approve of their dead bodies laying out on the prairies, as food for hogs and wild animals[.] I shall report this to Genl Curtis and ask that they be buried. We went into the house of Mrs Ragain which was situated where the conflict raged most severe. There were many marks of balls in the clapboards, and fences were completely razed to the ground. Mrs Ragain stated that all the men in the neighborhood were in the army and there being none but women and children at home was the reason the bodies of the dead had not been buried. The great number of dead horses and men, in the vicinity must produce sickness when the warm spring weather causes them to decompose.

The places where Marmaduke and Cabell were captured, and where their artillery was taken were pointed out to me and a correct map drawn of the place. Mr Miller also took a fine sketch in which the whole scene was most faithfully represented. Mrs Ragain told me many anecdotes of the fight which were new to me[.] Near the right of the rebel line was a family living and the rebels stationed there completely robbed the house of all that was in it. The lady saw Genl Marmaduke and asked that his men might be

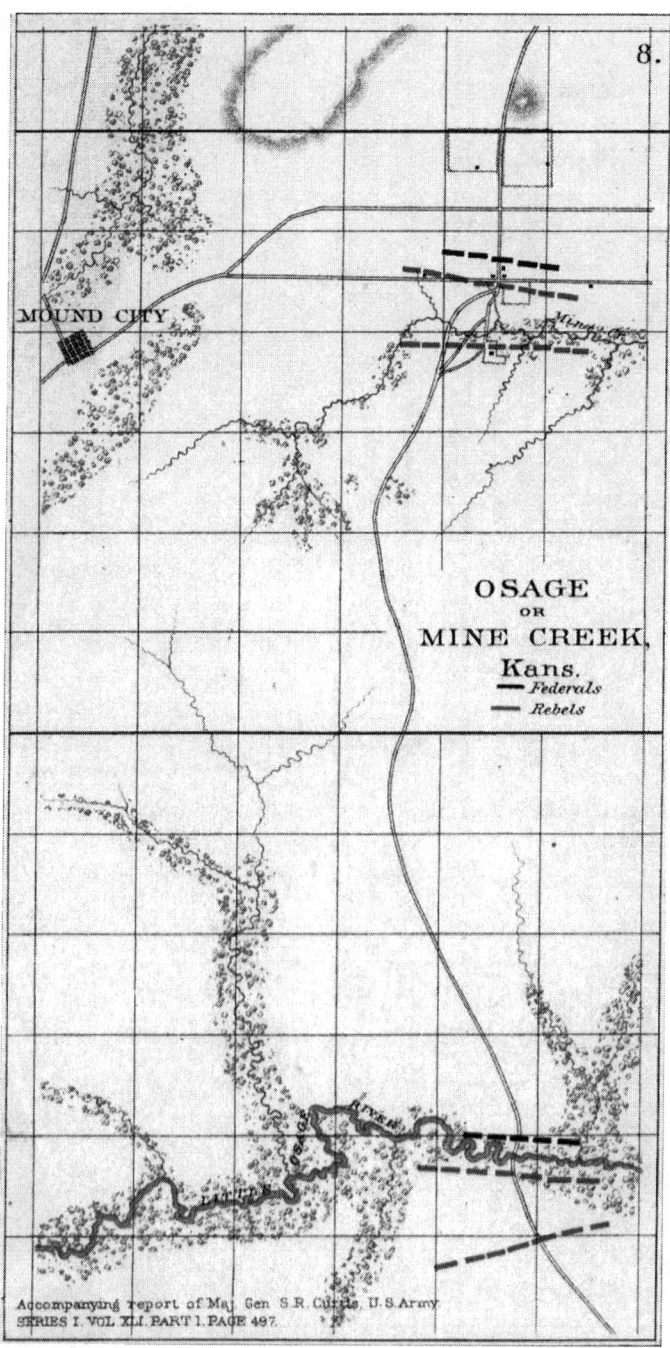

Osage or Mine Creek. Prepared by Lyman G. Bennett. From Davis, George B., Leslie J. Perry, and Joseph W. Kirkley, *Atlas to Accompany the Official Records of the Union and Confederate Armies* (Washington, DC: Government Printing Office, 1891–1895).

restrained and that her property be returned. The General in order to divert her attention to som[e]thing else went through with a series of circus performances with his mair and calling the woman's attention to his spirited animal told her that she was named Mis Mary Price and asked how she liked Mis Mary. After the tide of battle had rolled away from this portion of the field, the woman went to mrs Ragain's to recount her losses and to see what damages her neighbor had sustained. There she again saw Genl Marmaduke but this time a prisoner she immediately sang out to him "How are ye General? and how is Miss Mary Price now?" The rebel general could not see it[.][14]

An hour sufficed to complete the required sketches and then we returned to Mound City about 11 A.M. We immediately started for Fort Scott and reached the town about sundown[.]

Friday January 6" 1865

The big Irish chambermaid was bustling about my room almost before I had completed dressing and informed me that I had ocupied a bed used by a bride and groom the night before. I did not observe any particular sensations any more than if it had been ocupied by a dog instead of a blushing bride, being tired I slept soundly and sweetly. Col Blair[15] and I reconoirtred the country before noon and ground was selected for fortifications and quarters about a mile southeast of town on a height overlooking the town and country and contiguous to several springs which I was informed had never failed in the dryest seasons[.] I also observed coal cropping out in one or two places. An abundance of rock existed all over the tract, in fact it is nowhere covered by more than two feet of soil, but it was loose and easily excavated.

I also had a table made for the purpose of platting maps and sketches, and made arrangements with Mr Diamond at the hotel for boa[r]d for myself and family at $1.50 pr day a piece.

Saturday January 7" 1865

Was engaged in drafting plans of Battle fields at Big Blue, Westport, and Mine Creek. Both Miller and myself very busy. Received a letter from Mr Judson[16] in which he was very apprehensive I had never done anything towards getting Wilson's body and sending it home for burial. He has no reason to believe this. However I trust the body has arrived before this time and his fears relieved[.]

Sunday January 8" 1865

Procuring a horse of Mr Shannon of the Quarter-master's department, Miller and myself started about 9 A.M. for the camp of Co F 3d Wis Cavalry near Deerfield in Missouri for the purpose of sketching a plan of Charlots Battle Field which is near By. We had a pleasant ride over the hills and across a wide level prairie too, and across the Marmaton to the camp. It was near noon when we arrived and the Lieutenant in command urged us to remain until after dinner and he would go with us to the Battle ground. It was snowing and we were wet and cold and accepted his kind hospitalities and had our horses stabled out of the storm and was highly entertained with music from a melodian which one of the sergeants had purchased. Miller is a good musician and is at home with most any instrument.

After a plain camp dinner, our horses were brought out and we went to the battle ground on a gallop, the snow driving into our faces and in large flakes which made one feel like being pelted with snow balls. This battle

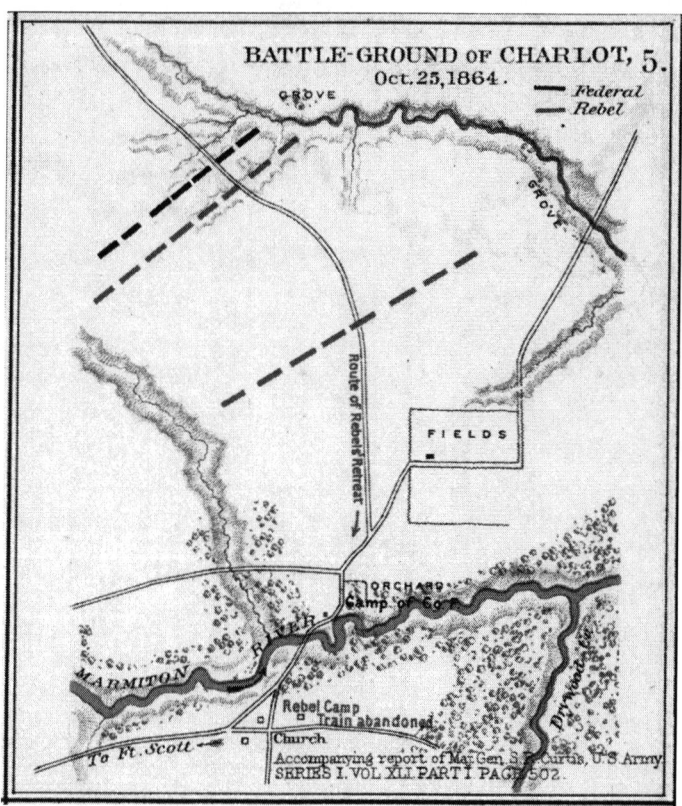

Battle-Ground of Charlot. Prepared by Lyman G. Bennett. From Davis, George B., Leslie J. Perry, and Joseph W. Kirkley, *Atlas to Accompany the Official Records of the Union and Confederate Armies* (Washington, DC: Government Printing Office, 1891–1895).

ground was on a smooth prairie with no fences to break the line of march or impede the charge. It was the last of a series of conflicts on the 25 of October and like the others resulted disastrously to the rebels. Their line of battle was here more than a mile in length, stretching across the prairie from the Marmaton timber to that of a small stream to the [e]ast. There center was partialy covered, by a depression in the prairie caused by a water course in stormy weather but which is dry four fifths of the year[.] This was not much of a shelter for the rebels and was a slight impediment to our victorious troops, flushed with three victories before during the day. The rebels must have been greatly demoralized, fatigued and disheartened with the loss of two or three generals and all but four of their cannon, and their path strewn with dead and wounded men[.][17]

The snow fast falling prevented my making a good sketch, but the small amount of topography, assisted by my memory required but little sketching. On our return we passed the place where the rebels camped after the battle and though more than two months had elapsed, yet hundreds of acres was still strewn with the debris of a defeated army. A hundred or more wagons was left and burned. Tons of ammunition, Cannon and guns of all kinds[.] Strewed over the fields were old iron, fragments of boxes, blank[ets] and cloth[e]s[.] More than half of Price's train and booty was here destroyed and abandoned. This was a sad days works for him, four defeats and the loss of men[,] officers[,] and material[.][18]

Monday January 9" 1865

Was all day busy on Plans of Battle fields scarcely took time to go to dinner[.] Col Blair gives us room to work in his own office and all with whom I have business to transact are very corteous and gentlemanly. Bright day and pleasant weather we are having[.]

Tuesday January 10" 1865

Completed Plans of Battle fields, and wrote a Report of my operations to Lt Robinson[.] Miller has drawn a horse and will start tomorrow or the next day to Newtonia to take sketches there while I push the survey about Ft Scott[.]

Wednesday Jan 11" 1865

Surveyed eighty acres being the North half of the North West Quarter of Section 32 Town[ship] 25 Range 25 S for government purposes. The day was bright and pleasant. We had no difficulty in finding such government corners as were standing[.] It required considerable chaining to fix the center of the section and the corners of the 80 selected and the orderlies who helped me were glad when the work was completed and they allowed to return to town. There is a range of heights to the east, south east south and southwest of the city with here and there elevations higher than the ridge with which they are connected. Such an elevation exists on the land selected and it not only commands the town but all its approaches. Water, stone and coal are other inducements in favor of the selection I have made.

Thursday January 12" 1865

Have prosecuted my surveys all day faithfully on foot and alone. Let me here describe the manner in which a topographical engineer makes his surveys[.] Having ascertained some well defined starting point such as a section corner, In his field book he notes all the objects he wishes to be shown on his map such as hills, fields[,] houses, streams[,] roads, ravines &c &c and notes their distances in his field books. He then follows some known line either a rode, a section line noting the intersection and distance and course of all the objects on his rout[e]. One leaf of his field book may include an 80 acre tract of a quarter section and when the surveyed corners are standing, it is but an easy job to pass around the section and through it if necessary, and get an almost mathematical[ly] correct map of the tract and when the several 80 acre tracts are put together the map is completed. A person wants considerable practice and a good knowledge of distances. In this manner a number of miles may be traveled and a wide scope of country surveyed in a day. I traveled on foot, had a late start and surveyed over two sections comprising considerable bluff land, a number of small streams as well as other objects[.] Miller left for Newtonia to day.

Friday January 13" 1865

A bright and pleasant day. Borrowing friend Hewitt[']s horse I rode to the Carter's on the Dry Wood Creek nine miles south of Ft Scott to look after Wilson Judson's horse. Mr Carter was gone and having no money to pay for her keeping I left her there. She was still very poor about in the same order she was when I brought her from Ft Smith. She is not worth ten dollars and will not cover her raw bones with flesh before next summer.

On my return I stoped at a coal bank where some men were digging coal. In this region of country coal exists everywhere, generaly near the surface, and is procured by taking off the rock and dirt overlaying it. This vein was thirteen inches thick. I have heard of none as yet more than two feet in thickness but should suppose that thicker veins existed deeper in the earth. The coal is most excellent, burns as readily as pitch & furnishes a large amount of heat. Sometimes whole acres of land has been dug up to get at the coal. This must soon bring railroads to Ft Scott. The scarcity of wood in Kansas must bring this coal in great demand, as an article for daily family use.

Saturday January 14" 1865

Somewhat cold, did not prosecute my surveys but commenced a map of the Environs of Ft Scott. Received a dispatch from Lt Robinson to come to Leavenworth[.]

Sunday January 15" 1865

Having much to do yet before the surveys were completed, and being urged by Robinson to proced to Leavenworth, I went out to day over the hills to the south and south west of town, and accomplished more than on any other day. The Sun came out bright and pleasant and my feelings were buoyant with the hope of seeing my family soon[.] As soon as Miller returns from Newtonia I shall start for Leavenworth[.]

Monday January 16" 1865

Surveyed the country north of the Marmaton river, and as there was but little to do then soon accomplished my task and have but little more traveling to do[.]

Tuesday January 17" 1865

Worked in the office at my map. At night the members of the Band was serrinading about town until after midnight[.] They took care to go to places where wines and liquors were to be had and was sure of being treated. A gang of idlers went with them and of course pitched in and helped themselves when the decanters were set out[.] The Band make very good music for new beginners. All of them are natural musicians and music on brass horns comes easy to them[.] Received a letter from Sister Carrie.[19]

Wednesday January 18" 1865

Lt Robinson, Commissary of the Post is my roommate and is a social gentlemanly companion. Received a letter from the Chief Engineer stating that he wished me to go to Ft Kearney[20] and attend to engineer matters there[.] Completed my surveys to day and am now ready to leave as soon as Miller returns. He should have been back to day[.]

Thursday January 19" 1865

Miller returned about noon. He was delighted with his trip. The weather was warm and pleasant and nothing of a disagreeable nature occurred on the rout[e]. He took a very good map of the Battle field at Newtonia,[21] and several sketches of landscapes there and on the way. There is nothing more to keep us here and tomorrow we return to Leavenworth[.] Received a letter from Melissa. She do[es] not know whether to stay where she is and wait for me or come to Ft Scott. I shall solve the problem for her on my return[.]

Friday January 20" 1865

It required some time to collect all our traps and get started. Once under way we made good progress over good roads. Ft Scott though situated on the extreme verge of civilized settlements in Kansas is quite a large town, and the stranger is astonished at the amount of business that is carried on. Government gives much of the business which is done here. From here all the army supplies for Forts Gibson[22] and Smith[23] is forwarded. This is also the center of much inland business and merchants are full of business. There is much capital here and many large buildings. The coal in this region must bring Railroads here in a short time. Along the Marmaton River is a belt of fine timber which is rapidly being cut, but as the country becomes filled up this will be wholy inadequate for the wants of the country. A south wind blew strong raising such a dust that it was unpleasant traveling. The roads were thronged with teams, and so thick was the dust that we would almost run upon them before being discovered. We reached Twin Springs before sundown and stopped for the night traveling about 35 miles[.]

Saturday January 21" 1865

A complete change occurred in the weather during the night. The wind veered around to the north, and we found the morning extremely cold. Soon after daylight the stage arrived and the passengers took breakfast when we stopped[.] Some were almost frozen in riding from Paoli[.] By running a portion of the way we managed to keep warm. At the Marias de Cygnes we found a fire and warmed ourselves by it. Arrived at Paoli at noon. Examined the works there and after dinner proceeded on our journey being joined by Mr Ramus the person in charge of construction at Paoli[.]

Soon it began to snow, the flakes were driven into our faces by the strong north wind and we suffered severely from cold. Reached Olathe about sundown and for the remander of the day hugged the stove until the frost was driven out of our systems. There had been a muster of militia there during the day and very many of the "lads" were drunk[.]

Sunday January 22nd 1865

A clear morning but very cold. Started from Olathe about 9 A.M. The ground covered with snow[.] Reached the Kansas river about noon, found it frozen over, but not solid enough to bear the team[.] In reconoitering our driver broke through the ice and got wet. We finaly procured axes and and poles and cut and broke up the ice for twenty rods to the ferry boat, then Miller and I mounted our horses and going ahead the mules followed after. The water about two feet deep and covered with cakes of ice. We were about two hours in getting over. After crossing we pushed on for Leavenworth as fast as we could urge the mules and suffered considerably from the cold. Arrived at Cousin Clara Bulkley[24] at 4 P.M. found supper ready and a warm fire burning Mellie and Minnie were in excellent spirits and again my little family was united[.]

Monday January 23"

Reported for duty at the Fort and was immediately set to platting up my notes preparatory to going to Fort Kearney[.] Genl Curtis suggested some alterations in maps of Battle fields[.] Cold and cheerless out doors[.]

Tuesday January 24" 1865

Worked at mapping all day. Looked for a boarding place in town but without success[.]

Wednesday January 25" 1865

It is a long tiresome walk to and from the Fort six miles at least. If on no other account I must leave Mrs Bulkleys. Saw Mrs Thompson, can get board at $85[.]oo pr month. Rather cold, worked on maps all day, rode some with Cousin Fernum in the evening[.]

Thursday January 26" 1865

On my return from the Fort found Melissa domaciled at Mrs Thompsons, we have a small but warm room with a pleasant prospect and good fair. Shall like all but the high price[.]

Friday January 27" 1865

To day has been like the others, and I am fatigued from a long walk and close work at the Fort. The weather continues clear but cold, the walking is good, from here to the fort and I do not mind it much. There is a sameness in the weather and in the country that is not found elsewhere[.]

Saturday January 28" 1865

A warm pleasant day. At work as usual at the Fort, and walk to town in the evening[.]

Sunday January 29" 1865

Have been to the Congregational Church after which took a short walk with Minnie. The sun has melted the frost and made it muddy in places. Stay home in the evening on account of mud[.]

Monday January 30" 1865

Rain has been the order of the day and now is every where. Talk of Illinois mud this Kansas institution waxes the lot.

Tuesday January 31"

Mud and maps! Am getting my work pretty well along.

Sunday February 12" 1865

Prepared for church expecting that Mellie would be down from Clara's earley and was bound she should not find me unprepared. Sat down to read, expecting her every minute, but she came not until a late hour, and when she did come, she was in the greatest hurry imaginable to get ready in time[.] No hen with a family of two dozen chicks to support never scratched harder to supply their wants, than she. She had been to the Methodist "Love Feast" and was now in for church. At rather a late hour we arrived, and found the Presiding Elder expatiating larg[e]ly on the Missionary cause. He was followed by Bro. Mitchel and then the donations commenced until $490 was raised, and Brothers Cloud and Mitchel were made Life members of the parent society. Col Cloud[25] was very popular among the pious breatheren and sisters, gave largely and was made general cock robin by the devoted ones. Perhaps an insight into his licentous course in Arkansas would have been viewed with holy horror by the saints. I learned that in the evening the missionary fund was raised to over $1000[.]oo[.] Went to the theatre, at night to hear the Revd Mr Kalloch preach. He is a Baptist and as that denomination have no house in Leavenworth, the theatre was hired for the occasion. Mr Kalloch is among the first pulpit orators of America. He preached a splendid sermon to an overflowing and attentive house. He has not entirely weaned himself from the world yet and is speculating largely in Kansas town sites[.][26]

Monday February 13" 1865

Went to the Fort expecting to start for Denvar, but learned that the expedition would not leave until Tuesday. Did a little work at mapping but most of the time was busy preparing for the trip. It rained in the afternoon, and I got completely wet while endeavoring to make an arrangement with the Signal Corps to proceed to Ft Kearney with them. But not being able to "consolidate" satisfactorily I have determined to get a team and go with my own conveyance. Rode to town in the rain. Found Mellie used up over the wash tub. Two weeks dirty duds had accumulated and it was rather a hard job for her.

Tuesday Feb 14" 1865

It has been a busy day for all concerned. Our whole outfit was to be drawn from the Quarter masters and prepared for the trip. It was a constant succession of storms of rain[,] snow[,] and sleet and in this drizzle our work had to be done[.] Commissary & Quartermaster supplies to draw teams, harness, forage, blankets &c &c but it was all accomplished before night, and I went splashing through the mud to town to pass the last night with my family before setting out on my long trip westward. Mellie bore up bravely and shed but few tears. I cheered her up as well as I could and we both planned for the future. May God be with me on this cold winter journey, preserve me from dangers and trials, and keep me from the hands of the savages.

18 | Fort Leavenworth, Kansas, to Fort Kearney, Nebraska Territory

February 15, 1865–March 3, 1865

The Platt is one of the most peculiar streams and the country bordering it is the most peculiar I ever saw[.]

—Lyman G. Bennett, March 1, 1865

On February 15, 1865, Bennett experienced an emotional departure from his family near Fort Leavenworth. He probably returned to his family sometime in April or May 1865 after making a nearly two thousand-mile round trip. It took Bennett's group fifteen days of struggle through rain, snowstorms, and mud to cover the nearly three hundred miles to Fort Kearney in the Nebraska Territory. First traveling northward close to Atchison, Kansas, the party then trekked along the Fort Leavenworth Road to Marysville where they crossed the Big Blue River.[1] The group then stayed at a series of stagecoach stations until they entered Nebraska along the Independence-St. Joe Road.[2] They picked up the Overland Trail and followed it to Fort Kearney. Bennett inspected the fortifications there and reported with humor on a skunk infestation and a rifle pit being used in a novel and unmilitary way. Although a marvelously detailed account of travel westward in the closing days of the Civil War, Bennett's racist perspective on Native Americans, discussed in the preface, will be jarring to readers.

Wednesday Feb 15" 1865

Three years of campaigning have learned me to require but little preparation in undertaking the most difficult and lengthy enterprises[.] "Few and short were the prayers we said" as I folded my little family in my arms and kissed them a long good by. The tears would come to Mellie's eyes in spite of her efforts to repress them. One more kiss and a grasp of the hand and I was away to the Fort. There I found that our raw mules had upset our wagon in the mud before setting out not only causing a delay of a hour but spilling all our stores into the street, but the fragments had been picked up and put together before I arrived and our teams sent out upon the road. It was terribly muddy, with about two inches of snow and water covering the ground[.] The water was rushing dow[n] the slopes, the rivulets were full, and gathering force and volumn as they went rushing madly on, soon become foaming torrents. The mud was universal, and our horses went plash, plash, all day long. From Fort out the country was rough and broken and the day was passed in slowly pulling up hill and then down, and then up and down again all day long. We could move but very slow and the mules were very much fatigued[.] we camped about sun down about a mile from Mount Pleasant a little one horse town 16 miles from the Fort. We thus traveled 15 miles[.] The 16" Regt[3] was camped about a mile further on. We thus had gone as far in one day as the Cavalry had in two. My business was to attend the horses and get the wood. Williams brought the water and got supper, while our "Jehue" took care of the mules[.] It was long after dark when we had finished our supper and was reposing upon the ground[.]

Thursday Feb 16" 1865

Mud! Mud! All day we have been slow plodding through the mud and nearly as liberal an allowance as yesterday. We halted an hour at the camp of the 16" but finding they were not to march we joined a heavily laden citizen's train and pushed on, but their movements were too slow for us and we soon left them and pushed ahead alone, and camped about 4 P.M. about four miles west of Atchison[.] The country passed over was well settled high rolling prairie with scarcely a stick of timber. Our camp was a mile and a half to the right of the road where a few scrubby hickory trees grew. We chopped our fore firewood cooked and eat our suppers and a storm of rain and snow coming on we went early to bed in the wagon. It grew cold and windy in the night but our warm blankets kept us from the effects of the storm. The cattle came around the wagon and disturbed us some, but there was little for them to get and we let them do about as they pleased. We traveled to day about 12 miles[.]

Friday February 17" 1865

It was very cold during the night and so froze the mud that for two or three hours it was tolerably decent traveling, but the sun soon thawed the ice and the remander of the day was but a repetition of our two days waiding through the mud. The country was less broken than that we had passed over and more sparsley settled[.] Much of the time we were entirely out of sight of timber—the prairies stretching out as far as we could see. We passed through Lancaster abot noon a little city of two or three houses ten miles from Atchison and where the overland stage Co has a station[.] About 3 P.M. we reached Grasshopper creek where was a narrow belt of brush and scrubby timber and being an excellent place to camp we determined to wait until the command come up. Grasshopper is forty miles from Leavenworth, we thus traveled 13 miles to day. Our driver shot a prairie chicken so we will have fresh meat for our dinner tomorrow[.]

Saturday February 18" 1865
Remained in camp all day, passed our time in reading an old paper which by accident came out with us. The cows were disposed to cultivate our acquaintance, and was all night perambulating about camp sharing the corn and hay with our horses and mules, rubbing at our heads and tasting of our rations[.] It was no use clubbing them away for they would soon return again with reinforcements[.] I was up about two dozen times during the night and threw billets of wood after them until cook said his wood was scattered over an acre of ground[.] The citizens train passed about 3 P.M. and will probably encamp five mile hence. We will overtake them tomorrow and unless they are two slow we will keep together. We are rather Jewish in our customs, resting Saturday and traveling Sundays[.] John brought in another prairie chicken to day. Wrote Mellie a letter and mailed it. Hope to hear from her at Kearney[.]

Sunday February 19" 1865
Resumed our journey, and found the traveling somewhat improved. I [illegible word] up slowly however in consequence of the ground being frozen a few inches below the surface and the moisture on top cannot penetrate below the ice. Where the grass is thick the ground is quite solid and the wheels cut in but little. We passed through a little town called Kimekuk [Kennekuk] near the edge of the Kikapoo reserve. Here we also passed the train which passed our place of encampment last evening. There are but few indians living on the Kikapoo Reserve, some it is said are with the wild indians making war on the whites[.] About 3 P.M. we halted at an indian farm to wait for the wagon to come up[.] Notwithstanding there were comfortable log houses built, yet the indians were living in a wigwam made of flags and rushes plaited together. We lifted an old bear skin which served as a door and within found four lazy indians and one squaw stretched out, around a little fire built on the ground in the center of the wigwam[.] Around it was a low platform built and mats made of rushes spread upon it. This constituted their bed[,] their floor

and seats. Besides the indians and the squaw was a cat, a pig with a bell on, and several dogs. The indians said not a word or made any ceremony when we entered. A few grunts in indian to each other was all that could be got out of them. One Indian amused himself by scratching the pig's back, ~~who~~ which appeared to like it. The wagon finaly came up and, it being some time before night we went two miles further to Big Creek and camped. It looked like rain and we all slept in the wagon[.]

Monday February 20" 1865

We rested finely in the wagon—was up early and disposed of our breakfast and was away[.] We had scarcely got under way when it began to rain. The roads had not more than half recovered from the recent wetting they had received and in an unaccountably short time was reduced to the sloppiest and muddiest kind of mud, and the mules had about all they could do to haul the wagon along. A five mile heat brought us to another Kansas city called Grenada and I was nearly disposed to stop. The wagon soon came up and we pushed on two miles further to a small log tavern and a good barn. Men[,] mules[,] and horses by this time were completely soaked, and we resolved to go no further through the rain. The roads are in a terrible condition. In no place did I ever see worse. How we are to get through to Kearney I hardly know. This is our sixth day on the road and we are not half way yet. I believe I will take the teams and wagon no further than Kearney, and pack our traps from then on[.]

Distance traveled 8 miles.

Tuesday February 21" 1865

Th[e] rain continued to fall in torrents until midnight almost forming a sea of water and mud[.]

Had the country been level, the whole surface would have been inundated. During the night the stage came to the creek near the house, and the bridge being torn up by the water, it stopped at the place we were, and remained until morning.

When we arose the storm had ceased and I found the roads firm and in better condition than before the storm. The bridge required mending but an hour's work set it to rights. The plank[s] that had floated away were found and replaced and about 9 A.M. we were away. Except in low places the roads were tolerable fair and we pushed on over the rolling prairies, rapidly. Now up a long slope, and then another, then down occasionly crossing a little stream, full of water from the recent rain. The Big Nemaha was very full, but the bridge had not washed away although threatened. The village of

Sinaca [Seneca] on the west of the river is a fine little town and has quite a business appearance. Along the Nemaha is quite a fine belt of timber which in Kansas is invaluable. We camped three miles west of Seneca on a little stream where was a little wood having traveled 18 miles[.]

Wednesday Feb 22nd 1865

We started shortly after sunrise, determined to push the teams to the utmost. The character of the country was like all we had passed over[.] High rolling prairie with a stream occasionly from five to ten and fifteen miles apart and a narrow thread of timber on the banks. A peculiarity of the timber is that it cannot be seen untill just upon it. The timber is generaly short and hid in the ravines.

About 2 P.M. we came to where the Oketa river branched from the Marysville road, and as the former was reported ten miles the shortest we started for Oketa, and had gone five miles when we learned that the Big Blue was up so high that it could not be crossed so we struck across the prairie in the direction of Marysville. We had gone about two miles when night came on and we were obliged to camp on the naked prairie. It was a lonly and desolate place, and as darkness drew its shroud around, My thoughts wandered to other & more pleasant associations[.] traveled 25 miles[.]

Thursday Feb 23d 1865

We left our prairie camp and pushed on to Marysville a distance judged to be 11 miles. Marysville is quite a fine little village for Kansas and a County seat. In consequence of the stage company changing their rout[e] via Oketo Much of the travel and business on the Marysville has disappeared and the tide of emigration now follows the stage by Oketa. A fine How[e] truss Bridge has been built across the Blue and we were able to cross without difficulty. The water was nearly up to the bridge and 23 feet in the channel. The water was black with the loose friable soil which had washed down from the hills. The Big Muddy would have been a more appropriate name at this time than the Big Blue. After crossing the river, for some distance the country was quite hilly, but the roads were generaly dry. Our driver pushed his mules along quite lively and we made 28 miles during the day. The prairie grass was tall and I fired it towards night. The fire spread rapidly and nearly all night lighted up the sky, until a slight rain in the night put out the blaze. We camped in a fine grove of timber 17 miles from Marysville, where pilgrims and soldiers from time immemorial had been in the habit of camping on account of wood and water.

Friday February 24" 1865.

We traveled 27 miles to day over a dreary prairie uninhabited except at stage stations which are established from 9 to 15 miles apart. We left Kansas before noon and to night are on Nebraska soil, at a ranch ocupied by one Hess. We got about fifty pounds of hay of him and was astonished at a bill of $2.50 for it. We will feast our animals on corn hereafter if this is the rate of charges in Nebraska[.] This is one of the most dreary countries I ever saw. The soil is rich enough, and the grass grows tall and rank, but there is nothing but interminable prairies to be seen all around. Weather cold and windy[.]

Saturday February 25" 1865

As we started across the prairie this morning the wind freshened up to a gail, and a storm of snow, set in square in our faces. The ground was frozen, and we walked after the wagon to keep warm. The storm increased at every step and soon the snow filled the air and whent whirling by or cutting directly into our faces. By the time we had gone five miles and reached the little Sandy creek we could not see ten rods ahead of us. This stream had also been much swollen and the bridge had been carried away, but the water had fallen and the ford shallow and we crossed without difficulty. Five miles further we toiled against the wind and reached the Big Sandy[.] Here two bridges had been carried away and there had as yet been no crossing by the ford[.] In driving the mules into the water they became entangled in their harness, and we had considerable difficulty in crossing. Horses, mules and men were more or less wet, and suffered from cold in consequence but by running over the prairies we became warmed up. The mules however shook with the cold and it required considerable of an effort to keep them along. At length I became tired and disgusted with their slow movements and galloped ahead to the next stage station 12 miles from the Big Sandy, and succeeded in getting my horse in a warm stable, and thawed myself out by a warm fire. In about three hours to [sic] mule driver and Thomas came up with but one of the mules, and reported that the others with the wagon was two miles back in a freezing condition, and could not pull the wagon another inch. I saddled my horse and went back to where the mules were and detaching them from the wagon, succeeded in driving them to the station, but it required the free use of the lash to make them move. A warm stable and a little rubbing warmed them up, and we had the satisfaction of knowing that they would not freeze to death. The storm continued until in the night and then the sky cleared up, but the cold was intense and our blankets were not sufficient to keep us warm. Our bill here amounted to nearly eight dollars. A man must be the owner of a gold mine to live in this country[.] Traveled 20 miles.

Sunday February 26" 1865

The sky was clear to day, but the weather was intensly cold. The sun shown in a sickly melancholy manner, imparting no warmth to the prairies around. Nine miles travel brought us to the Little Blue where we watered our animals and then pushed on nine miles further up the east side of the Little Blue[.] The roads were hard, and smooth and had it not been for our late start in the morning we would have went 25 miles, as it was we only accomplished 18 miles. There are some settlements along the Blue, and the country looks a little as if made for white folks. There is a very narrow belt of cotton wood timber along the stream, but does not amount to much. The animals suffered from cold during the night, standing out doors.

Monday February 27" 1865

Our road still led us along the Blue over good roads. To day the cold relaxed a little and what little snow was left and had not been blown away, thawed. In the road there was none left as the wind had swept it as as clean as mother's kitchen. Passed many deserted ranches the buildings of which had been burned by the indians last summer. We encamped where had been considerable of a farm, but the buildings had been burned and all was desolate and lonly. We found plenty of dry wood from the remains of an old stable and built up a rousing fire, the first good camp fire we have had on the trip[.] Williams cooked a mess of beans. Our pickles, beans, pork and biscuit, coffee &c furnished a most excellent meal and we shall remember the luxuries and comfort of this camp, which contrasts widely with all our former experiences in camping out[.] We traveled 22 miles.

Tuesday February 28" 1865

Our mule driver pushed his team to the utmost and kept them on the trot most of the time[.] The road led over a level prairie, with no hills to do down or up, by dint of loud hollowing much whipping and considerable swearing he drove his team 34 miles, making the best time since we started. We camped at a stage station 28 miles from Ft Kearney. The night was cold, and we had a poor night's rest, in the "Pilgrim house" attatched to the station. The term "Pilgrim" is new to me & is applied to the overland emigrants to the gold regions, Utah and California[.] A number of wolf pelts was stretched over the walls of our pilgrim shanty, and were taken from the carcases of some of the varmints that had been poisoned. A well was being dry [drilled?] here and had already reached the debth of 75 feet with no indication of water[.] The ocupants of the Station were the most gentlemanly of any we had passed, and we were not confronted in the morning with an exhorbitant bill[.]

Wednesday March 1st 1865

I pushed on ahead of the team and after two hours ride found myself among the low conical bluffs which skirt the Platt river valley, and from between them I obtained a glimpse of the river timber in the distance. The Platt is one of the most peculiar streams and the country bordering it is the most peculiar I ever saw[.] The valley from the low bluffs on each side is from three to ten miles wide and as level as a house floor, with just enough sand to keep the roads always dry. There are no banks to the river, you suddenly come to a thread of water meandering among grassy islands and sand bars, with a sluggish current, and but a few inches in debth[.] Here and there are cotton wood trees bordering the water and some of the smaller islands are covered with willows. It is a strange country to me. The level plain stretches away until earth and sky meets in the mirage far in the distance. At a little dog town nine miles from Ft Kearney I stopped for a drink of water, and finding a Missouri Democrat of a recent date, I passed an hour in reading the news. In the 15 days I have been on the road most important events have occured[.] Charleston and Wilmington[4] have been captured from the rebels and our armies are everywhere succesful and the rebellion must soon succumb before the vigorous blows that is being dealt on every hand. I reached Kearney about 2 P.M. two hours ahead of my team, and made arrangements for quarters for self and party[.]

I had an introduction to Genl Mitchell[5] and Col Livingstone[6] who were very courteous and gentlemanly. I was a little amused with Capt Gillett.[7] I was in his office making arrangements with his Adjutant for quarters when the Capt came in and judging from my rusty appearance that I was a soldier and in the way he was on the point of ordering me out of the office when a hint from the Adjutant restrained him. When he found who I was he apologised and we both had a hearty laugh at my expense. I expect I do look as rough and rusty as the highest private in the rearest rank[.]

Thursday March 2nd, 1865

Visited the fortifications and works about the place and found them in an unfinished condition, and work entirely suspended in consequence of snow and cold weather[.] as planned by the old German who had them in charge the intrenchments were to be from 14 to 16 feet high. The bastions at the corners were round, resembling minerats on a mehometan mosque, and so small that the recoil of a large gun when fired would have skattered and knocked them down. I discharged the dutchman and put Capt Gillett, commander of the Post in charge of the works. I also cut down the height of the intrenchment to eight feet on the revetment and ordered the bastions to be enlarged. A redern [redan] of earth that was designed to extend across the eastern side of the Post I changed to a rifle pit and laid out a Lunette South West of the Post so that all the approaches could be covered by cannon[.]

On the south was a crooked ditch called a rifle pit, which the soldiers used as a privey until it was filled with excrement. I had to laugh when this was pointed out as a rifle pit. It would not afford protection to a snow bird[.]

I was busy all day in drafting my plans and giving instructions for the completion of the works[.]

I must not forget to mention that a letter from Mellie, gave me much pleasure and was as sunshine on the wintry scene around. I wrote a letter to her and one to father[.]

Among the institutions of Ft Kearney are skunks. They live and raise their families under the buildings of the Post and I am told that under every house is a family of them. The family under my room were rather quarrelsome and several times during the day and evening I could listen to their dissensions. During the night the quarrel broke out with fury on both sides and bottles of night blooming Cereus was empt[i]ed in liberal quantities[.] The stench awoke us and we found ourselv[e]s nearly suffocated. To put our heads under the bed cloth[e]s would do no good, we could not stand it except by holding the nose and breathing through the mouth, and the tast was sickening. In this way we spent an hour, alternately pukeing and then scolding the skunks and threatening all kinds of vengence. Lt Moore[8] declared it was thick enough to cut like cheese. I was never before in so close proximety to them without being able to retreat or help myself in some way. The whole place smells of skunks and at two bits apiece for their pelts a person could make an independent fortune in gathering skunk fur[.]

Friday March 3d 1865

Have made my preparations for leaving here tomorrow. Made a Report of operations here and written letters to Mellie and Lt. Robinson. There was considerable snow fell during the night and it has been cold and wintry to day. I shall have an unpleasant trip I fear, but the roads are good and I shall make good progress. Williams has taken a sketch of the place, and will finish it up when he has time[.] The officers here are gentlemanly and courteous, and I am very favorably impressed with Col Livingston and Capt Gillett.

19 | Fort Kearney to Denver

March 4, 1865–March 22, 1865

I had dreamed of indians during the night and every flap of the canvass wagon cover in the wind would awaken me, and I would listen for the steathy steps of indians.

—Lyman G. Bennett, March 6, 1865

In a journey of fifteen days, Lyman Bennett traveled 391 miles from Fort Kearney to Denver City. These entries revealed his usual interest in his surroundings and his notions of what beautiful landscapes entailed. Dismissively he wrote on March 4, "I doubt if there are many spots on earth calculated to excite emotions less pleasureable than the region along Platte river[.]" Writing a day later, he revealed that the Great Plains intrigued him, such as the fact that "distances here are very deceptive, one would imagine a bluff ten miles away, but two. In stretching our eyes ahead the mirage would cause every object in the distance to assume fanciful shapes and appear as large again as they realy were. It is strange how these mirages will deceive[.] One not accastomed to them would declair that lakes and forests existed in the distance and yet after traveling all day, one would be no nearer these imaginary lakes and forests than in the morning. This is a strange country[.]" Bison interested him, but the herd spotted on March 11 roamed "at least ten mile distant," which was a disappointment. Five days later, he saw "the back bone of the continent, the Rocky mountains." When he caught his first sight of them that day his "feelings were as boyant and I felt as spryhtly [sprightly] as a young and happy child[.]"

Intermixed with his excitement at seeing new landscapes were deep feelings of dread and anxiety at the possibility of encountering Native Americans. Just months before on November 29, 1864, Colonel John M. Chivington led a force of approximately seven hundred Colorado soldiers in an attack on a Cheyenne and Arapaho camp along Sand Creek in the Colorado Territory. Although Chivington's force suffered fifteen men killed and fifty wounded, over 150 Native

Americans were killed in this one-sided encounter with many being women and children whose bodies were mutilated.[1] The Sand Creek Massacre shifted power to militant Cheyenne and Arapaho,[2] and they launched a furious series of raids in January and February 1865 over a more than eighty-mile stretch of trail from western Nebraska to near Denver.[3] These winter attacks stopped mail, supplies, and telegrams from reaching Denver for a month.[4] Bennett was among the first travelers to head west to Denver after these retaliatory strikes.

Commenting regularly on destruction that he observed, Bennett was particularly unimpressed by Julesburg, Colorado, which was nevertheless an important stagecoach stop. Cheyenne and Arapaho had systematically looted the town on January 7 and soon returned to loot it again on February 2. During their second attack, they set the town's buildings on fire.[5] Altogether, approximately fifty people were killed along the trail, property damages amounted to "several thousand dollars, and [there was] a few weeks' disruption of commerce and travel."[6] Major General Samuel R. Curtis, Bennett's mentor, proved ineffective at keeping the trails open.[7] Authorities merged the Department of Missouri with Curtis's Department of Kansas, and Major General Grenville M. Dodge, who Bennett had met at Rolla in November 1861, took command on January 30, 1865.[8] Working energetically, Dodge arranged for reinforcements, ordered all westbound wagon trains to be accompanied by one hundred armed men, and ordered cavalry to protect the Overland Mail.[9] Bennett arrived safely in Denver City on March 18 and took up lodging at a ranch owned by John Coryell, a relative.

Saturday March 4th 1865

I doubt if there are many spots on earth calculated to excite emotions less pleasureable than the region along Platte river[.] Few traces of life or civilization is visible for miles upon miles, and many that are left are deserted by their owners in consequence of the dreaded indian. The eye stretches over a vast extent of level monotonous prairie and lights upon no living forms except his companions. There are a few scattered ranches, but danger and apprehension have forced the tenants to heard together at the Posts protected by military forces, and the country is left to indians and wild beasts. I wandered listless up the road, surrounded by boundless prairies, and a clear blue sky above and pinched with cold. Winter still reigns in this wild region and is not disposed to relax his grasp. At night I try to rest in the wagon beside horses and mules and by day I hurry along the well beaten track with no comforts except those carried with me, and but few companions except my thoughts[.] But these are enough. They spring up with every mile we travel, my very lonliness affords food for thought[.]

We did not get a very early start and the long day's march before us made us move in a hurry. I forgot my blanket and went back two miles after it. Kearney City, two miles west of the Fort is the first object that meets the eye. To my mind it is a wretched congregation of adobe hovels and calculated to extinguish all aspirations in a civilized being to reside in the country. Adobe houses are made of sod, cut with a plow the desired width and thickness, and then with a spade the required length, after which it is put up in regular mason work style[.] The walls of buildings are usualy two feet thick and eight or ten feet high and are dark and dingy both without and within[.] The alkaline soils in this country forms a kind of cement and it is said that these adobe or sod houses have been known to stand fifteen and twenty years. I judge them to be the horror of a neat house wife[.] Some of them are whitewashed within and are neat and clean. In speaking of ranches an adobe or sod house is understood.

After passing Kearney city a few ranches are scattered along the road, mostly unoccupied and they present a forlorn appearance. A few trees fringe the Platt river, but are of a stunted growth. Otherwise there is nothing to relieve the monotony of the scene, except the low bluffs on either side of the river and usualy ten miles away[.] The river bottom or vally is from fifteen to twenty five miles wide and as level as a house floor. The roads are hard as a pavement and the traveling delightful. We met a portion of the 11" Ohio Cav[10] about noon, who were on their way to Omaha to be mustered out[.] They were a healthy looking set of fellows indicating that the elevated regions about Ft Laramie agreed with them, about every man in the outfit inquired the distance to Ft Kearney until I was completely tired of constantly singing out "seventeen miles" and [g]alloped by them to escape their endless questioning. We reached Plum Creek Post about 4 P.M. having made an excelent drive[.] Capt Majors[11] the commander provided for our wants, provided fuel for the boys and invited me to share his quarters and table[.] I examined the works in process of construction, and found them adapted for the defense of the place against indians and made no alterations in the plan but directed the captain to complete them as soon as possible. The officers quarters are of wood and very convenient and comfortable. The mens are of adobe, and very warm, in fact too warm, and are to[o] close and dark I directed that they be ventilated and more light be provided for. The enclosure was of sod and about half completed. Little can be done until the frost is out of the ground. I passed a pleasant evening with the captain and formed the acquaintance of one Williams the telegraph operator and formerly a resident of Kendall County[,] Ill[.]

Sunday March 5"
Resumed our westward pilgrimage early, and was furnished an escort of five men[.] The country. [*sic*] The country was similar in every respect to that passed over yesterday.

First was Millaley's Ranch[12] 16 miles distant and ocupied by a guard of only ten men from Plum Creek. They occupied a portion of the adobe houses belonging to the Stage station, which was a sufficient protection against indian attack without any additions[.] The small guard here would be incapable of defending more extensive works.

At Miller's Ranch I found a captain and a small company ocupying the buildings which were ample for their accommodation. No other works had been erected. This is not of much importance as a military station, still it would be well to construct a sod wall around the ranch and I suggested this to the captain[.]

Ten miles from Miller's is Dan Smith's Ranch and stage station. Here was a sargent and ten men quartered in comfortable houses of cedar logs. The stable was large and made of sod. The defences were not good, but ten men could not defend more extensive works and consequently I gave no orders for any to be erected[.]

We had traveled 35 miles since morning and was 70 miles from Ft Kearney and encamped for the night. The day was bright and pleasant and nothing occurred to impede our progress. The mules were picketed out half a mile from the station to get grass and the horses put in the stable. The bluffs were higher than those of the day before but still from five to ten miles from the road. The distances here are very deceptive, one would imagine a bluff ten miles away, but two. In stretching our eyes ahead the mirage would cause every object in the distance to assume fanciful shapes and appear as large again as they realy were. It is strange how these mirages will deceive[.] One not accastomed to them would declair that lakes and forests existed in the distance and yet after traveling all day, one would be no nearer these imaginary lakes and forests than in the morning. This is a strange country[.]

Monday March 6" 1865
Awoke at sunrise and found ourselves and stock safe, no indian daring to molest or make afraid. I had dreamed of indians during the night and every

flap of the canvass wagon cover in the wind would awaken me, and I would listen for the steathy steps of indians. We were off rather early and reached Gillman's station at 11 A.M. Capt Porter of the 7" Iowa[13] with his company ocupied the Post. He had built a high adobe wall around his stables and wagon yard and his command ocupied the houses on the ranch which were ample for the command[.] The road ran uncomfortably near his quarters and I directed that an adobe wall be built across the road and in front of the quarters thus compelling the travel to go several yards from the houses and also affording a complete defense in front[.] Williams took a sketch of the place and we went on our way. After going about five miles I discovered objects on the prairie and foot of the bluffs which I could not make out. My imagination lead me to believe they were indians and I ordered the team to stop which I reconortred. I rode a mile or two towards them and found they were not savages but four footed beasts of some kind, and I motioned for the team to come on. Thinking it a heard of buffalo I approached and prepared for a chase, and an onslought among the animals after going three miles farther I found they were cattle and I returned to the road and joined the wagon. These cattle belonged to ranch men near Cottonwood and had fead [fled] more than six mile away from their owners. Reached Cottonwood at 2 P.M. and was cordialy welcomed by Major O'Brian[14] commanding the Post. I was soon provided for, and was invited by the Major to accept of the hospitalities of his table and bed. I made an inspection of the post & its surroundings with a view to continue my journey the next day but rumors of indians about coming in I decided to remain here one day. A man up one of the canons four miles from the Post was chopping wood when his attention was called to the fireing of guns in the distance. Peering about to discover the cause he saw to indians a mile or more away, coming towards him[.] His fears lent speed to his legs and he came in to the Post and reported his discovery. The appearance of a fire among the canons corroborated his suspicions and we all believe that indians are prowling in the neighborhood and I shall wait here for further developments[.] It has been a bright day and we have traveled 27 miles making 97 in three days since leaving Ft Kearney.

Tuesday March 7" 1865

Cold Clear and windy! After confering with Maj O'Brian I have made several alterations in the Post at this place[.] The south line of stockade I find within rifle range of the bluffs and easy of attack. I have directed that the pickets be pulled up and set 200 feet further north and that the ground be extended 400 feet on the other side and 250 on the west. Thus affording room for hay yards corrall hospital &c & within the stockade, also two block houses at opposite corners of the enclosure, the whole to be 900 X 954 feet[.] The defences here is a stockade of cedar. All the quarters stables &c are of cedar, and are very comfortable and convenient[.] The situation is fine and everything is clean and in good order. Fine springs are near at hand and I doubt if there is a finer location on the whole route. The bluffs here are quite high and much nearer the river than anywhere below, being not more than a mile and a half from the river. The stage station is 3 miles above the garrison and I have given written orders for it to be removed to within one half a mile of the post.

Have been very busy all day drafting plans for the works here and in making a report to Lt Robinson[.][15]

I find it necessary to remove the road several hundred feet to the north of where it now runs, so as not to annoy the garrison with the dust which arises from the immense travel in summer. It is also necessary that a bridge be built across a deep ravine here which is difficult for loaded teams to pass. Mr Gillman will build it for the privelege of collecting toll for a series of years. Passed the day very buisily yet at the same time very pleasurably with Maj O'Brian[.]

Wednesday March 8" 1865

On awaking this morning the snow had whitened the ground and was eddying in gusts around the quarters. It was intensly cold and the wind keen and cutting from the Northwest[.] It was cruel for men and teams to face the storm and I resolved to lay over another day. Have drafted plan of bridge, written up my journal read Moors poetical works, written to Mellie and the storm abating after noon I have resolved to push on tomorrow. Williams has taken a view of the post and though it is freezing cold out doors I shall endeavor to push on tomorrow. Capt Majors and Porter arrived from below just at night with about 80 men for Julesburg, some of the men were badly frozen. I[f] not to[o] cold they go on to morrow and I will go with them.

Maj O'Brian has given me considerable information in relation to the existence of gold mines in the Black hills and as I am to visit that neighborhood it may be well to look after the "gold[.]"[16]

Thursday March 9" 1865

We were rather late upon the road, owing to the escort being behind time, but once under way we made rapid progress. The wind cut into our faces and was very cold raising clouds of sand and dust that was blinding. The high bluffs in some places almost overhung the road, affording facilities for indian attack, had they been on the alert in the cañons. But sometimes indians are found on the plains. A story is told of a dutchman, that was traveling across the plains, and the fear of indians gradualy wearing off in consequence of the non appearance of indians, he sometimes would straggle from his escort. One day he discovered a fine buffalo robe laying upon the prairie[.] Approaching to secure his prize, he was frightened at the appearance of an indian under it, in the attitude of attack[.] The poor dutchman cried out in affright, "Some one had pitter come here quick, or ire gone up[.]" About ten miles above cottonwood I observed a kind of rock exposed upon the sides of the bluff. This is the first rock I had observed on the rout[e]. It was peculiar, being soft and clayey, like marl and is used for wells and for building purposes. Twelve miles from Cottonwood we passed Jack Morrow's ranch, a large two story timber building, the best I have seen on the rout[e].[17] At several points flocks of wild geese were along the road picking up the corn dropped from the waggons. I tried my revolver at them long range at them, but the only effect was to scare them.

Thirty three miles we passed Fremont's Spring a fine fountain close by the road, and near the stage station. I found the Post at O'Fallons Bluff, in a bad situation commanded by bluffs and approached by ravines, affording cover for attack. I directed the Post removed to the springs two miles below and drew a plan for their construction. The Lt in command here was very courteous and gentlemanly and freely tendered the hospitalities of his table and bed.[18] Distance traveled 35 miles[.]

Friday March 10" 1865

The road from the Post leads over the bluff affording fine views of the surrounding country. The day was more warm and pleasant than yesterday and I enjoyed the ride and the prospect from the bluff, the descent of which into the river valey was somewhat steep and more like hills than any place we had traveled over this side of the Little Blue. We reached Alkali station about two oclock. Capt Murphy 7" Iowa Cav[19] is stationed here and has displayed much industry and energy in constructing quarters and fortifications. I added a little to his plans and have no doubt he will complete his works early[.] Capt Murphy is much of a gentleman and I shall long remember the good dinner and breakfast I received. Here the stage station is one mile from the post. I ordered its removal nearer the station. Maj Majors arrived with his command and I shall go in his escort. Traveled 25 miles[.] The soil about Alkali is filled with alkeline matter which accumulates in large quantities in the grass on the surface of the ground and I am told is a good substitute for salasatus [?] or baking powders. The waters I understand is death to man and beast.

Saturday March 11" 1865

Started early with Maj. Major's command. It was a clear beautiful day. The Prairie dogs as we passed through their towns were scampering from hole to hole in full enjoyment of the bright sunshine, and genial warmth that was shed around. They are a strange little animal smaller than our Illinois rabbits and live in deep holes they dig in the prairie, the entrance to their habitations being surrounded by a mound of earth[.] On the approach of any one each scampers to his home and sits in a position to dart in and barks and chatters until approaching nearer they dart into their holes still scolding and barking under ground[.] Watch their houses for awhile and first one and another and another they stick their noses out to see if the co[a]st is clear, and gathering confidence will venture to a neighbor and will canvas together as to the nature of the monster that has dared to invade their city[.] Their flesh is said to be excellent food. We pass through many of their towns going up the Platt[e], some of them ocupying many acres.

    Away to the right we discover objects in the distance which Lt Tolbot[20] after a peep through the glass pronounced to be buffalo. They were at least ten mile distant, and entirely too far for me to devote a page to the natural history of the buffalo.

    In passing near the Platt[e] river we started up several grouse and Lt Talbot had the good fortune to shoot one. They were a little smaller & a little different from the prairie chicken[.] Reached Borear's station[21] about 3 P.M. Making 25 miles travel[.]

Sunday March 12" 1865

A Company of the 7" Iowa is stationed at Beauvois, commanded by Capt [no name supplied][.][22] But little had been done here beyond erecting quarters for officers and men. I drew up a plan for works, and Capt [extra spacing] promised to carry them out.

    It was a fine day, more cold than yesterday and somewhat cloudy. All along the road was the marks of indian depradations in the form of deserted and ruined ranches. The walls were crumbling and no human being was left except at one ranch not far from Julesburg. The bluffs here came near the river and the sandy beds of dry water courses formed deep cañons that were not easily crossed. A view from the hills was romantic indeed. There was considerable sand and coarse gravel in many places. Discovering moving objects near the bluffs ahead Maj Majors & I galloped forward to reconoitre not knowing but they were indians[.] We found Capt O'Brian[23] and several officers hunting Jack rabits but on our arrival they gave up the chase[.] The burning of the stage station had filled the few adobe houses at the post with men[,] officers[,] soldiers &c and one could scarcely find standing room[.] Capt Murphy is a genial good hearted fellow and we were perfectly at home with him[.] Some of the officers played poker all night[.] Traveled 30 miles[.]

Monday March 13" 1865

I had imagined Julesburg to be a sort of town, and was surprised to find nothing but a few quarters for soldiers and about a mile below the ruins of the stage station around these blackened ruins was the remains of scores of half burned wagons and a large amount of other property[.][24]

This is an important point for it is from here the travel up the North Platt[e] and to Laramie leaves the main rout[e] & also the telegraph here changes to the north[.] I saw Mr Reynols the stage agent. He was full of complaints against the soldiers for on account of depridations and found more fault with them than the indians. And yet if these soldiers were withdrawn complaints would be redoubled.

We were on the road about noon, which in many places was hilly and rough. We passed the Wisconsin Ranch where a train had been captured and destroyed by indians[.] Thousands of dollars of property was still scattered about. Hundreds of pounds of nails codfish and the broken and charred fragments of every species of property. The ranch was in ruins, and had been swep[t]ed by fire[.][25]

Near here we saw moving objects at a distance up the river. My glass revealed them to be human beings but whether white or red men I could not tell. While crossing to a high point to get a better view we scared up a jack rabit the first I had seen[.] He was soon out of sight his long ears erect and longer in proportion than the ears of a mule. We reached Buffalo Springs[26] about 5 P.M. Here was Co B Denvar Militia[27] under Lt Hayns, who treated me very corteously[.]

A snow storm set in and at dark it was several inches in debth[.] I took the dimensions of the ranch and gave such directions as I thought necessary for defense. The place is surrounded and commanded by hills and is not desirable for a Post[.] Traveled 18 miles[.]

Tuesday March 14"

The snow was about six inches deep and impeded our progress considerably, but the sun coming out soon thawed it. The sandy soil absorbed the water as fast as the snow melted and we were not bothered with mud[.] After going about four miles I found that I had forgotten my blankets and went back after them and it was 10 oclock before I overtook the escort. We passed several ruined ranches, the fragments of broken furnature and utensils were scattered all around.

Reached Lillian springs[28] about noon. Here Capt Cozzens with a company of militi[a] was stationed who gave me an escort to Valley Station[.][29] The road lead over some large sand hills and it was exceedingly difficult for the teams to haul the wagon alone. Added to this one of the mules was taken sick and wanted to lay down, but by the vigorous use of the whip we kept her moving, and towards evening she began to be better____ We met a train going to the states, with a large number of men and women, who perhapse had come here with high expectations of sudden wealth and their hopes had not been realized____ Washington Ranch, three miles from valley Station had not been destroyed[.] There had been some fighting but the indians had been driven away.[30]

We reached Valley Station about sundown[.] Major Tabbot received and treated us very courteously. Here I took the dimensions of the buildings and gave directions for works to be erected[.] There had been considerable fighting here but 20 men had kept the indians at bay and saved the station[.] Traveled 32 miles[.][31]

Wednesday March 15" 1865

Started early with a train that was bound for Denver under the charge of Mr Chivington[32] a son of Col Chivington of the 1st Colorado.[33] But their movements were to[o] slow for us and we pushed ahead. I found that my horse had but little or nothing to eat during the night and stopped at a ruined Ranch and fed a mess of corn. The snow which had fallen the day before was all gone, and the road was as dry and hard as at any time during the trip. About noon we reached the American Ranch which had also been visited by indians and destroyed. At this ranch was four men & the wife and children of the proprietor[.] The indians surrounded the ranch by hundreds, the four brave men within fought them for hours and killed large numbers of their savage foe, but it is supposed that their amunition gave out & the ranch fell into the hands of the savages[.] The bodies of four men were afterward found and buried[.] The family was gone and no trace has been obtained of them. It is supposed the woman and children are prisoners and that her husband has been reserved to be tortured[.] This was a large ranch, the finest one on the whole rout[e] and the indians obtained much booty and set the place on fire. They always carry off their dead, but two dead indians were afterwards found left behind it is supposed their comrades did not see them[.] One was badly burned and the other lay upon the prairie. The burned indian had been set up against the ranch and was still there as we passed. The other had been scalped and mutilated by those passing along until his head was entirely peeled and his body disfigured and eaten by dogs[.][34]

Two miles further along, was Godfrey's Ranch[.] The indians at the time of the destruction of the American ranch, had made demonstrations here but Mr Godfrey with four others had kept them at bay and driven them off, killing several.[35] At Godfrey's was several women, who dressed in mens cloths and during the fight was in sight of the indians who took them to be men and supposed there was a much larger garrison than there realy were. [text is missing in the original] in the hottest of the fight when the prairie was on fire and had nearly reached Godfrey's hay stack, he sallied out and extinguished the flames, while the bullets was falling around him like hail. Had his hay caught on fire the whole ranch would have been destroyed.

Shortly after leaving Godfrey's I observed occasional patches of old snow, and in some places the road was muddy from its me[l]ting thus showing that we was in a more elevated region and in the neighborhood of the mountains[.] On every side however was a level plain with a range of low bluffs on either side. Sometimes the bluffs came near the river and the road either skirted or passed over them.

Reached Beaver Creek about 4 P.M. & put up for the night. Here was a company of Colorado militi[a] encamped and I was soon at home. I here met Henry Stafford formerly of Oswego, now married and a resident of Denver. At Beaver Creek is a bridge erected by the stage Co and tole collected[.] Traveled 32 miles.

Thursday March 16" 1865.

It was a beautiful morning, the sun was bright and it was delightful to travel in the thin pure air of the plains. After an hour or two the distant clouds in the west cleared away, and for the first time I caught a glimpse of the mountains. The first I was able to distinguish was Pike's Peak and afterwards Long's Peak. They were low in the horizon and difficult to distinguish from clouds.

For a day or two I had been on the lookout for the mountains, and every fragment of cloud that I could discover peering up from the western horizon I would watch intently to see if it was not a mountain peak, but it would gradualy change its appearance or fade entirely away, so that it was sometime before I could be convinced that it was mountains and not cloud that I saw. When satisfied as to what they were, my feelings were as boyant and I felt as spryhtly [sprightly] as a young and happy child[.] We were near the end of our trip and I was soon to see the back bone of the continent, the Rocky mountains.

We reached Junction station about noon[.] Here we were to leave the Platt and strike across the country to Denver, on what is called the Cut off. The river was raising from the snows in the mountains and at their base melting and we sometimes stopped to see the floating fields of ice as it moved solemnly down stream until striking an island or some impediment the ice would break into larger cakes and then gorge. Pieces of ice would be driven on the land ~~and~~ or breaking into fragments would finaly be carried down stream. We struck off on to the Cut off after partaking of refreshments with the officer commanding the Post[.] For eight or ten miles the roads were smooth and dry, but we soon entered sand hills & it was extremely slow and difficult traveling for about five miles. The wheels of our wagon sinking into the sand sometimes a foot.

We finaly reached the Bijean a small stream crossing the plains and emptying into the platt[.] Here we saw trees, the first in about 200 miles of travel. They were cottonwood and short and scrubby and resembled an old orchard[.] Bijean was pretty well up from the melting snow which here covered the plain. We experienced no difficulty in crossing and soon reached the station and stopped for the night[.] The station was a miserable dirty hovel, full of filth and lice and we chose to sleep in our wagon to laying in this miserable den. Traveled 38 miles.

Friday March 17" 1865

From Bijean the country raises with more of a grade than along the Platt and eventualy becomes hilly, as we advanced the snow was found in large quantities and covered the whole surface of the ground like a sheet[.] The roads were muddy and as bad as any we had seen in Kansas[.] We made very slow progress over the hills. I was nearly sick[.] My bones ached and I was glad to ride in the wagon[.] I saw a large drove of antelope bounding over the plains and learned that they were in large numbers in this region[.] This was the 2nd drove I had seen[.]

We reached Living Springs a stage station 24 miles from Bijean and stopped for the night[.] Here was a company of the 2nd Colorado Cav[36] stationed and the officer in charge, a German Lieutenant was very gentlemanly and corteous[.] Very soon Friend Stafford and a militi[a] Lieut came along and were going to Kiowa creek 11 miles further and stop for the night[.] I saddled my horse and joined them leaving the boys and team to come after the next day. The road was good to the Kiowa and we reached the station shortly after dark. The Kiowa creek was up but we forded without difficulty[.] I traveled 35 miles to day and the team 24[.]

Saturday March 18" 1865

A bank of cloud had hung over the mountains so that I had caught only occasional glimpses of a few of the snow covered peaks[.] But this morning not a cloud was to be seen and stretching to the north and south as far as the eye could see was the rocky mounts in all their grandeur and beauty. The sun was tipping the highest points with gold and the scene was truely delightful[.] We went ten miles to Box elder and took breakfast after which we pushed on again over almost intolerable rodes[.] The snow was near a foot deep and thawing fast, and our horses went splashing through the slush not very rapidly[.] In spots the ground was bare but here the mud was worse than the snow[.]

At the Toal gate 10 miles from denver was a well strongly impregnated with saltpeter. Very much if not all the water on the plains are tinctured with some mineral substance or alkali[.] We reached Denver City about 1 P.M. My first inquiry was for the Post Office and was gratified by receiving a letter from Mellie with cheering news from her. I then made my way to Coryell's ranch three miles up the Plate from Denver, and was glad to meet with friends once more. A season of refreshment and rest is now before me[.] Traveled 34 miles[.]

Sunday March 19"

The team arrived about noon and I soon made arrangements for their keeping and then returned to John H Coryell's[.][37]

Monday March 20" 1865

Proceeded to town and held an interview with Genl Moonlight[38] in relation to the location of the Posts in his District. The following post were fixed upon[:] Living Springs on the Cut off 42 miles from Denver[,] Junction 40 from the Latter. American Ranch Valley station Spring Hill and Julesburg. Returned in the afternoon to John Coryell's[.]

Tuesday March 21" 1865

The Commissary at Denver kindly offered me a place in his office to make the required plans for the Posts and the Quartermaster furnished me with a table and I proceded to my work. Williams also compleated one or two views of posts in the way. Libbie Coryell[39] came to town with me on horse back and I returned with her before night[.] She complained much of the churndasher gait of her horse, and for an experienced rider made considerable complaints of having hurt[?] [bust?] the saddle. More than Mellie did the first riding she ever did.

Wednesday March 22nd 1865

Completed my plans to day and submitted them to Col Moonlight who approved them and forwarded them to the different posts with orders for the troops stationed there to proceed at once with their construction[.] Looked about the city and the business of the town. Cherry Creek is quite high with a large stream of water caused by melting snow. A foot bridge is being constructed across it. There scarcely had even been any water in Cherry creek and large brick buildings had been built in its bed, when the freshet last April carried everything away, destroying much property and carrying houses away[.][40] On this creek were the first gold discoveries but never in paying quantities. On the east bank of the Platt are is still seen the old diggings of early miners. Acres of the ground is perfectly honeycombed with the holes dug by miners.

20 | A Tour of Colorado Gold Mines

March 23, 1865–March 24, 1865

> *Our road threaded the defiles of the grandest mountains of the continent. At last I was at the "ultima thule" of my school boy dreams, "the Rocky Mountains." Huge ramparts of rock towered skyward and toppled over the path.*
>
> —Lyman G. Bennett, March 23, 1865

The Panic of 1857 resulted in an economic depression that displaced thousands of Americans.[1] The discovery of gold in the Rocky Mountains the following year ignited a rush and provided a destination for more than one hundred thousand people.[2] With the arrival of these gold seekers, mining towns quickly developed, and the Colorado Territory was born through an act of Congress in February 1861.[3]

Ever inquisitive, Lyman Bennett desired to see some of the gold mines for himself. Traveling on horseback, Bennett and John Coryell proceeded west from Denver into the mining country. The two men ventured to Central City where they toured two of the Gregory mines, named for John H. Gregory, who made one of the key discoveries there in 1859.[4] Central City had attracted thousands of people intent on instant wealth in 1859, but when Bennett and Coryell toured the town, it had already begun to decline in its influence.[5] Black Hawk, a mile east of Central City, was home to many stamping mills that Bennett clearly described.[6] For someone as mechanically minded as Bennett, it is no wonder that he had "seldom passed two more interesting days than those among the mountains in the gold mines[.]"

Thursday March 23d 1865

John Coryell and I ate an early breakfast and then left for the mountains determined to see something of Rocky Mountain scenery and gold mining before returning to the states. We went by the way of Denver in order to get William's horse for John. His riding animal was away down on the Platt river. Crossing the platt we were soon in a rapid pace across the plains which lay at the foot of the mountains. To all appearances it was not more than three miles but after a two hours' ride I was surprised to learn that we had gone 12 miles and had only reached the foot of the first hills. All along the mountains is a rocky range from 200 to 500 feet in height called the hog back with occasional breaks where streams have forced their way through from the mountains. This Hog back is a sort of outer enclosure, confining the more lofty crags and peaks within. Our road lead through an opening into quite an extensive valley laying between the hog back and mountains proper. Nestling in this valley on the banks of Clear creek and surrounded by towering mountains on every side was Golden city, 15 miles from Denver and now the capitol of Colorado. I[t] was a small dull town, with no business except that derived from those passing too and fro between Denver and the gold mines[.] In this valley it was almost like summer, the sun shown out brightly and no wind could scale the surrounding ramparts of rock. This town and its surroundings recalled to mind Irving's sleepy Hollow. This must be the sleepy hollow of Colorado.

A mile or two further on we entered a narrow defile, and here the greed for speculation was again shown, by an attempt to build another city. The few tumble down houses and cluster of sheds here collected in this narrow pass was called Golden Gate. The person must be possessed of unparallelled stupidity who would make his home here, or imagine a city would be built here. We had a good road to travel over. Much work had been required to put it in order but it was in good condition and with no very heavy grade. The windings were interminable and twisted in labarynthian mazes around the granite crags which every where reared their bald rocky peaks heavenward. A brawling muddy stream was flowing down the valley swelled by the melting snows[.]

Our road threaded the defiles of the grandest mountains of the continent. At last I was at the "ultima thule" of my school boy dreams, "the Rocky Mountains." Huge ramparts of rock towered skyward and toppled over the path. The stratas being generaly in a vertical position or slightly tilted. How they come [to be] set up on edge, is a theme for geological study. Little streams leaped in beautiful cascades from ledge to ledge and tumbled along their rocky channels and finaly plunged into the muddy stream in the valley. The waters of the smaller streams exceeded exceed anything I had ever seen of limpid purity. It may be poetical to talk of "liquid crystals," but no crystal has the perfection and transparency of these mountain streams. But the larger stream in the valley, swelled by the numberless streams leaping down from the snowy tops of the mountains was muddied by the loose and friable soil of the valley, which was easily washed. The mountain sides were fortified with stupendous walls of rock rising in successive masses one above another: with now and then a scragley pine moaning in the breeze and stoutly maintaining a foothold in the rocky cliffs. The snowy range could be seen through the opening chasms cutting the sky ~~with~~ in sharp profiles with images of castles, towers buttresses and battlements of the whitest marble, for these distant peaks were yet covered with snow, in many places eternal[.] It must be delightful to ramble among these hills in summer time, when the grass and flowers have sprang into life, and breezes cool from the snow capped peaks is wafted through the defiles, and all around pictures of green and white and gray, is mirrored viticaly before you.

We dined at the Hake House a log accommodation for travelers at the foot of the first ridge we came to called the Guy hill. Our fair was good the price stiff, but braced by a full meal, we overcome the hill with ease, considerable work has been expended on the roads and the hill is not difficult to overcome.

On the summit of this ridge, was presented a distinct view of the plains, stretching away for leages like the ocean. We passed down the other side of the ridge, and plunged into a valley more narrow, crooked and romantic than the other but with the same general characteristics except a heavy growth of timber clothing the hill sides in many places[.] The timber consisted of small pines, hemlock and spruce, all evergreens and was ~~some~~ a pleasing relief to the gray rocks around. Passing over another ridge and down another crooked cañon we came to North Clear Creek, a considerable stream, on which ~~in~~ the principle mining operations ~~are~~ is carried on[.] We soon came to old flumes ~~and~~ abandoned sluices, and broken machinery of the rudest construction, being the debris of early mining operations before the quartz leads were opened. Shortly we came to houses some of logs and others neat and tidy, portraying the good tasts of the ocupants. These shelters in some nook or under some beetling crag, formed a romantic and pleasing picture. Clear creek was muddy as a swamp ditch in fact was fairly thick with the pulverized rock washed from the quartz mills and denommated [denominated] tailings. Next was the busy town of Black Hawk pinched up in a narrow ravine scarcely wide enough for the creek to flow freely along. The houses were found every where, perched upon rocks, dug into the side of the hills, crowding into the streets and stretching along the sides of the gulches which radiated from the main vally like the limbs of the forest trees. All was business, bustle and mud. The clang of machinery, the puff of steam the dash of water, the increasing tramp, tramp tramp of quartz mills, gave the narrow valley the appearance of an ant hill teaming with life[.] A new white church sat upon a hill, its clanspie [?] pointing heavenward, and was an object of note here in this ~~fountain head seat~~ throne of mammon. A mile further on was Central City but I could not discover a cesation of houses and mills and machinery to indicate the line of demarcation between the two town[s]. What I had observed at Black Hawk existed at Central City only on a somewhat more extensive scale. Central is the larger of the two places. The sun was yet an hour above the horizon as we took care of our horses after a 40 mile ride and not withstanding the fatigues of the trip we climbed one of the mountains, peered down abandoned shafts, looked up hidden curiosities and amused ourselves by tumbling rocks down the deep shafts and listened to the dull echoes as it bounded from side to side into the cavernous deabths below. An earley bed and a good night's rest prepared us for the labors as well as the wonders of [word missing][.]

Friday March 24" 1865

As soon as we had taken our breakfast we sallied out upon the hills to the gold mines. A few shafts we[re] just being opened some were not sunk more than ten feet others 40, 50 & 100, a vein of quartz is struck, and at once operations commence until it is ascertained if their is gold enough to pay expences. Scores have been abandoned on account of poor pay, while others are down 400 feet and s[t]ill being pushed downward. The quartz veins are from two to six feet in thickness and perhaps miles in length and no one knows how deep. These leads usualy run out at the debth of from 60 to 100 feet and nothing but the solid granite is left but by perserveing and pushing the shaft from 20 to 100 feet further down the quartz vein is struck again and richer than before. The granite over this last vein is called wall rock.

We came to a shaft in which the windless was run by steam and learned that the mine was known as Gregory No 2. This was sunk about 200 feet, and was reached by ladders from platforms about 20 feet below each other. Lighting a candle each, we proceeded down! down! down! untill we began to think the bottom had fallen out, but finaly came to the end where the miners were at work[.] The quartz is blasted and then placed in buckets and hauled to the top by windlass and then taken to the mill. Our candles gave a dim sickly light, but enough to show bright and glittering gems all around and overhead. Most of the gold is contained in pyrites of iron which glitters like crystals in the candle light. We gathered a few specimens and them [then] commenced our toilsome ascent. Oh it was hard getting to the top of earth again. A week's labor in the harvest field would not have been more fatiguing. At the height we were, the air was very thin and rair and it was difficult to take in sufficient breath[.] We at length emerged into daylight perfectly exhausted, our knees and joints trembling and unable to proceed on our explorations until after a half hour's rest. Some of the shafts have struck veins of water and are wet and disagreeable to work. Pumps are kept running night and day to drain them. At the Old Gregory lead enough water is obtained from the shaft for the whole operation of running the quartz mill. We went to the Old Gregory mine, where new and extensive machinery is in operation and more being put in operation. The quartz is pulverized as fine as flouer by being placed in cast iron boxes and then crushed by iron stamps[.] These weigh from 100 to 400 pounds and are raised by machinery about two feet and then let fall on the rock. Of course the hardest stone is soon pummeled to attoms. At this mill 20 stamps are in operation and 20 more are being put up. A stream of water is let into the pounding boxes, and flows off again carrying the crushed quartz in the form of muddy water. This flows over copper sheets, galvanized with quick silver. The particles of gold adhears to these plates and forms what is called amalgum, a thin coating of gold and quicksilver. The crushed rock and iron pyrites flow off and is carried away by the currant of water[.] Each day the machinery is stopped, the amalgum cleaned from the copper plates, and retorted, more quicksilver is put on and the mill set in motion for another 24 hours run. The mills run night and day. In this locality are about 200 mills. Some estimate can be formed of the work accomplished and the amount of machinery in operation[.]

At the star mills we saw them retorting the gold. This is done by placing the amalgum in an air tight iron vessel and placing it in the fire. The quicksilver when heated escapes through a pipe in the form of a fluid and being precipitated in water becomes liquid and is saved with little loss while the pure gold remains in the retort. At the star Mills only 12 stamps were in operation and yet the yield for one day was 54 ounces of pure gold[.] In the territory are 300 mills at 50 ounces per day for each mill, makes an aggregate of 15000 ounces or about or between $300 000 & $400.000 per day. And as the source of supply is inexhaustible & increasing every day some approximation can be formed of the mineral wealth of the rocky mountains. We went to the Bob tail lead to[o], the Black hawk and pretty generaly around the mines and returned to the hotel at dinner time, exhausted, but loaded with fine quartz specamins. We had seen enough to satisfy my curiosity and as our expences was running up pretty fast we concluded to leave and stop at some house on the road over night. Our bills for one day amounted to $18[.]oo which was paid and we mounted our steeds and steared for Denver. On getting to the Michigan house where we had thought to halt it was so early in the afternoon that we pushed on, reaching Golden City shortly after sun down, and being only 15 miles from home, we concluded after supper to go on again. The night was dark and we wandered over the prairies many miles out of our way and reached home about 11 oclock, tired and ready for bed. I have seldom passed two more interesting days than those among the mountains in the gold mines[.]

21 | To Fort Laramie and Back to Denver

March 25, 1865–April 15, 1865

Here was the remains of chimneys and barracks where Old Camp Waback was situated. The graves of many soldiers still existed on the top of a hill overlooking the camp. Now all was desolate and silent and nothing but the chilling wind or the bounding antelope disturbed the silence which prevails around.

—Lyman G. Bennett, March 30, 1865

Bennett did not remain long in Denver. After a stay of eight days, he continued inspection work by journeying 230 miles northward to Fort Laramie, situated in what would become Wyoming Territory during 1868. After reporting on defensive conditions at Fort Laramie, Lyman next made his way to Julesburg, Colorado, which took him southeast on the Overland Trail by the well-known landmarks of Scotts Bluff, Chimney Rock, and Courthouse Rock. Like many previous travelers, Bennett remarked on these impressive natural formations. Along the trail he observed damage from the Native American attacks earlier in the year, but also commented that the telegraph line had been temporarily rebuilt. In Julesburg, thanks to the reconstructed line, he learned of the surrender of General Robert E. Lee's army at Appomattox on the very day that it occurred. Bennett may have planned to travel east from Julesburg to visit his family at Leavenworth, Kansas, but instead Major General Patrick E. Connor, the commander of the District of the Plains, ordered him to return to Denver.

After enduring an uncomfortable stagecoach trip to Denver, Bennett probably returned to Leavenworth sometime in April or May. In July, he began writing again to recount participation in his last adventure of the Civil War years. The final part of his diary has entries from July 3 through October 4, 1865, that described his experiences in the Powder River Expedition. This last part of his diary was published in David E. Wagner's *Powder River Odyssey: Nelson Cole's Western Campaign of 1865: The Journals of Lyman G. Bennett and Other Eyewitness Accounts* (2009) and is not included here.

Major General John Pope and Major General Grenville M. Dodge planned the Powder River campaign that was designed to punish

Cheyenne, Arapaho, and Sioux tribal members for their recent attacks.[1] Three columns were deployed with Colonel Nelson D. Cole in command of the eastern, or right, column.[2] Bennett was employed as Cole's engineer during the campaign. Leaving Omaha City, Nebraska, on July 1, the column marched into the Dakota Territory and eventually to the Powder River in Montana Territory. Several skirmishes were fought against Native Americans, but supply problems, punctuated by a lack of preparation time, contributed to the ineffectiveness of Cole's column. Overall, none of the columns achieved great success.[3] Colonel Cole recognized Bennett's contribution in his report of the expedition by stating: "To the untiring energy and devotion to his profession of Mr. L. G. Bennett, my engineer officer, is due the accurate record of the valuable information of the region passed over. Constantly moving, he by observation made himself thoroughly familiar with the topography of the country, and gleaned much in relation to the mineralogy along the route."[4]

Saturday March 25" 1865

Proceeded to Denver on horse back, and turned over one span of mules, our harnes[s] and wagon to the Quartermaster and prepared to pack our things on a mule the rest of the trip[.] The wagon is to[o] big and clumsy. Saw Cousine Brant Coryell.[5] He went to John Coryell's and all day long we had an insessent string of stories to tell. Williams took a sketch of John's house and as it was cold and stormy staid over night with me. Indeed this has been a very stormy cold day, but the time was well put in in visiting with Brant, and the others.

Sunday March 26" 1865

Another cold day, which was devoted in doors to social chat and good cheer and to writing letters. John Coryell is a noble fellow, free hearted and generous to a fault. I wish him well in all his undertakings[.]

Monday March 27" 1865

The sun after an absence of two days showed his face and looked as smiling and cheerful as a lover. Ate an early breakfast, said good by all around and was off to town. John accompanied me, and assisted in packing our mule. It was 10 oclock before we were ready and off. Taking a near cut and striking Cherry creek below the regular crossing, our horses immediately sank in the quick sand and John's was not fortionate enough to get out but floundered around and was soon buried except his head, we tied a rope about his neck and hauled him out. In his struggles he cut one leg severely[.] Quicksand exists in all the streams in the country and sometimes swallows up whole teams.

Damages repaired we proceeded to the regular crossing and was off. We found an excellent road and every few miles a settlement of farmers along the creeks running from the mountains into the Platt[.] It was a fine day—the snow disappeared and with the mountains close at hand on our left, we enjoyed our ride and the scenery. We crosed Clear Creek, Boulder Creek Dry creek, Left Hand Creek and the St Vrains. Just before sundown we stopped at a little town on the St Vrains called Burlington and passed the night under a straw stack. The straw furnished an excellent bed. In the morning Traveled 35 miles[.]

Tuesday March 28

In the morning it was snowing and lay to the debth of four inches on our blankets. Of course we slept under the snow sheet as warm as toast. I do not know when a feather bed was more comfortable and warm. We dug our traps from under the snow and was soon off, the storm continuing at intervals until after noon, but thawing faster than it fell and at night when we reached Camp Collins no snow was remaining[.] Capt Evans[6] commanding the post showed me all that was to be seen at his post & entertained me very hospitably[.] Camp Collins is on the Cache La Poudre creek a few miles from the foot of the mountains. The stream is very fine and clear and full of trout. Camp Collins though new is in good condition and the quarters are excellent[.] Traveled 35 miles[.]

Wednesday March 29"

Found Lt Triggs[7] and Lt Lewis at Collins, who were about to start for Laramie, thus obviating the necessity of an escort. Our party consisted of 17 men, well armed, and thought ourselves sufficient for any number of red skins who might meet us. All except Williams [and] Lt Lewis were on the road two hours before us, and it was 10 oclock before we overhauled the remander of the party[.]

It was a fine bright morning. The mountains were near and every ravine and elevation could be seen distinctly. The white top of Long's Peak reaching to the clouds, appeared through the thin air but a few miles distant, when at least fifty intervened. Our road skirted along the foot of the mountains, which reared in dark masses to the clouds and on the other hand the prairies stretched away like the ocean[.]

Soon we our road commenced winding among the hills and up their gradual slopes until we found ourselves in a cold region with patches of snow, on the northern slopes of the hill sides[.] The wind was keen and cutting, and in passing up ravines it came sweeping through and penetrated our thickest clothing. Up some [?] hollow was some strange natural scenery[.] On either side the wall of conglomerate rock was worn and perforated by the wind, and presented a varied and pleasing scenery. But it was to[o] cold to admire the fantastic shapes of the rocks. A few of us pushed on ahead of the wagon over hills and across ravines over a continuous field of snow which in many places was drifted several feet. In a deep hollow 38 miles from camp we found a spring and small stream of water, called Lone Tice creek. Here we built a fire of willow brush and waited several hours until after dark for the wagon and remander of the party to come up[.] We had brought wood from Camp Collins and hastily cooked our supper and went to bed to keep warm. The wind shrieked down the ravine and no amount of covering could keep us warm[.] 38 Miles

Thursday March 30" 1865
Our march was resumed over bleak snow covered hills, and many were the devious windings to avoid the drifts of snow. The day was clear, not a cloud in the sky, but at the same time freezing cold. We passed the Chayanne Pass about noon. This is said to be one of the best passes through the mountains, but always cold, and with a a tornado of wind pouring down through it. Near where our road passed was a fine spring of clear cold water[.] Lodge Pole creek issues from this pass and runs into the Platte near Julesburg. Here was the remains of chimneys and barracks where Old Camp Waback[8] was situated. The graves of many soldiers still existed on the top of a hill overlooking the camp. Now all was desolate and silent and nothing but the chilling wind or the bounding antelope disturbed the silence which prevails around. From here the road gradualy descended and when we had reached Horse Creek the snow had mostly disappeared. The ragged mountains to the west and the loose crumbly hills near the road, formed a strange appearance and arrested the attention of the most stoical traveler. Scarcely a tree could be found during the whole day. Sand hills and rocks, and the naked plain, prevaded the whole rout[e][.] We camped near the head of the Chug Water creek, or at the point where it issues from the mountains. Here was plenty of wood and good water, and a blazing fire a hearty supper and warm blankets lead us to forget the fatigues of the day.

Traveled 48 miles[.]

Friday March 31" 1865

Our rout[e] to day was down the valley of the Chug winding among the hills which rose on either side of the valley sometimes perpendicularly like colums in some old ruins[.] The scenery was truely romantic and interesting[.] Early in the morning I noticed two parallell walls of rock which ran ~~parallell~~ in sight of the road for miles forming a natural causway between the perpendicular walls. Thus it ran over hills and mountains and was lost in the distance[.] This was undoubtedly caused by some convulsion of nature in which two neighboring strata of rock were tilted into a verticle position[.]

Shortly after noon we came to a stack of hay, which some heardsman had cut, but was undoubtedly forced to leave in consequence of indian difficulties. A perpendicular column of rock, or rather clayey substance which abounds all through the country and by the winds are cut into fanciful shapes rose up from the valley for 200 feet. This was called Chimney Rock of the Chug, [probably near Chugwater, Wyoming] and I had williams take a sketch of it. We camped again on the Chug and found an abundance of wood. Lt Taggs [Triggs] was quite unwell and went on to Laramie[.]

Traveled 49 miles[.]

Saturday April 1st 1865

The camp was aroused at 4 oclock and the party prepared to move before sunrise[.] Lt Lewis galloped ahead, leaving the wagons and soldiers to come after. Just as we were leaving the Chug and turning easterly over the ridge we came upon an encampment of indians hearding their stock of ponies and cattle. They were friendly sioux and dependent upon Ft Laramie. The scenery near the Chug springs was as grand as any thing I had yet seen, resembling the dismantled towers the crumbling walls and battlements of some grand old ruin. We had no time or I should have had a sketch taken.

Twenty five miles was quickly passed and before 11 A.M. we reached Ft Laramie[.] I was hospitably entertained by Major McKay[9] the commandant, and found several old acquaintances, among the officers of the indian expedition among which was Col Banner [Bonner?] Capt [Thomas J.] Majors & Lt [John?] Talbot, who all remembered that it was the 1st day of April, and many were the amusing incidents of the day. It was a bright warm day, the squaws were listlessly wandering about the garrison or slily squatted in some corner. I was astonished to learn that nearly every officer at the post and many of the men had their squaw wives and cohabited with them the same as married people, notwithstanding many of them had wives and children in the States. One Capt with a family in Ohio and himself a minister of the gospel had a black filthy looking squaw as his wife[.] Strange what creatures human beings will become. Passed the night at the Mr Bullock Post sutler and was furnished with a fine clean bed, and the best of fair.

Sunday April 2nd 1865

Passed the day writing letters and bringing up my journal which has been sadly neglected of late. There were indications of rain and the sky was dark and lowering, and a sort of mist was swept by the wind over the ground but it did not make out to rain.

In conversation with Mr Bullock and others I learned of rich gold discoveries in the Black Hills, in the neighborhood where the hostile indians are congregated. It is said that the first accounts of a posative nature, was that a band of indians in removing their encampment, had occasion to cross a stream. Their lodge poles were fastened to ponies and the ends dragged on the ground behind. In crosing the stream a lodge pole uncovered a glittering object which attracted the attention of a squaw coming up behind and she picked it up which proved to be about $15[.]00 worth of gold. Indians in coming to the Fort were in the habit of exchanging pieces of gold with the soldiers for tobacco and trinkets but would not tell from whence it came for fear the whites would overrun their country, but it is now pretty generaly known that the streams among the Black hills sparkle with gold, and that indians pick up considerable quantities by hunting along the bed of streams, thus showing the region to be particularly rich in gold. Mr Bullock likewise showed me many specimens of fossils and petrifications found among these hills at a place known as the "bad Lands" among which are petrified snakes, turtles and even eagles quils turned to stone[.] In the rocks are many shells in a fine state of preservation and unlike any fossils I ever before saw, and which are usualy found in lime rocks.[10]

Made some examination in relation to the location of Lunetts for covering the place[.] Laramie is in a bad place for defense being surrounded with hills which command it from all sides and ravines that would cover an approaching party. Added to this the buildings are scattered over much ground and would require 500 men for an efficient defense, unless lunetts or earthworks of some kind are thrown up[.]

Called upon Lt Trigg[s] who has his wife at the Post, and they have had the fortune to find a boy in this out of the way region where one would suspect nothing but half breeds could be produced.

Monday April 3d 1864 [1865]

A party of three men started with an ambulance for Julesburg, with the mail[.] To these I attached my party, making six in all. The genial warmth of the sun added beauty to the varied scenery which met our view as we wound among or clambered over the hills which in many places approached very near the Platt, and at other times receded forming broad level flats bordering the river. But one house was passed and that was a ranch five miles from Laramie[.] The lazy indians and half breed children lounging about showed the mixture of the races that tenanted the place.

At times the hills rising almost perpendicularly from the plain, to the height of several hundred feet presented a magnificent sight. Not a tree or living shoot of any kind except a thin growth of Cottonwood along the Platt could any where be seen. Camped near the Indian Agency 32 miles from Laramie[.] I set fire to the tall grass on a small island in the river. The flames leaping in the air lighted up the heavens and surrounding hills and flats and presented a wild and romantic scene. The upward mail party also a squad of the 11" Kansas[11] encamped at the same place, altogether forming a party too strong of [sic] any ordinary band of indians[.]

Tuesday April 4th 1865

A cold south east wind sat in during the night, making it disagreeable traviling to face it. We saw a number of wild geese and had a few shots, but missed them. At Horse creek we saw a large number of them, also wolv[e]s and cyota. We met Col Plum[12] with the 11" Kansas cavalry on their way to join the expedition against the indians[.]

We reached Camp Mitchell about noon[.] This is a fine post about three miles from Scotts bluffs and in admirable condition[.] It is built after the Mexican fashion with high walls and loopholes and a high look out[.]

The materials are sod and plastered over with alkaline soil and lime[.] The rooms and men's quarters are clean and airy and kept in as good order as any I ever saw. Capt Shuman 11th Ohio cavalry[13] is in command of the post and has exhibited much tast and labor in the construction of the Post. No amount of indians can affect anything here.

Scotts Bluff near by are singular and attractive formation of a sort of dry hard clay that at a distance resemble rock[.] It rises up from the plain in irregular masses several hundred feet and a person at their base is made giddy in looking up the dizzy height[.] A few dwarf cedars on the summit imparts a little of life to the otherwise ragged and lifeless scene. They resemble some grand old ruin with piles on piles of rugged and irregular masonry thrown together with here and there a broken column or tottering turret once fashioned by the hand of man. Long and curiously I gazed at this mass of sun dried clay, which the winds and storms at some period will reduce to a level with the surrounding plain[.] After dinner and resting our stock we proceeded fifteen miles to Ficklins Ranch, a small dirty log house, with no floors and filth and vermin on the walls and strewed over the ground in this unattractive pen we were to pass the night, after a ride of 35 miles[.]

Wednesday April 5"

On awakening the wind was howling about our cabin, and the snow flying in eddying and hurried gusts across the plain. The cold was intense which with the snow prevented our pushing forward. All day it snowed and the wind swept down the valley the river froze over, and we passed the day in telling yarns, eating rations and teasing John the mule driver. A more stormy bad day I have not experienced. I cannot get accustomed to these long winters of fifteen month[s] in the year.

Thursday April 6"

It was a cold morning but clear, and chimney rock loomed up plainly, about fifteen miles distant. This singular formation is well worthy the inspection of the tourist[.] Our road led us to within four or five miles but the cold prevented my going to it for about 150 feet the rock or clay is connicle in shap[e], and can be climbed with some difficulty but from the top of the cone a perpendicular shaft shoots up for 150 feet[.] Tis said that within ten years 100 feet of this shaft has crumbled and fallen down[.] The existence of this isolated column of clay, standing in the midst of a prairie proves the dryness and purity of the atmosphere, for were there rains of any account this would have been dissolved and leveled with the plain ages ago. There is the most undoubted evidence that this whole region was once a disordered mass of dry clay, but frost and wind have reduced them to the scattered and wired ruins we see around us. Chimney Rock is worthy an elaborate description.

Passing on about noon we came to Court House rock another curiousity of the same nature and material as Chimney rock[.] As our road passed but a few hundred yards away, I rode to its base, and was surprised at its giddy height, reaching skyward it seemed almost to the clouds. A sunrise view from its summit must be grand[.]

We made our way through drifts of snow crossing Lawrences Fork and Pumpkin creek occasionly walking to keep up the circulation until about 4 P.M. when we reached Mud Spring Station, a poor tumble down log house a telegraph station and guarded by 15 soldiers from Camp Mitchell.[14] Into this dungeon we immersed ourself until morning. The news of the capture of Richmond was taken from the wires and we all felt happy at the announcement.[15] Here was a long battle fought with the indians during the winter first with the little garrison but afterwards with Col Collins and 100 men[.][16]

Friday April 7"

The mail party and my own boys were slow to get ready, and I pushed on ahead, over a rough broken country, the ravines being filled with snow drifts exceedingly difficult to get through. From the tops of the hills were magnificent views of the surrounding country. Court House and Chimney rocks loomed up in the distance, and the sun appearing made the day and the scene somewhat interesting. The day passed without anything of interest occurring, save a wolf now and then scampering over the prairie, or flocks of wild fowl flying about. Many miles of the telegraph had been tourn down by the indians and was only temporaly put up on poles that were only high enough to keep the wires from the ground, and nales for insolators[.]

We camped on the Pole Creek, a fine stream of clear cold water. W[e] had but little wood barely sufficient to boil our coffee. By getting under the lea [lee] of a sand bank we managed to sleep comfortably, but the night was cold and raw[.]

Saturday April 8" 1865

A cold blustering morning and disagreeable riding. I pushed on ahead of the team and about 1 P.M. reached the South Platt, opposite Julesburg. Ice covered about two thirds of the river, but not stout enough to bear my horse and it was extremely difficult crossing. His feet would break through, and the water was extremly cold, with considerable current, and in places belly deep. When nearly across, a large floe of ice came rushing down and nearly came upsetting both myself and horse.

On reaching Julesburg I was cold, wet and hungry, and about drilled out, and felt more like going to bed than anything else[.] The team did not get in untill nearly night. Received a telegraph dispatch from Lt [George T.] Robinson ordering me to report to Genl Connor[17] and a dispatch from the latter for me to come to Denver immediately. I am fearful I shall not be able to go home as soon as I anticipated.

Sunday April 9th

Passed the day in reading newspapers, and writing letters. Weather warm and damp and snow disappearing. In the evening a number of white soldiers and indians had a regular battle with snow balls. All went in with a whoop and yell, and it was hard to tell who were the masters. News of the surrender of Lee and his army all feeling good[.]

Monday April 10"

The stage for Denver was crowded, and it was impossible for me to get a seat, and I am obliged to lay over, with nothing on my hands to pass away the time.

My plan is to leave Williams and John here until I can go to Denver and confer with Genl Connor, and if required to remain, send for them but if permitted to go to Leavenworth telegraph for them to proceed and I follow after[.]

Tuesday April 11"
I left Julesburg on the stage about 8 A.M. there being 9 passengers in all. One a lady, Mrs La Fair, going to meet her husband, at Black Hawk in the mountains. Three of the men were on the way for Salt Lake, some said they had wives and families there. All were rather hard cases, told their stories and used considerable profane and vulgar language. I am sorry for the lady passenger.

At Buffalo Springs there were no stage horses fit to go out and Capt Clark[18] gave us his own horse and fitted us up with a team to proceed to Lillian Springs—At Lillian there were no horses to put in and we stood a fair chance of laying over 36 hours[.] I finaly found a man with seven yoke of cattle and pressed three of them to haul the stage to Valley Station 18 miles. When we started the buglar at the Post sounded the march and we star set out with many a cheer and hurah.

Wednesday April 12" 1865
We reached Valley Station about 4 A.M. and after considerable of an effort succeeded in arousing the stage drivers and the ranch keeper but none would start out until morning and we were obliged to retire to the coach and doze the remainder of the night away.

After breakfast we set out on our journey with a poor lot of stock and made slow time to Godfreys 18 miles, when we again changed horses for a team that was in good order and we were pushed through in a hurry to Junction Station which we reached at 8 oclock P.M. In changing horses the team that was to be attached to the coach ran away and it was not until 11 oclock that we got started. Bad roads made the trip in the night slow indeed and the 18 miles to the Bijou took us until nearly morning. The coach was our bed room as well as our parlor and kitchen. I felt sorry for our single lady passenger, crowded up between rough and vulger men. She is a good woman and of good principles[.]

Thursday April 13" 1865
It was a bright warm day, but mud was everywhere. The poor teams we had

could scarcely pull the empty coach. The passengers footed it more than half the time until we reached ~~Lillian~~ Living Springs where a fresh team was procured and we pushed on more rapidly and without walking except at one bad place in the road which we reached about midnight. It was dark and I stumbled into the mud nearly knee deep. We were all pretty well fatigued and slept notwithstanding the swinging of the coach.

Friday April 14" 1865

Reached Denver about 7 A.M. glad that a long tedtious and disagreeable trip was ended. Had just five dollars in my pocket and in need of $100 to get along properly[.] After breakfast walked up the Platt to John H Coryell's ranch[.] Found him absent but the remander of the family at home[.] Libie lent me 6 dollars & on my return to town I purchased a hat to shield from the store & sun for $5[.]oo[.]

Saturday April 15" 1865

[no entry]

22 | After the War

Lyman Bennett's participation in the Powder River Expedition ended with his arrival at Fort Laramie on September 30, 1865. Upon arrival, he cleaned up, wrote letters, and bemoaned the fact that no letters were awaiting his arrival: "It is cruel not to hear from home for weeks & months." Bennett mentioned meeting the famous mountain man, Jim Bridger, in his last diary entry on October 4. Traveling east, Bennett reunited with his family at Leavenworth and left the service for good in March 1866.[1] He had served in the military or as a civilian employee for nearly five years.

The Bennett family soon left Leavenworth and returned to Kendall County, Illinois, where they settled again near Yorkville.[2] The couple's second child, Edgar (Eddie) Agaz Bennett, was born on September 17, 1866.[3] A year later, a third child, William, was also born, but he died on April 1, 1868.[4] Their fourth and last child, Carrie, was born on Christmas Day of 1868.[5]

To support his growing family, the ever-energetic Bennett worked at various tasks. From at least 1869 to 1872, he served as Kendall County surveyor.[6] Bennett took on a new role as circuit clerk for Kendall County in 1872 when the incumbent, Albert M. Hobbs, died. Hobbs and Bennett had served together in Company E of the Thirty-Sixth Illinois Infantry, with Hobbs rising to eventual command of the company. The Thirty-Sixth Illinois went on to see heavy combat in the Western Theater, and Hobbs had been wounded in the knee during combat at Chickamauga.[7] When he returned to Kendall County, voters elected him as circuit clerk, and he remained in that position until his death.[8] Apparently troubled by his old wound, Hobbs agreed to an operation, but he died as a result of it in January 1872.[9] Bennett completed the remainder of his former comrade's term and then was elected in his own right in 1876.[10] As a part of his tenure, Bennett and Melissa hosted Senator John A. Logan, a former Civil War general, in 1876 during the senator's reelection bid and treated him to a dinner at their house.[11] Bennett also earned income by drawing an attractive panoramic map of Winona, Minnesota, that was sold as a lithograph, and he further contributed to an atlas of Kendall County.[12] In addition, Bennett probably engaged in farming as the 1870 census listed him as owning $600 in real estate. He likely helped his father work his farm as well.[13]

In 1871, Bennett's parents relocated to Greene County, Missouri, along with some of his younger siblings. His younger brother, Charles, lived with his parents and helped them with the farm, eventually becoming the manager of the family's farm.[14] Bennett's parents, Charles and Louisa, had built a prosperous farm in Illinois with a real estate value of $16,000, but for unknown reasons they decided to migrate to Missouri.[15] Springfield is the county seat of Greene County, where Lyman had spent some time during the war, and he had liked the area, so perhaps he recommended Greene County to his parents. In the 1870s, Lyman probably wrote his "Recruiting in Dixie" piece, which was part of a larger work illustrated by him. His detailed diaries also became the basis for the first thirteen chapters of the *History of the Thirty-Sixth Regiment Illinois Volunteers, during the War of the Rebellion* (1876), co-authored with William M. Haigh. Although Bennett's association with the regiment ceased after Pea Ridge, he also wrote the narrative for the regiment's operations through the battle of Perryville.[16] Meanwhile his extended family thrived in Greene County, which led Lyman and Melissa to follow them there in December 1880, when his circuit clerk term ended.[17] In Greene County, Lyman, who at one point during the war had difficulties paying a bill for five dollars, purchased land and "erected a handsome residence" for his family.[18] He plunged into community organizations including the Masons, the Odd Fellows, the Good Templars, to which he had long belonged, and the Grand Army of the Republic.[19] In politics, he continued as a supporter of the Republican Party.[20] The family joined the Calvary Presbyterian Church, where "the family are highly respected by all who know them."[21] His brother, Charles, married Sarah Smith from Kendall County in 1881[22] and ended up farming the rich tract of land near Springfield that his parents acquired a decade earlier.[23]

The 1880s, though, were also a time of loss. Lyman's father, Charles, died on July 29, 1882, and his mother, Louisa, passed away on April 14, 1886. Initially buried in New York, Lyman reinterred his father in Springfield's Hazelwood Cemetery in 1888, where his mother had been buried.[24] Sadly, Lyman and Melissa's daughter, Carrie, died on July 28, 1888 at the age of only nineteen.[25]

Financially speaking, Lyman "tilled [the land] with good success" and felt stable enough to express willingness to join with his brother Charles and commit $1,000 each to establish a sugar refinery.[26] Whether this venture was actually established is unknown. Amid all these activities, Lyman continued to work as a surveyor. He platted the Pickwick Place Addition (now on the National Register of Historic Places) in Springfield in 1890,[27] conducted surveys in "public lands" in the Oklahoma Territory,[28] and surveyed some proposed railroad lines in Missouri.[29]

In December 1897, though, Bennett applied for a veteran's invalid pension. The medical examination revealed rheumatism in his left knee, an

Taken near Springfield, Missouri, in the mid to late 1890s, this Bennett family photograph includes several people mentioned in the book. Lyman Bennett stands fifth from the right, and his wife, Melissa, sits directly in front of him. Standing next to Lyman is his brother, Charles. Another brother, Martin (Matt), sits third from the left and holds a cat in his lap.

extensive network of varicose veins in the left leg, a "tender & stiffened" left hip, chronic catarrh, and a slightly "impaired" left wrist due to a bullet wound from his military service. Bennett blamed his knee problems on an injury sustained when a horse fell on it during the War. Officials approved a pension payment of eight dollars per month due to "partial inability to earn a Support by Manual Labor."[30] On December 8, 1899, Bennett, who resided "about one mile south of the end of the Pickwick street car line, was seriously hurt by the fall of his horse about 8 o'clock last night. The man's collar bone was broken and two or three ribs fractured." Bennett commented that the old horse fell due to "a little fright," although he admitted that "the shock was so great that his memory is not entirely clear as to what happened at that time."[31] In July 1903 health problems had increased to such a degree that Lyman applied for a full invalid pension. The examining board now noted that although Bennett's mental condition was "sound," he looked "aged & feeble" due to the onset of heart disease and an enlarged prostate gland in addition to the health issues documented in 1897. Examiners recommended a full invalid pension of twelve dollars per month, which was approved by officials.[32] Lyman passed away only a few months later on February 24, 1904, at age seventy-one. Members of the Captain John Matthews Post of the Grand Army of the Republic aided during the burial service at Hazelwood Cemetery.[33] His wife, Melissa, who had battled pneumonia for a week, died one week after Lyman.[34] The obituary read that she and Lyman had "contributed largely to the growth and upbuilding of the city. She was a lady respected among women and her passing creates a void in a home and in the community."[35] There were two surviving children: Minnie, now Mrs. C. E. Phillips, and Edgar.[36]

Notes

Preface
1. Lyman G. Bennett, diary, 1 January 1865, Lyman G. Bennett Papers, R0274, The State Historical Society of Missouri Research Center–Rolla.
2. Bennett, diary, 23 April and 4 July 1857, Bennett Papers.

1. Before the War
1. Lyman G. Bennett, "Personal Reminiscences" (MS, ca. 1875), 2, The Huntington Library.
2. Bennett, "Personal Reminiscences," 43–44.
3. Ibid., 14, 38.
4. Ibid., 18–20.
5. Ibid., 42–43.
6. Ibid., 2.
7. Ibid., 4–6.
8. R. I. Holcombe, ed., *History of Greene County, Missouri, Written and Compiled from the Most Authentic Official and Private Sources* (St. Louis: Western Historical Co., 1883), 880; Bennett, "Personal Reminiscences," 41–42.
9. Bennett, "Personal Reminiscences," 41–42.
10. *Seventh Census of the United States*, 1850, Kendall County, Ill., 247.
11. Ibid.
12. E. W. Hicks, *History of Kendall County Illinois, from the Earliest Discoveries to the Present Time* (Aurora, IL: Knickerbocker & Hodder, 1877), 267.
13. Newton Bateman and Paul Selby, eds., *Encyclopedia of Illinois and History of Kendall County* (Chicago: Munsell Publishing Co., 1914), 1:850.
14. Bateman and Selby, *Encyclopedia of Illinois*, 1:799, 803.
15. Dan Patterson and Clinton Terry, *Surveying in Early America: The Point of Beginning* (Cincinnati: Univ. of Cincinnati Press, 2020), 16.
16. Lyman G. Bennett, diary, 19 January and 16 August 1857, Lyman G. Bennett Papers, R0274, The State Historical Society of Missouri Research Center-Rolla.
17. Bennett, diary, 20 January 1857, Bennett Papers.
18. Ibid., 8 January, 6 February 1857.
19. Ibid., 6 January, 14 February 1857.
20. Ibid., 14, 16, 18 February 1857.

21. Ibid., 14 February 1857.
22. Ibid., 16 February 1857.
23. William Marvel, *Lincoln's Mercenaries: Economic Motivation among Union Soldiers during the Civil War* (Baton Rouge: Louisiana State Univ. Press, 2018), 13.
24. Marvel, *Lincoln's Mercenaries,* 13.
25. Ibid.
26. Ibid., 15.
27. David M. Fahey, *Temperance & Racism: John Bull, Johnny Reb, and the Good Templars* (Lexington: The Univ. Press of Kentucky, 1996), 6; Jack S. Blocker, Jr., *American Temperance Movements: Cycles of Reform* (Boston: Twayne Publishers, 1989), 50.
28. Fahey, *Temperance & Racism,* 9, 11.
29. Fahey, *Temperance & Racism,* 11; Blocker, *American Temperance,* 51.
30. Fahey, *Temperance & Racism,* 11.
31. Bennett, diary, 4 May 1857, Bennett Papers.
32. Ibid., 4 May, 17 August, 27 January 1857.
33. Ibid., 4 February, 6–7 April 1857.
34. Ibid., 6–7 April, 9–10 April 1857.
35. Ibid., 9–10 April 1857.
36. Ibid., 20 January, 9 March, 22 October 1857.
37. Ibid., 28 April 1857.
38. Ibid., 8 May 1857.
39. Ibid., 19 August, 16 May 1857.
40. Ibid., 21 May 1857.
41. Ibid., 19 August 1857.
42. Ibid., 24 May 1857.
43. Ibid., 26–27 May 1857.
44. Ibid., 30 May, 3–4 June 1857.
45. Ibid., 26–27 May, 14 June 1857.
46. Ibid., 20 July, 13 September 1857. Bennett's drawing of a mosquito appears in one of these entries.
47. Ibid., 4 July 1857.
48. Ibid.
49. Ibid., 20, 27, 31 July, 3 August 1857.
50. Ibid., 12 August 1857.
51. Ibid., 26 August, 15 September 1857 and subsequent entries.
52. Ibid., 4 October 1857.
53. Ibid., 9 November, 1, 5, 8 December 1857.
54. Ibid., 1, 5 December 1857.
55. Ibid., 23, 24 December 1857.

56. Ibid., 10 May 1857.
57. Ibid., 8 August 1857.
58. Ibid., 31 October, 4 November 1857.
59. Ibid., 4 November 1857.
60. Ibid., 5 November 1857.
61. Bateman and Selby, *Encyclopedia of Illinois*, 1:760.
62. Illinois, US Compiled Marriages, 1851–1900 (Provo, UT: Ancestry.com Operations, Inc., 2005); Pension records on Lyman G. Bennett, RG T-288, Invalid Record Application 1200162, National Archives.
63. *Eighth Census of the United States*, 1860, Kendall County, Ill., 239.
64. "A School Teacher: A Letter Sent by Lyman Bennett," *White River Valley Historical Quarterly* 4, no. 1 (Fall 1970): 14. Note that the article title is in error, as this letter was sent by James Rawles, not by Lyman Bennett.
65. "A School Teacher," 15.
66. *Eighth Census*, 1860, Kendall County, Ill., 239.
67. *Eighth Census*, 1860, Kendall County, Ill., 271.
68. "Minnie L. Bennett Phillips," Find a Grave (website), accessed December 3, 2020, http://www.findagrave.com/memorial/38800666/minnie-l-phillips; Pension records, Lyman G. Bennett.
69. *General Land Office Records, 1776–2015* (Provo, UT: Ancestry.com Operations, Inc., 2008).

2. Off to War

1. James M. McPherson, *For Cause & Comrades: Why Men Fought in the Civil War* (New York: Oxford Univ. Press, 1997), 16.
2. William Marvel, *Lincoln's Mercenaries: Economic Motivation among Union Soldiers during the Civil War* (Baton Rouge: Louisiana State Univ. Press, 2018), 19.
3. Newton Bateman and Paul Selby, eds., *Encyclopedia of Illinois and History of Kendall County* (Chicago: Munsell Publishing Co., 1914), 1:736; E. W. Hicks, *History of Kendall County Illinois, from the Earliest Discoveries to the Present Time* (Aurora, IL: Knickerbocker & Hodder, 1877), 302.
4. Hicks, *History of Kendall County Illinois,* 302.
5. Ibid.
6. McPherson, *For Cause & Comrades,* 18–19.
7. Marvel, *Lincoln's Mercenaries*, 18–19, 22, 24.
8. *Eighth Census of the United States*, 1860, Kendall County, Ill., 239.
9. Marvel, *Lincoln's Mercenaries,* 46.
10. L. G. Bennett and William M. Haigh, *History of the Thirty-Sixth Regiment Illinois Volunteers, during the War of the Rebellion* (1876; repr., Marengo, IL: Prairie State Press, Inc., 1999), 11–12.

11. Bennett and Haigh, *Thirty-Sixth Regiment Illinois*, 12.
12. It was unusual for an infantry regiment to have mounted companies, but the regimental history listed ten companies of infantry and separate A and B cavalry companies. The cavalry companies were eventually detached from the regiment. Bennett and Haigh, *Thirty-Sixth Regiment Illinois*, 52–53; *The Union Army: A History of Military Affairs in the Loyal States 1861–1865* (1908; repr., Wilmington, NC: Broadfoot Publishing Co., 1998), 3:276–77; Janet B. Hewitt., ed., *Supplement to the Official Records of the Union and Confederate Armies* (Wilmington, NC: Broadfoot Publishing Co., 1995), 10:578–672.
13. Bennett and Haigh, *Thirty-Sixth Regiment Illinois*, 13.
14. *The Union Army*, 3:276; Bennett and Haigh, *Thirty-Sixth Regiment Illinois*, 17.
15. Bennett Compiled Service Record, Thirty-Sixth Illinois Infantry, National Archives; Bennett and Haigh, *Thirty-Sixth Regiment Illinois*, 44.
16. Lyman G. Bennett, diary, 28 August 1861, Lyman G. Bennett Papers, R0274, The State Historical Society of Missouri Research Center–Rolla.
17. A reference to the Confederate victory at the First Battle of Bull Run, or Battle of First Manassas, fought on July 21, 1861.
18. Born in 1817 in Bavaria, Nicholas Greusel immigrated to New York City with his family when he was sixteen years old. He worked in a variety of odd jobs and served in the Mexican American War. He was an employee of the Chicago, Burlington and Quincy Railroad when the war began. The 1860 census listed him as a conductor with only one hundred dollars in real estate and $300 in personal estate. After much active service as colonel of the Thirty-Sixth Illinois, Greusel left the service due to rheumatism in 1864. Larry J. Daniel, *Battle of Stones River: The Forgotten Conflict between the Confederate Army of Tennessee and the Union Army of the Cumberland* (Baton Rouge: Louisiana State Univ. Press, 2012) 107, 214; Bennett and Haigh, *Thirty-Sixth Regiment Illinois*, 13–16; *Eighth Census of the United States*, 1860, Kane County, Ill., 143.
19. Major General John C. Fremont commanded the Department of the West at the time. Born in 1813, Fremont achieved fame due to his prewar western expeditions and his 1856 presidential nomination from the newly created Republican Party. A controversial figure, Fremont left the army in 1862. Ezra J. Warner, *Generals in Blue: Lives of the Union Commanders* (1964; repr., Baton Rouge: Louisiana State Univ. Press, 1992), 160–61.
20. The Young America Guards, commanded by Elias B. Baldwin, became Company C of the Thirty-Sixth Illinois Infantry. Bennett and Haigh, *Thirty-Sixth Regiment Illinois*, 18, 38.
21. The Bristol Company was designated as Company E of the Thirty-Sixth Illinois. Ibid., 18; Hewitt, *Supplement to the Official Records*, 10:629.

22. Charles D. Fish commanded Company E, and he served until he resigned his commission in Mississippi in the summer of 1862. Bennett and Haigh, *Thirty-Sixth Regiment Illinois*, 221.
23. Albert M. Hobbs helped recruit and then became captain of Company E following the resignation of Captain Fish. Captured at the Battle of Stones River on December 31, 1862, Hobbs spent five months in Libby Prison before being exchanged. He was wounded in the knee at the Battle of Chickamauga, which ended his military service. Upon returning to Kendall County, Illinois, voters chose him as circuit clerk, and he remained in that position until his death in January 1872 following an operation on his knee. Lyman Bennett filled out his term as circuit clerk. Hicks, *History of Kendall County Illinois*, 304; Bennett and Haigh, *Thirty-Sixth Regiment Illinois*, 402, 482, 489; Bateman and Selby, *Encyclopedia of Illinois*, 1:821.
24. William H. Clark gained the reputation as the "regimental wag," or jokester. He was wounded in the arm at the Battle of Perryville on October 8, 1862. He was unable to secure a leave of absence to see his ill wife, who died in April 1863. Clark was promoted to regimental adjutant on March 12, 1863. Bennett and Haigh, *Thirty-Sixth Regiment Illinois*, 60, 278, 426.
25. Bennett's dating is off on this entry. Bennett's company arrived on Tuesday, August 20 according to the *Thirty-Sixth Regiment Illinois*, which he cowrote.
26. A reference to Company D of the Thirty-Sixth Illinois.
27. Bennett wrote two entries each for August 28 and August 29. Regrettably, it is impossible to determine the correct dates for these. However, his diary entries at least match the 1861 calendar beginning with his Friday, August 30 entry.
28. Colonel John B. Wyman commanded the Thirteenth Illinois Infantry. Wyman brought the Thirteenth to a high state of drill expertise. Wyman was killed on December 28, 1862, during an assault on Chickasaw Bluffs. *The Union Army*, 3:252; United States War Department, *War of the Rebellion: A Compilation of the Official Records of the Union and Confederate Armies*, 128 vols. (1880–1901; repr., Harrisburg, PA: National Historical Society, 1971), ser. 1, 17, pt. 1:654.
29. The Sons of Temperance, a secret brotherhood, spread quickly through the United States in the 1840s. It eventually had "over 5,000 chapters in the U. S., England, Ireland, and Australia." The organization also functioned as a mutual insurance company. David J. Hanson, "Sons of Temperance: A Temperance Brotherhood & Insurance Company," Alcohol Problems and Solutions (website), accessed May 17, 2021, https://www.alcoholproblemsand solutions.org/sons-of-temperance-a-temperance-brotherhood/.
30. There was no Captain Price in the Thirty-Sixth Illinois. This is possibly a reference to Captain William P. Pierce.

31. Captain Albert Gallatin Brackett mustered the regiment on September 23, 1861. Hewitt, *Supplement to the Official Records*, 10:578.
32. An attorney and a native of New York, Lieutenant Colonel Edward S. Joslyn resigned his commission in 1862. According to the regimental history co-authored by Bennett, Joslyn was "Fearless and outspoken, none who knew him doubted his patriotism or courage." Bennett and Haigh, *Thirty-Sixth Regiment Illinois*, 16, 221.
33. A reference to Orville B. Merrill, appointed first lieutenant of Company I. He became captain of the company sometime in 1862 after the resignation of Captain Samuel C. Camp. Wounded and captured at the Battle of Stones River, Merrill returned after his exchange several months later. Ibid., 49, 186, 394, 432.
34. Bennett mentioned Lieutenant William Walker with less sympathy in the regimental history where his profession as an auctioneer meant that "his duties were performed in 'just a going, gentlemen,—going—going—gone' sort of a way, exciting the laughter of some and the disgust of others." Ibid., 34.
35. John VanPelt served as first lieutenant of Company D, and Joseph Whitham served in the same company. Private Whitham was wounded in the abdomen at the Battle of Kennesaw Mountain on June 27, 1864. Ibid., 43–44, 611.
36. C. P. Drake, his wife, and his two children lived near Lyman Bennett's family. A native of Connecticut, C. P. Drake was a thirty-seven-year-old railroad agent in 1860. *Eighth Census of the United States*, 1860, Kendall County, Ill., 239.
37. Captain Samuel C. Camp commanded Company I but resigned sometime in 1862. Bennett and Haigh, *Thirty-Sixth Regiment Illinois*, 49, 186.
38. Sergeant Gustav Voss served in Company I. Ibid., 49.

3. Trip to Rolla, Missouri

1. Patricia L. Faust, *Historical Times Illustrated Encyclopedia of the Civil War* (New York: Harper & Row, 1986), 501.
2. Louis S. Gerteis, *The Civil War in Missouri: A Military History* (Columbia: Univ. of Missouri Press, 2012), 53.
3. William Garrett Piston and Richard W. Hatcher, III, *Wilson's Creek: The Second Battle of the Civil War and the Men Who Fought It* (Chapel Hill: The Univ. of North Carolina Press, 2000), 28, 33, 36.
4. Gerteis, *Civil War in Missouri*, 4.
5. Ibid., 12, 30–31.
6. Ibid., 30–31.
7. Ibid., 30–31.

8. Ibid., 31.
9. Ibid., 33.
10. Ibid., 34, 36.
11. Ezra J. Warner, *Generals in Blue: Lives of the Union Commanders* (1964; repr., Baton Rouge: Louisiana State Univ. Press, 1992), 286, 448.
12. Faust, *Historical Times Illustrated Encyclopedia*, 833–34.
13. Ibid., 435–36.
14. William Garrett Piston, "Struggle for the Trans-Mississippi," *North & South* 11, no. 5 (October 2009): 16.
15. Piston, "Struggle," 16.
16. Marvin R. Cain and John F. Bradbury, Jr., "Union Troops and the Civil War in Southwestern Missouri and Northwestern Arkansas," *Missouri Historical Review* 88 (October 1993): 30.
17. Stephen Z. Starr, "The Grand Old Regiment," *The Wisconsin Magazine of History* 48, no. 1 (Autumn 1964): 25.
18. The conflict continued between Lieutenant Walker and Colonel Nicholas Greusel, a saga that had not yet quite played out.
19. Louisa and Caroline Bennett, sixteen and eighteen years old respectively in 1860, were younger sisters of Lyman's. *Eighth Census of the United States*, 1860, Kendall County, Ill., 239.
20. A reference to either Captain Melvin B. Baldwin of Company A or Captain Elias Baldwin of Company C of the Thirty-Sixth Illinois. L. G. Bennett and William M. Haigh, *History of the Thirty-Sixth Regiment Illinois Volunteers, during the War of the Rebellion* (1876; repr., Marengo, IL: Prairie State Press, Inc., 1999), 38, 41.
21. Colonel James A. Mulligan raised the Twenty-Third Illinois Infantry, known as the Irish Brigade. Mulligan commanded Union forces, including the Twenty-Third Illinois, at Lexington, Missouri, and surrendered on September 20, 1861, after holding out three days against the Missouri State Guard. Most of the men were exchanged after a relatively brief time. Mulligan died of wounds received at the Battle of Winchester, Virginia, in July 1864. Mark M. Boatner, III, *The Civil War Dictionary* (New York: David McKay Co., Inc., 1959), 574; Larry Wood, *The Siege of Lexington Missouri: The Battle of the Hemp Bales* (Charleston, SC: The History Press, 2014), 105, 122–23.
22. The St. Louis Arsenal was a "three-story building on the banks of the Mississippi" along with a number of outlying buildings housing a large quantity of arms and ammunition. Piston and Hatcher, *Wilson's Creek*, 28; National Geospatial-Intelligence Agency, *Wilson's Creek Staff Ride: Fieldbook*, 3rd ed. (St. Louis: NGA, 2012), 2–5.
23. Companies A and B indeed received "Minie and Enfield rifles and the other

companies remodeled Springfield muskets, caliber 69." *The Union Army: A History of Military Affairs in the Loyal States 1861–1865* (1908; repr., Wilmington, NC: Broadfoot Publishing Co., 1998), 3:276.

24. Authorities organized the Third Missouri Infantry in September 1861 in St. Louis. A three-year regiment, the Third served at the Battle of Pea Ridge and then campaigned in the western theater for the remainder of the war. Frederick H. Dyer, comp., *A Compendium of the War of the Rebellion* (1908; repr., Dayton, OH: Morningside Bookshop, 1978), 2:1323.

25. A reference to the Thirteenth Illinois Infantry.

4. Camp Life in Rolla

1. John F. Bradbury, Jr., "'Good Water & Wood but the Country is a Miserable Botch': Flatland Soldiers Confront the Ozarks," *Missouri Historical Review* 90 (January 1996): 169.
2. Bradbury, "Good Water," 169.
3. William Garrett Piston, "Struggle for the Trans-Mississippi," *North & South* 11, no. 5 (October 2009): 16.
4. Piston, "Struggle," 18.
5. Ibid., 15, 18.
6. Ibid., 18.
7. William L. Shea and Earl J. Hess, *Pea Ridge: Civil War Campaign in the West* (Chapel Hill: The Univ. of North Carolina Press, 1992), 3.
8. Twenty-six-year-old native New Yorker Sergeant William J. Willett lived with his parents near Bristol in Kendall County, Illinois, according to the 1860 census. Willett served in Company E and was killed at the Battle of Chickamauga. L. G. Bennett and William M. Haigh, *History of the Thirty-Sixth Regiment Illinois Volunteers, during the War of the Rebellion* (1876; repr., Marengo, IL: Prairie State Press, Inc., 1999), 44, 482; *Eighth Census of the United States*, 1860, Kendall County, Ill., 117.
9. Following the Pea Ridge campaign, the Fourth Iowa was transferred east of the Mississippi River and served for the remainder of the war in the Western Theater. Frederick H. Dyer, comp., *A Compendium of the War of the Rebellion*, (1908; repr., Dayton, OH: Morningside Bookshop, 1978), 2:1166.
10. George W. Walker enlisted in Oswego on May 24, 1861, in Company H of the Thirteenth Illinois Infantry. Illinois GenWeb (website), accessed February 16, 2021, https://civilwar.illinoisgenweb.org/r050/013-h-in.html.
11. John Martin enlisted in Oswego on May 24, 1861, in Company H of the Thirteenth Illinois Infantry. He re-enlisted in January 1864 and transferred to Company I of the Fifty-Sixth Illinois Infantry. Illinois GenWeb (website), accessed February 16, 2021, https://civilwar.illinoisgenweb.org/r050/013-h-in.html.

12. The Thirteenth Illinois served in the Pea Ridge campaign and then was transferred east of the Mississippi River and spent the remainder of their service in the Western Theater. Dyer, *Compendium*, 2:1050.
13. The *Chicago Daily Tribune* was established in 1847 and then "transformed by the arrival in 1855 of editor and co-owner Joseph Medill, who turned the paper into one of the leading voices of the new Republican Party." Mark R. Wilson, "Chicago Tribune," in *Encyclopedia of Chicago* (website), accessed February 16, 2021, http://www.encyclopedia.chicagohistory.org/pages/275.html.
14. Captain William P. Pierce commanded Company D of the Thirty-Sixth Illinois and then became assistant surgeon of the regiment. In January 1863 Pierce became surgeon of the Eighty-Eighth Illinois Infantry, "a most worthy and credible promotion." Bennett and Haigh, *Thirty-Sixth Regiment Illinois*, 43, 412.
15. Possibly a reference to wounded from the Battle of Wilson's Creek fought on August 10, 1861, near Springfield, Missouri.
16. The Fifteenth Illinois Infantry served in Fremont's army in Missouri in the fall of 1861 and in early 1862 transferred east of the Mississippi River where they were part of the army that captured Fort Donelson, Tennessee. Almost their entire remaining service took place in the Western Theater. Dyer, *Compendium*, 2:1050–51.
17. Private Judson W. Hanson served in Company E of the Thirty-Sixth Illinois. The nineteen-year-old native of New York enlisted on August 24, 1861 in Bristol. According to the regimental history he re-enlisted in 1864, but records indicate that he mustered out on October 8, 1864. Bennett and Haigh, *Thirty-Sixth Regiment Illinois*, 44, 559; Historical Data Systems, Inc., American Civil War Research Database, accessed on February 16, 2021, http://www.civilwardata.com/active/hdsquery.dll?SoldierHistory?U&504756.
18. Possibly a reference to a skirmish at Springfield on October 25, 1861, although the Seventh Missouri did not fight there. Dyer, *Compendium*, 1:799.
19. Major William D. Bowen commanded the First Battalion Missouri Cavalry. After service in the Pea Ridge campaign, the battalion was transferred east of the Mississippi River where they were eventually assigned to the Tenth Missouri Cavalry. Shea and Hess, *Pea Ridge*, 334; Dyer, *Compendium*, 2:1312.
20. Probably a reference to one of the two cavalry companies that were part of the Thirty-Sixth Illinois at this time.
21. Woodstock, Illinois, is in McHenry County, and Company H was from that county. The company's lieutenants were First Lieutenant Alfred H. Sellers and Second Lieutenant Charles F. Dyke. Bennett and Haigh, *Thirty-Sixth Regiment Illinois*, 48.
22. Major General David Hunter became commander of the Department of

the West in early November 1861 after Major General John C. Fremont was relieved from that position after emancipating Missouri slaves without authorization. Hunter was already on duty in Missouri, and "he took over operations against the Missouri State Guard which continued to retreat toward Pineville, near the Arkansas border." William Garrett Piston and Richard W. Hatcher, III, *Wilson's Creek: The Second Battle of the Civil War and the Men Who Fought It* (Chapel Hill: The Univ. of North Carolina Press, 2000), 315.

23. Sterling Price, a former Missouri governor, commanded the Missouri State Guard. Ezra J. Warner, *Generals in Gray: Lives of the Confederate Commanders* (1959; repr., Baton Rouge: Louisiana State Univ. Press, 1987), 247.

5. Engineering Work in Rolla

1. John F. Bradbury, Jr., "Fort Wyman and the Defenses of Rolla," Pioneer Times 7, no. 3 (July 1983): 247–48.
2. Bradbury, "Fort Wyman," 249.
3. Ibid.
4. Ibid.
5. Ibid., 250.
6. Ibid., 251.
7. John F. Bradbury, Jr., "Fort Wyman, Fort Dette and the Defense of the Railhead," *PCHS [Phelps County Historical Society] Newsletter*, new series no. 44 (October 2011): 5.
8. Bradbury, "Fort Wyman," 251.
9. Ibid.
10. Ibid.
11. John F. Bradbury, Jr., "'Buckwheat Cake Philanthropy': Refugees and the Union Army in the Ozarks," *Arkansas Historical Quarterly* 57, no. 3 (Autumn 1998): 236.
12. Bradbury, "'Buckwheat Cake,'" 240–41.
13. Ibid., 236.
14. Ibid., 236, 238.
15. Ibid., 237, 241.
16. Ibid., 239.
17. Vera Cruz, Missouri, was the county seat of Douglas County at that time. Bruce Nichols, *Guerrilla Warfare in Civil War Missouri* (Jefferson, NC: McFarland and Co., Inc., 2012), 1:76.
18. This skirmish is undocumented in the *Official Records*.
19. Possibly a reference to James Haggin McBride, who "was appointed brigadier general of the 7th Division of the Missouri State Guard" at the beginning of the Civil War. Bruce S. Allardice, *More Generals in Gray* (Baton Rouge: Louisiana State Univ. Press, 1995), 155.

20. Solomon Collins resided in Ozark County, Missouri, in 1850. A married, illiterate farmer, Collins had served in two Tennessee militia companies during the War of 1812. His age was listed as sixty-one in the 1850 census, which made him about seventy-one in 1861. *Seventh Census of the United States*, 1850, Ozark County, Mo., 24; Solomon Collins, War of 1812 pension application, accessed February 17, 2021, Fold3.com.
21. Born in New York in 1830, Marcus LaRue Harrison lived in Nashville, Illinois, in 1850. At the time he was married to Rebecca Axley, who bore two sons before dying in 1861. Harrison worked for the Chicago, Burlington and Quincy Railroad and the Burlington and Missouri River Railroad before the conflict. He was the "master of buildings and car repairs" for the latter railroad by 1861. Harrison enlisted as a private in September 1861 in the Thirty-Sixth Illinois and was "an acting lieutenant of engineers" at Fort Wyman. In July 1862 he was appointed as "chief engineer for the district of Southwest Missouri." That summer he raised the First Arkansas Cavalry, a regiment he commanded for the rest of the war. Harrison developed the idea of farm colonies in northwest Arkansas later in the war, which proved controversial. After the war, he became a railroad promoter in Arkansas, dabbled in politics, engaged in surveying work in north central Arkansas, and was appointed to the post office department. Richard A. Bland, "Marcus LaRue Harrison (1830–1890)," *Encyclopedia of Arkansas* (website), accessed February 21, 2021, https://encyclopediaofarkansas.net/entries/marcus-larue-harrison-1665/; L. G. Bennett and William M. Haigh, *History of the Thirty-Sixth Regiment Illinois Volunteers, during the War of the Rebellion* (1876; repr., Marengo, IL: Prairie State Press, Inc., 1999), 92; *History of Benton, Washington, Carroll, Madison, Crawford, Franklin, and Sebastian Counties, Arkansas* (Chicago: Goodspeed Publishing Co., 1889), 220.
22. Fought on October 13, 1861, the skirmish at Wet Glaze, also known as Dutch or Monday Hollow, took place about twenty miles from Lebanon, Missouri. Estimated enemy losses ranged from twenty-nine to more than one hundred men. The only man killed in the Union force was a cavalryman in the Fremont Battalion. United States War Department, *War of the Rebellion: A Compilation of the Official Records of the Union and Confederate Armies*, 128 vols. (1880–1901; repr., Harrisburg, PA: National Historical Society, 1971), ser. 1, 3:236–41.
23. Major William D. Bowen commanded the First Battalion Missouri Cavalry. *Official Records*, ser. 1, 3:240–41.
24. First Lieutenant Edward S. Chappel of Company A of the Thirty-Sixth Illinois. Bennett and Haigh, *Thirty-Sixth Regiment Illinois*, 38.
25. Following a court-martial called in January 1862, only one witness was available, and he testified that Lieutenant William Walker "was not present"

during the mustering in of Company I. This ended the matter, and Walker resigned his commission from Company I on February 28, 1862. However, "the Regiment was not rid of dissension." Bennett and Haigh, *Thirty-Sixth Regiment Illinois*, 96–98; Illinois GenWeb (website), accessed February 17, 2021, https://civilwar.illinoisgenweb.org/r050/036-i-in.html.

26. A reference to Company H. Illinois GenWeb (website), accessed February 17, 2021, https://civilwar.illinoisgenweb.org/civilwar/reg_html/036_reg.html.
27. A reference to Company I. Illinois GenWeb (website), accessed February 17, 2021, https://civilwar.illinoisgenweb.org/civilwar/reg_html/036_reg.html.
28. "His body was placed in a coffin, draped with the National flag, and forwarded to friends in Elgin, for burial. The Regiment was drawn up in two lines, in open order, extending from camp to railroad station, between which Company A. with reversed arms followed the coffin, which was preceded by the band, playing a funeral dirge, the solemn cadences of which added a mournful solemnity to the sadness of the hour. Captain Baldwin took charge of the body and proceeded with it to Elgin." Bennett and Haigh, *Thirty-Sixth Regiment Illinois*, 73.
29. Bennett's 1857 diary mentioned McDowell but never provided a first name.
30. Captain Bacon Montgomery commanded a company in the Fremont Battalion (Missouri) Cavalry. *Official Records*, ser. 1, 3:239, 791.
31. A reference again to the skirmish at Wet Glaze, which occurred on October 13, 1861.
32. A reference to the Thirteenth Illinois Infantry.
33. The abbreviation "inst" here is short for "instant," which Bennett uses to mean the same month.
34. Major Clark Wright commanded the Fremont Battalion (Missouri) Cavalry. *Official Records*, ser. 1, 3:238–240.
35. Turner is not mentioned in the reports of this skirmish. A Colonel William W. Summers was captured at Wet Glaze. Ibid., 3:238, 241.
36. Captain Theodore A. Switzler commanded a company in the Fremont Battalion (Missouri) Cavalry. Ibid., 3:239, 805.
37. Possibly a reference to Captain S. N. Wood, who commanded the Kansas Rangers—a unit that campaigned actively in southwest Missouri in the fall of 1861. *Official Records*, ser. 1, 8:366.
38. The "darkey" was not mentioned in the reports of the skirmish at Wet Glaze. *Official Records*, ser. 1, 3:236–41.
39. In 1838, Colonel John S. Phelps, a native of Connecticut, "moved to Springfield, Missouri, where he at once became not only financially successful but politically prominent." Prior to the Civil War, he served for eighteen years in Congress. In 1860, Phelps labeled himself a "retired lawyer" and had

accumulated $65,000 worth of real estate and $12,000 worth of personal estate. Ten slaves comprised part of his wealth. At Rolla, Phelps commanded Phelps' Independent Missouri Regiment, which served in the Pea Ridge campaign. In July 1862, President Abraham Lincoln appointed him military governor of Arkansas, and in 1876 voters elected him governor of Missouri. Ezra J. Warner, *Generals in Blue: Lives of the Union Commanders* (1964; repr., Baton Rouge: Louisiana State Univ. Press, 1992), 367–68; William L. Shea and Earl J. Hess, *Pea Ridge: Civil War Campaign in the West* (Chapel Hill: The Univ. of North Carolina Press, 1992), 334; *Eighth Census of the United States*, 1860, Greene County, Mo., 135; *Eighth Census of the United States*, Slave Schedule, Greene County, Mo., 12.

40. A younger brother, Guy (also known as George) Bennett was born in approximately 1835 in New York. His residence in 1860 is unknown, but after the war he lived in Lawrence, Kansas, and was married to Sadie Drew, "a daughter of Governor Drew of Arkansas." Lyman G. Bennett, "Personal Reminiscences" (MS, ca. 1875), 43, The Huntington Library.
41. George G. Lyon served as chaplain of the Thirty-Sixth Illinois. Bennett and Haigh, *Thirty-Sixth Regiment Illinois*, 38.
42. The *Kendall County Free Press* began publication in 1857, but its ending date is unknown. Only a handful of issues survived from the Civil War years. Library of Congress, Chronicling America (website), accessed February 17, 2021.
43. First Lieutenant George F. Stonax served in Company F of the Thirty-Sixth Illinois. Bennett and Haigh, *Thirty-Sixth Regiment Illinois*, 45.
44. Corporal Hiram Wagner of Company E of the Thirty-Sixth Illinois Infantry. Bennett and Haigh, *Thirty-Sixth Regiment Illinois*, 44.
45. Corporal David G. Cromwell of Company E of the Thirty-Sixth Illinois Infantry. Ibid., 44.
46. Sample Orr was a forty-three-year-old farmer and lawyer, a native of Tennessee, and a resident of Springfield, Missouri, in 1860. *Eighth Census of the United States*, 1860, Greene County, Mo., 170.
47. Here Lyman references an interesting incident that he did not mention in the regimental history.
48. Captain John S. Coleman commanded Company A of Phelps' Independent Missouri Regiment. The thirty-one-year-old enlisted at Rolla, Missouri, on August 28, 1861. Coleman Compiled Service Record, Phelps' Independent Missouri Regiment, Fold3.com (website), accessed February 24, 2021.
49. Captain Josephus G. Rich commanded Company B in Phelps' Regiment. He was thirty-three years old and mustered in on October 3, 1861, in Rolla. Rich Compiled Service Record, Phelps' Independent Missouri Regiment, Fold3.com (website), accessed February 18, 2021.

6. Surveying Work in Rolla

1. Ezra J. Warner, *Generals in Blue: Lives of the Union Commanders* (1964; repr., Baton Rouge: Louisiana State Univ. Press, 1992), 127–28.
2. Earl B. McElfresh, *Maps and Mapmakers of the Civil War* (New York: Harry N. Abrams, Inc., 1999), 37.
3. McElfresh, *Maps and Mapmakers*, 64.
4. Ibid., 65.
5. Ibid., 24.
6. Ibid., 23–24.
7. Ibid., 23, 150.
8. Ibid., 31.
9. Ibid., 31–32.
10. Ibid., 32.
11. Orville B. Shelburne, *From Presidio to the Pecos River: Surveying the United States-Mexico Boundary along the Rio Grande, 1852 and 1853* (Norman: Univ. of Oklahoma Press, 2020), 37.
12. Andro Linklater, *Measuring America: How the United States Was Shaped by the Greatest Land Sale in History* (New York: Plume Books, 2003), 16–17.
13. Linklater, *Measuring America*, 77–78.
14. McElfresh, *Maps and Mapmakers*, 26–27.
15. Ibid., 14–15.
16. Ibid., 40, 46, 60.
17. Ibid., 68.
18. Ibid., 54.
19. Lyman G. Bennett, diary, 12 December 1861, Lyman G. Bennett Papers, R0274, The State Historical Society of Missouri Research Center–Rolla.
20. Alonzo H. Barry served as major of the Thirty-Sixth Illinois Infantry. He resigned his commission during the hot summer of 1862 in Mississippi. L. G. Bennett and William M. Haigh, *History of the Thirty-Sixth Regiment Illinois Volunteers, during the War of the Rebellion* (1876; repr., Marengo, IL: Prairie State Press, Inc., 1999), 38, 215.
21. Promoted to brigadier general after the Battle of Pea Ridge, Dodge mostly served in the Western Theater afterwards and was promoted to major general in June 1864. After the war, he became chief engineer of the Union Pacific Railroad, one of the two railroads that constructed the first transcontinental railroad. Later in life he continued his work with railroads and surveying. He died in Council Bluffs, Iowa, in 1916. Warner, *Generals in Blue*, 127–28.
22. Likely a reference to the *Kendall County Free Press*.
23. Bennett is possibly referring to Sherer's Independent Cavalry Company, which was organized in Aurora, Illinois, and became the Thirty-Sixth

Illinois Infantry's original Company B. Eventually detached from the Thirty-Sixth Illinois, it campaigned as an independent cavalry company in the Western Theater. Frederick H. Dyer, comp., *A Compendium of the War of the Rebellion*, 2 vols. (1908; repr., Dayton, OH: Morningside Bookshop, 1978), 2:1034; *The Union Army: A History of Military Affairs in the Loyal States 1861–1865* (1908; repr., Wilmington, NC: Broadfoot Publishing Co., 1998), 3:276–77.

24. The Thirteenth Illinois Infantry.
25. The Twelfth Missouri Infantry was organized in St. Louis in August 1861. After fighting in the Pea Ridge campaign, the Twelfth Missouri campaigned actively in the Western Theater, losing 112 men killed or mortally wounded. Dyer, *Compendium*, 2:1328.
26. A reference to an expedition consisting of elements of the Thirty-Sixth Illinois, the Fourth Iowa, and Wood's Kansas Rangers against a Confederate named Freeman. According to a report made by Colonel Grenville M. Dodge, the expedition marched to Texas County, Missouri, and brought back "a large amount of property, stock, and several prominent rebel prisoners" on November 9, 1861, rather than November 8 as Bennett stated. United States War Department, *War of the Rebellion: A Compilation of the Official Records of the Union and Confederate Armies*, 128 vols. (1880–1901; repr., Harrisburg, PA: National Historical Society, 1971), ser. 1, 3:255.
27. Delos W. Young served as surgeon of the Thirty-Sixth Illinois Infantry and was promoted to brigade surgeon by the fall of 1862. "Dr. Young was in many respects a remarkable man; his ability as a Surgeon was of a high order, and with him nothing seemed too much to do for his friends." Bennett and Haigh, *Thirty-Sixth Regiment Illinois*, 38, 289, 331, 416.
28. Captain Henry A. Smith commanded Cavalry Company B of the Thirty-Sixth Illinois Infantry. Bennett did not refer to this incident in the regimental history. Ibid., 53.
29. Captain Albert Jenks commanded Cavalry Company A of the Thirty-Sixth Illinois Infantry. Ibid., 52.
30. A twenty-four-year-old married brother of Melissa Bennett's, Edgar Lyon was working as a clerk in Kendall County, Illinois. *Eighth Census of the United States*, 1860, Kendall County, Ill., 271.
31. Second Lieutenant William F. Sutherland served in Company I of the Thirty-Sixth Illinois Infantry. Bennett and Haigh, *Thirty-Sixth Regiment Illinois*, 49.
32. Referred to in the regiment's postwar history as "always constitutionally thirsty," William Todd was a musician in Company E of the Thirty-Sixth Illinois Infantry and became the drum major of the regiment. Bennett and Haigh, *Thirty-Sixth Regiment Illinois*, 44, 196.

33. Franz Sigel emigrated from Germany in 1852 after participating in the Revolutions of 1848. He taught school first in New York City and then in St. Louis, the home of many German immigrants, where "by 1861 he was director of schools." The Lincoln administration appointed him brigadier general in August 1861 followed by a promotion to major general in March 1862. Warner, *Generals in Blue*, 447–48.
34. May refer to Thomas Roe Freeman, who initially commanded a regiment in the Seventh Division of the Missouri State Guard. He later recruited several companies into a cavalry battalion "that operated on the Arkansas-Missouri border by mid-1863." James E. McGhee, *Guide to Missouri Confederate Units, 1861–1865* (Fayetteville: Univ. of Arkansas Press, 2008), 111.
35. Another reference to the expedition that took place in early November 1861. Dodge reported that the detachment returned with "a large amount of property, stock, and several prominent rebel prisoners. They drove Freeman from Texas County, and Captain Wood, in command of cavalry, is still in pursuit of him." *Official Records*, ser. 1, 3:255.
36. On July 5, 1861, Sigel led an attack against a stronger force of Missouri State Guardsmen under the command of Governor Claiborne Fox Jackson. The Guardsmen soon forced Sigel's men to retreat, which they did in an orderly fashion for a number of miles. Louis S. Gerteis, *The Civil War in Missouri: A Military History* (Columbia: Univ. of Missouri Press, 2012), 44–48.
37. Sigel led an unsuccessful flanking maneuver at the Battle of Wilson's Creek on August 10, 1861. Patricia L. Faust, *Historical Times Illustrated Encyclopedia of the Civil War* (New York: Harper & Row, 1986), 833–34.
38. A reference to either James Roseman or Wilbur F. Roseman, both of Company G of the Thirty-Sixth Illinois. Private James Roseman later served as a lieutenant in the First Arkansas Cavalry commanded by Colonel M. LaRue Harrison. Private Wilbur Roseman was wounded in action on December 31, 1862, at the Battle of Stones River. Bennett and Haigh, *Thirty-Sixth Regiment Illinois*, 47, 222, 396.
39. Alexander Sandor Asboth, a native of Hungary, immigrated to the United States after his involvement in the Hungarian Revolution of 1848. When the war started, he became General John C. Fremont's chief of staff and then was appointed brigadier general in March 1862. He commanded a division at the Battle of Pea Ridge, where he was wounded. In 1864, he was wounded seriously at the Battle of Marinna, Florida, while commanding the District of West Florida. He died in 1868 after serving as "U. S. minister to the Argentine Republic and Uruguay." Warner, *Generals in Blue*, 11.
40. The 1860 census for Greene County, Missouri, listed a Thomas Greene, a forty-seven-year-old married farmer. He and his wife, Almira, had seven

children and lived in Taylor Township near Springfield. Six of these children were daughters ranging in age from an infant to nineteen-year-old Parella. *Eighth Census of the United States*, 1860, Greene County, Mo., 107.

41. Colonel Charles Knobelsdorff commanded the Forty-Fourth Illinois Infantry. After fighting at the Battle of Pea Ridge, the regiment served in the Western Theater, where they lost 135 officers and enlisted men killed and mortally wounded. William L. Shea and Earl J. Hess, *Pea Ridge: Civil War Campaign in the West* (Chapel Hill: The Univ. of North Carolina Press, 1992), 331; Dyer, *Compendium*, 2:1065–66.

42. Born in Tennessee in 1811, Benjamin McCulloch migrated to Texas, fought at the Battle of San Jacinto, and then went on to serve during the Mexican American War. Upon his return he became a US marshal for six years in Texas. He became a brigadier general in May 1861 and commanded regiments from Arkansas, Louisiana, and Texas during the Battle of Wilson's Creek. McCulloch was killed at the Battle of Pea Ridge. Ezra J. Warner, *Generals in Gray: Lives of the Confederate Commanders* (1959; repr., Baton Rouge: Louisiana State Univ. Press, 1987), 200–201.

43. A Jacob Bauman, a forty-six-year-old married farmer and carpenter, resided in Greene County, Missouri, in 1860 with a large family. A native of Pennsylvania, Bauman had neighbors primarily from Tennessee, Virginia, and North Carolina. *Eighth Census of the United States*, 1860, Greene County, Mo., 110.

44. Bennett included this story in the regimental history. Bennett and Haigh, *Thirty-Sixth Regiment Illinois*, 92.

45. Unfortunately, it is not possible to identify the cave that Bennett and his friends visited near Rolla.

46. Isaac N. Buck served as quartermaster of the Thirty-Sixth Illinois Infantry. Bennett and Haigh, *Thirty-Sixth Regiment Illinois*, 38.

47. A detachment of 120 men under the command of Major William D. Bowen of the First Battalion Missouri Cavalry marched to Salem, Missouri, "to bring in some witnesses in the case of some prisoners he [Grenville Dodge] has now in the fort." Attacked on December 3, 1861, "by 300 rebels, under command of Colonels Freeman and Turner" near Salem, the Federal detachment suffered losses of four men killed or mortally wounded and eight wounded. *Official Records*, ser. 1, 8:33–34.

48. Mrs. Jane Greusel, Colonel Greusel's wife, was a native of Canada. *Eighth Census of the United States*, 1860, Kane County, Ill., 143.

49. Henry Wager Halleck was born in New York in 1815 and graduated from the United States Military Academy in 1839. After various assignments, he resigned his commission in 1854 and turned to various ventures in California that made him quite prosperous. Upon the recommendation of General

Winfield Scott, Halleck was appointed major general in August 1861. Halleck commanded the Department of Missouri when Bennett wrote to him. Warner, *Generals in Blue*, 195–96.

50. Major William D. Bowen reported that the rebel loss was sixteen killed and mortally wounded and twenty wounded. Bowen did not report taking any prisoners. *Official Records*, ser. 1, 8:34.
51. Harrison's poem was written in the style of Edgar Allan Poe's popular, and often parodied, poem, "The Raven." Edgar Allan Poe, *Poetry and Tales*, The Library of America (New York: Literary Classics of the United States, 1984), 81–86.
52. Louisa Canfield Bennett.
53. Major William D. Bowen and his First Battalion Missouri Cavalry marched from Salem, Missouri, on December 5, 1861, in search of the enemy. Bowen's men "followed them from Jack's Forks to Spring Valley, scouring the country in every direction as we advanced." He reported that "20 prisoners and some 35 horses" were captured. *Official Records*, ser. 1, 8:36–37.
54. A classmate of Henry W. Halleck's, Colonel George Thom, an "aide-de-camp and chief of Topographical Engineers," was serving at the time on a board to examine the St. Louis defenses. A native of New Hampshire, Thom graduated seventh in his United States Military Academy class of 1839 and was breveted second lieutenant of topographical engineers. His Civil War assignments were diverse ones mostly in the Western and Eastern Theaters. At the end of the war he was breveted brigadier general in the US Army. *Official Records*, ser. 1, 8:398, 401; George W. Cullum, *Biographical Register of the Officers and Graduates of the U. S. Military Academy at West Point, N. Y.*, 2nd ed. (New York: D. Van Nostrand, 1868), 1:575–76.
55. Lieutenant Clark survived this illness.

7. St. Louis

1. William Garrett Piston and Thomas P. Sweeney, *Portraits of Conflict: A Photographic History of Missouri in the Civil War* (Fayetteville: Univ. of Arkansas Press, 2009), 205.
2. Piston and Thomas P. Sweeney, *Portraits of Conflict*, 204.
3. Ibid.
4. Ibid., 205.
5. Ibid., 205.
6. Ibid., 203.
7. "Thomas Biddle," Find a Grave (website), accessed March 4, 2021, https://www.findagrave.com/memorial/9349/thomas-biddle.
8. "Catharine C Biddle," Find a Grave (website), accessed March 4, 2021, https://www.findagrave.com/memorial/46844638/catharine-c-biddle.

9. According to the 1860 census, Mary Phelps was a native of Maryland and forty-six years old. *Eighth Census of the United States*, 1860, Greene County, Mo., 135.
10. Brigadier General George Washington Cullum, a native of New York, served as Major General Henry W. Halleck's chief of staff. Cullum is most famous for his *Biographical Register of the Officers and Graduates of the U. S. Military Academy* first published in 1868. Ezra J. Warner, *Generals in Blue: Lives of the Union Commanders* (1964; repr., Baton Rouge: Louisiana State Univ. Press, 1992), 105–6.
11. John C. Kelton graduated from the United States Military Academy in 1851 and was breveted second lieutenant in the Sixth US Infantry and stationed at Fort Snelling, Minnesota. In September 1861 he was appointed colonel of the Ninth Missouri Infantry but served as assistant adjutant general of the Department of the West in St. Louis from November 24, 1861, to March 11, 1862. George W. Cullum, *Biographical Register of the Officers and Graduates of the U. S. Military Academy at West Point, N. Y.*, 3rd ed. (Houghton, Mifflin, and Co., 1891), 2:459.
12. Major General John C. Fremont established Benton Barracks, north of St. Louis, Missouri, as a training facility in 1861. The Civil War Muse (website), accessed March 2, 2021, http://www.thecivilwarmuse.com/index.php?page=benton-barracks.
13. Major James Totten, an 1841 graduate of the United States Military Academy. He commanded the Little Rock Arsenal when the war started and then fought in several of the early military actions in Missouri, most notably Wilson's Creek, where he commanded an artillery battery. He served as an assistant inspector general on Halleck's staff from November 23, 1861, to February 19, 1862. Totten continued in the military until he was dismissed in 1870 on a variety of charges. Cullum, *Biographical Register*, 3rd ed. (1891), 2:89–90.
14. Private Eugene Lake was a member of the Second Independent Battery, Ohio Light Artillery, part of Brigadier General Alexander S. Asboth's division. After experiencing combat at Pea Ridge, the battery saw action during the Vicksburg Campaign and then was stationed in Louisiana. *Soldiers and Sailors Database*, accessed March 2, 2021, https://www.nps.gov/civilwar/search-soldiers-detail.htm?soldierId=519A48B1-DC7A-DF11-BF36-B8AC6F5D926A; William L. Shea and Earl J. Hess, *Pea Ridge: Civil War Campaign in the West* (Chapel Hill: The Univ. of North Carolina Press, 1992), 332; Frederick H. Dyer, comp., *A Compendium of the War of the Rebellion*, 2 vols. (1908; repr., Dayton, OH: Morningside Bookshop, 1978), 2:1488.
15. According to Bennett's December 22, 1861, entry, Frank Briggs served in Dodson's Kane County Independent Company.

16. Marcus LaRue Harrison's first wife, Rebecca Axley, died sometime in 1861. Harrison married Medora Bigby sometime during the war and divorced her in 1873. Richard A. Bland, "Marcus LaRue Harrison (1830–1890)," *Encyclopedia of Arkansas* (website), accessed February 21, 2021, https://encyclopediaofarkansas.net/entries/marcus-larue-harrison-1665/.
17. Organized in September 1861, Dodson's Kane County Independent Company Cavalry campaigned in Missouri and Arkansas and then became Company H of the Fifteenth Illinois Cavalry in December 1862. Dyer, *Compendium*, 2:1033.
18. Peter Schryver served as a musician in Company E of the Thirty-Sixth Illinois Infantry. L. G. Bennett and William M. Haigh, *History of the Thirty-Sixth Regiment Illinois Volunteers, during the War of the Rebellion* (1876; repr., Marengo, IL: Prairie State Press, Inc., 1999), 44.
19. A reference to Charles M. Bennett mentioned in chapter one.
20. Jaundice was a fairly common illness during the Civil War and, fortunately, rarely fatal. Most cases "were probably due to viral hepatitis, a relatively benign disease." Alfred Jay Bollet, *Civil War Medicine: Challenges and Triumphs* (Tucson, AZ: Galen Press, Ltd., 2002), 287.
21. Based on standards created by Dr. Benjamin Rush (1745–1813), many physicians still practiced his "heroic therapy" of bleeding, blistering, and purging. Rush believed that fluids needed to be drawn off to counteract an overactive vascular system. Ira M. Rutkow, *Bleeding Blue and Gray: Civil War Surgery and the Evolution of American Medicine* (New York: Random House, 2005), 45.
22. John 6:37 (King James Version): "All that the Father giveth me shall come to me; and him that cometh to me I will in no wise cast out."
23. Mary Biddle Vandervoort (1812–1901) married Peter Vandervoort, but he died in 1851. The couple had eight children. Catharine C. Biddle (1816–1912) started philanthropic work in 1856 along with her two younger sisters. Hannah Stokes Biddle (1820–1907) and Elizabeth Newbold Biddle, along with their sister Catharine, were referred to as "The Three Graces." All conducted nursing in hospitals in Philadelphia during the war, and all, along with Mary, were buried in Laurel Hill Cemetery in Philadelphia. Their father, Charles Biddle (1787–1836) was a brother of Nicholas Biddle. "Catharine C Biddle," Find a Grave (website), accessed March 4, 2021, https://www.findagrave.com/memorial/46844638/catharine-c-biddle.
24. Private Dwight Follett served in Company D of the Twenty-Second Ohio Infantry, a regiment organized at Benton Barracks in the fall of 1861. According to the 1860 census, he was an Ohio native and twenty-one years old. *Soldiers and Sailors Database*, accessed March 2, 2021, https://www.nps.gov/civilwar/search-soldiers-detail.htm?soldierId=E3CCB49D-DC7A-DF11-BF3

6-B8AC6F5D926A; Dyer, *Compendium*, 2:1506; *Eighth Census of the United States*, 1860, Licking County, Oh., 10.

25. Bennett quoted from the 1854 poem, "Life's Sunny Spots," by Eulalie, a pseudonym for Mary Eulalie Shannon. Eulalie, *Poems* (Cincinnati: Moore, Wilstach & Keys, 1854), 6.
26. Dr. Joseph K. Barnes was appointed in November 1861 as medical director on Major General David Hunter's staff. Barnes became surgeon general in 1863. United States War Department, *War of the Rebellion: A Compilation of the Official Records of the Union and Confederate Armies*, 128 vols. (1880–1901; repr., Harrisburg, PA: National Historical Society, 1971), ser. 1, 8:370; Shauna Devine, *Learning from the Wounded: The Civil War and the Rise of American Medical Science* (Chapel Hill: The Univ. of North Carolina Press, 2014), 36.
27. Erysipelas is a painful and contagious infection caused by streptococcal bacteria. It is sometimes called "St. Anthony's Fire" or "the Rose." Rutkow, *Bleeding Blue and Gray*, 26.
28. Catharine Follett was a native of Pennsylvania and listed as forty-eight years old in the 1860 census. Her son, Dwight, was the oldest of seven children in the household. *Eighth Census of the United States*, 1860, Licking County, Oh., 10.
29. No soldier named B. F. Wells is listed on the roster of Birge's Western Sharpshooters.
30. Colonel John W. Birge organized Birge's Western Sharpshooters in September and October 1861. Made up of enlistees from seven different states, the regiment mostly campaigned in the Western Theater. Dyer, *Compendium*, 2:1076; *The Union Army: A History of Military Affairs in the Loyal States 1861–1865* (1908; repr., Wilmington, NC: Broadfoot Publishing Co., 1998), 3:300–301.
31. George A. Wells (Company A), Edwin Wells (Company B), and Henry G. Wells (Company G) enlisted in the First Minnesota Infantry that compiled an outstanding combat record that stretched from First Bull Run to Appomattox. Dyer, *Compendium*, 2:1296; Return I. Holcombe, *History of The First Regiment Minnesota Volunteer Infantry, 1861–1864* (1916; repr., Ann Arbor, MI: St. Croix Valley Civil War Round Table, 2006), 462, 466, 496.
32. Wilson Judson was twenty years old in 1860 and enlisted one year later in Company K of the Thirteenth Illinois Infantry. Judson left the regiment to accept a promotion as first lieutenant and adjutant of the Fifty-Fourth United States Colored Troops in 1863. He died of typhoid at Fort Gibson, Indian Territory, on October 7, 1864. Eighth Census of the United States, 1860, Kendall County, Ill., 284; Judson Compiled Service Record, Fifty-Fourth Regiment United States Colored Troops, Fold3.com (website), accessed March 30, 2021.

33. A reference to Brigadier General Nathaniel Lyon, who was killed at the Battle of Wilson's Creek. Warner, *Generals in Blue*, 286.
34. A native of New York, Samuel Ryan Curtis grew up in Ohio and graduated from the United States Military Academy in 1831. He resigned a year later and became a civil engineer and an attorney. He commanded the Second Ohio Volunteers during the Mexican American War, and when the Civil War started, he became colonel of the Second Iowa Infantry. The Lincoln administration appointed him brigadier general in May 1861. He commanded the army during the Battle of Pea Ridge and received a promotion to major general after that victory. Warner, *Generals in Blue*, 107–8.
35. Matthew 7:12 (King James Version): "Therefore all things whatsoever ye would that men should do to you, do ye even so to them: for this is the law and the prophets."

8. Work in St. Louis

1. Probably the "friend McDowell" referred to in his 1857 diary.
2. Peter Joseph Osterhaus, a native of Prussia, participated in the 1848 revolutions and fled to the United States a year later. He settled in Illinois and then moved to St. Louis, the home of many German immigrants. Osterhaus entered the war as a major of a Missouri battalion that fought at Wilson's Creek and then commanded the Twelfth Missouri Infantry. After commanding a division at the Battle of Pea Ridge, he was promoted to brigadier general and afterwards served in the Western Theater. Ezra J. Warner, *Generals in Blue: Lives of the Union Commanders* (1964; repr., Baton Rouge: Louisiana State Univ. Press, 1992), 353.
3. Probably a reference to the Second Ohio Cavalry that began campaigning in Missouri in late January 1862. The Second Ohio Cavalry campaigned in all three theaters of the conflict by the time they mustered out in St. Louis in October 1865. Frederick H. Dyer, comp., *A Compendium of the War of the Rebellion*, 2 vols. (1908; repr., Dayton, OH: Morningside Bookshop, 1978), 2:1473–74.
4. Neuralgia medicine was used to treat "paroxysmal pain that extends along the course of one or more nerves." Jack D. Welsh, *Medical Histories of Union Generals* (Kent, OH; Kent State Univ. Press, 1996), 407.
5. Thomas à Kempis wrote *The Imitation of Christ* in the early fifteenth century.
6. Probably a reference to Lieutenant Colonel James B. McPherson, who served as an "aide-de-camp and assistant to chief of engineers" at the time. McPherson, an 1853 graduate of the United States Military Academy, rose quickly in the ranks and went on to command the Army of the Tennessee. McPherson was killed at the Battle of Atlanta on July 22, 1864. United States War

Department, *War of the Rebellion: A Compilation of the Official Records of the Union and Confederate Armies*, 128 vols. (1880–1901; repr., Harrisburg, PA: National Historical Society, 1971), ser. 1, 8:398; Patricia L. Faust, *Historical Times Illustrated Encyclopedia of the Civil War* (New York: Harper & Row, 1986), 466; Warner, *Generals in Blue*, 307–8.

7. The building that had housed the McDowell Medical College was converted into the Gratiot Street Prison, which held Confederate prisoners of war. Louis S. Gerteis, *Civil War St. Louis*, Modern War Studies (Lawrence: Univ. Press of Kansas, 2001), 170.

9. Furlough

1. A native of Massachusetts, T. Sanderson was listed in the 1860 census as a sixty-eight-year-old farmer. His wife, Ann, age sixty-three, was a Vermont native. *Eighth Census of the United States*, 1860, Kendall County, Ill., 247.
2. A forty-six-year-old court clerk in Oswego, J. J. Cole was a native of Rhode Island and a wealthy man with a personal estate valued at $30,000 and $10,000 of real estate. His wife, Mary, age fifty, was also a native of Rhode Island. *Eighth Census of the United States*, 1860, Kendall County, Ill., 289.
3. There were several Hollenbacks in Kendall County, and Mr. Hollenback could not be specifically identified.
4. L. C. Gorton was a fifty-two-year-old farmer and a native of New York. Lyman's brother, Charles, lived with the Gorton family. *Eighth Census of the United States*, 1860, Kendall County, Ill., 240.
5. Possibly a reference to E. Bibbins, a married, forty-four-year-old farmer and native of Pennsylvania. *Eighth Census of the United States*, 1860, Kendall County, Ill., 187.

10. Return to St. Louis and Rolla

1. James R. Knight, *The Battle of Pea Ridge: The Civil War Fight for the Ozarks* (Charleston, SC: The History Press, 2012), 39.
2. William L. Shea and Earl J. Hess, *Pea Ridge: Civil War Campaign in the West* (Chapel Hill: The Univ. of North Carolina Press, 1992), 5.
3. Shea and Hess, *Pea Ridge*, 7.
4. Ibid.
5. Ibid., 10.
6. Ibid., 12.
7. Ibid., 14.
8. Ibid., 14, 331.
9. Knight, *Battle of Pea Ridge*, 42.
10. Justus G. Ketchum from Bristol enlisted on May 24, 1861, in Company H of

the Thirteenth Illinois Infantry. Private Ketchum mustered out on June 7, 1865. Illinois GenWeb (website), accessed March 8, 2021, https://civilwar.illinoisgenweb.org/r050/013-h-in.html.
11. A reference to the *Chicago Tribune.*
12. Corporal Orrin Dickey served in Company I of the Thirty-Sixth Illinois Infantry. L. G. Bennett and William M. Haigh, *History of the Thirty-Sixth Regiment Illinois Volunteers, during the War of the Rebellion* (1876; repr., Marengo, IL: Prairie State Press, Inc., 1999), 50.
13. Federal forces under the command of Brigadier General Samuel R. Curtis reoccupied Springfield, Missouri, on February 13, 1862, after a short skirmish near the town. United States War Department, *War of the Rebellion: A Compilation of the Official Records of the Union and Confederate Armies*, 128 vols. (1880–1901; repr., Harrisburg, PA: National Historical Society, 1971), ser. 1, 8:59; John F. Bradbury, Jr., "'Buckwheat Cake Philanthropy': Refugees and the Union Army in the Ozarks," *Arkansas Historical Quarterly* 57, no. 3 (Autumn 1998): 239.

11. To Springfield

1. William L. Shea and Earl J. Hess, *Pea Ridge: Civil War Campaign in the West* (Chapel Hill: The Univ. of North Carolina Press, 1992), 27–28.
2. A puke was a northern label for "poor southern white trash." Michael Fellman, *Inside War: The Guerrilla Conflict in Missouri during the American Civil War* (New York: Oxford Univ. Press, 1989), 13.
3. Lieutenant Colonel Samuel N. Wood served in the Sixth Missouri Cavalry, which campaigned in southwest Missouri at this time. He is probably the same man as Captain S. N. Wood, mentioned in chapter four as commanding the Kansas Rangers. The Sixth Missouri Cavalry went on to serve in the Vicksburg Campaign and then in many smaller actions in Louisiana and Mississippi. United States War Department, *War of the Rebellion: A Compilation of the Official Records of the Union and Confederate Armies*, 128 vols. (1880–1901; repr., Harrisburg, PA: National Historical Society, 1971), ser. 1, 8:336; Frederick H. Dyer, comp., *A Compendium of the War of the Rebellion*, 2 vols. (1908; repr., Dayton, OH: Morningside Bookshop, 1978), 2:1306–7.
4. Besides processing grains, some gristmills in the region "carded wools, ginned cotton, and wove cloth." In the fall of 1862, some men in the Army of the Frontier operated gristmills to produce flour in northwest Arkansas. Michael A. Hughes, "Wartime Gristmill Destruction in Northwest Arkansas and Military Farm Colonies," in *Civil War Arkansas*, 32; M. Jane Johansson, *Albert C. Ellithorpe, The First Indian Home Guards and the Civil War on the Trans-Mississippi Frontier* (Baton Rouge: Louisiana State Univ. Press, 2016), 47.

5. Private James S. Hatch was "the youngest and tallest of the one hundred who first joined" Company E of the Thirty-Sixth Illinois Infantry. He was wounded in action at the Battles of Stones River and Chickamauga. After re-enlisting he went missing in action at the Battle of Kennesaw Mountain. L. G. Bennett and William M. Haigh, *History of the Thirty-Sixth Regiment Illinois Volunteers, during the War of the Rebellion* (1876; repr., Marengo, IL: Prairie State Press, Inc., 1999), 45, 395, 482, 559, 607, 609.
6. More than likely this is a reference to the skirmish that took place near Springfield, Missouri, on February 13, 1862, that resulted in the Federal army's capture of the town. *Official Records,* ser. 1, 8:59.
7. Men of the Missouri State Guard apparently used these to distinguish between regiments. At the Battle of Wilson's Creek, the men of the Fifth Missouri State Guard wore red badges on their left shoulders. William Garrett Piston and Richard W. Hatcher, III, *Wilson's Creek: The Second Battle of the Civil War and the Men Who Fought It* (Chapel Hill: The Univ. of North Carolina Press, 2000), 243.

12. From Springfield to Pea Ridge

1. William L. Shea and Earl J. Hess, *Pea Ridge: Civil War Campaign in the West* (Chapel Hill: The Univ. of North Carolina Press, 1992), 30.
2. Shea and Hess, *Pea Ridge*, 32–33.
3. Ibid.
4. Ibid., 34.
5. Ibid., 38.
6. Ibid., 41–43.
7. Ibid., 45.
8. Ibid., 48.
9. Ibid., 49.
10. Ibid., 52.
11. Ibid., 52–53.
12. Ibid., 11.
13. Ibid., 12.
14. Earl J. Hess, Richard W. Hatcher, III, William Garrett Piston, and William L. Shea, *Wilson's Creek, Pea Ridge & Prairie Grove: A Battlefield Guide with a Section on Wire Road* (Lincoln: Univ. of Nebraska Press, 2006), 230.
15. Shea and Hess, *Pea Ridge*, 12.
16. United States War Department, *War of the Rebellion: A Compilation of the Official Records of the Union and Confederate Armies*, 128 vols. (1880–1901; repr., Harrisburg, PA: National Historical Society, 1971), ser. 1, 8:192; Shea and Hess, *Pea Ridge*, 45.

17. A Private Jacob Horne served in Company C of Colonel John S. Phelps' Independent Missouri Regiment. *Soldiers and Sailors Database*, accessed March 15, 2021, https://www.nps.gov/civilwar/search-soldiers-detail.htm?soldierId=5F34E9A9-DC7A-DF11-BF36-B8AC6F5D926A.
18. A reference to a sinkhole on "Bloody Hill" where a number of Union soldiers were buried. Kip A. Lindberg and Thomas P. Sweeney, *"A Scene of Horrors": A Medical and Surgical History of the Battle of Wilson's Creek, August 10, 1861* (Cassville, MO: Litho Printers and Bindery, 2013), 90.
19. Corporal Daniel Whitney served in Company E of the Thirty-Sixth Illinois Infantry. L. G. Bennett and William M. Haigh, *History of the Thirty-Sixth Regiment Illinois Volunteers, during the War of the Rebellion* (1876; repr., Marengo, IL: Prairie State Press, Inc., 1999), 44.
20. The Dug Springs skirmish occurred on August 2, 1861, eight days before the Battle of Wilson's Creek. William Garrett Piston and Richard W. Hatcher, III, *Wilson's Creek: The Second Battle of the Civil War and the Men Who Fought It* (Chapel Hill: The Univ. of North Carolina Press, 2000), 139–40.
21. The skirmish at Crane Creek, Missouri, occurred on February 14, 1862. *Official Records*, ser. 1, 8:59.
22. Colonel Calvin A. Ellis commanded the First Missouri Cavalry. Shea and Hess, *Pea Ridge*, 333.
23. Bennett's assessment was incorrect. Brigadier General Samuel R. Curtis performed well throughout the Pea Ridge campaign. Brigadier General Franz Sigel had an excellent day on March 8, 1862, but otherwise his performance was "erratic." Ibid., 310–11.
24. Lieutenant Colonel Colley B. Holland was mustered into Colonel Phelps's regiment on October 17, 1861, and promoted from captain to lieutenant colonel on December 19, 1861. Holland Compiled Service Record, Phelps' Independent Missouri Regiment, Fold3.com (website), accessed March 15, 2021.
25. Second Lieutenant David K. Moore, age twenty-five, mustered in on October 3, 1861, in Rolla in a unit that became part of Phelps' Missouri Regiment. On detached service in Cassville, Missouri, in January and February 1862, he died of disease on March 23, 1862. Moore Compiled Service Record, Phelps' Independent Missouri Regiment, Fold3.com (website), accessed March 15, 2021.
26. Bennett's account was mostly accurate. According to reports by Brigadier General Samuel R. Curtis and Colonel Clark Wright of the Sixth Missouri Cavalry, about five hundred "Texas Rangers" surprised Captain Montgomery's men at Keetsville on February 25, 1862, destroying five wagons, seizing sixty to seventy horses, and killing two men. *Official Records*, ser. 1, 8:74–75.

27. Mathew Ritchey, a former state legislator and county judge, founded Newtonia in 1854. A Unionist and slaveholder, he and his business partner, Samuel P. Cloud, owned a gristmill. Newtonia became the site of two battles: one in September 1862 and the other in October 1864. The Ritchey Mansion still stands today. Larry Wood, *The Two Civil War Battles of Newtonia* (Charleston, SC: The History Press, 2010), 13–15, 148.
28. The stone barn and the stone fence surrounding it became a Confederate defensive point during the first Battle of Newtonia on September 30, 1862. Wood, *Two Civil War Battles*, 63–67.
29. The Sixth Missouri Cavalry, commanded by Colonel Clark Wright, was in Cassville on February 27, 1862. *Official Records*, ser. 1, 8:74.
30. Brigadier General Jefferson C. Davis, a Mexican War veteran, commanded a division made up of units from his home state of Indiana as well as Illinois. Davis achieved notoriety on September 29, 1862, when he murdered Major General William Nelson at the Galt House in Louisville, Kentucky. Ezra J. Warner, *Generals in Blue: Lives of the Union Commanders* (1964; repr., Baton Rouge: Louisiana State Univ. Press, 1992), 115–16.
31. Private Clark W. Edwards of Company D of the Thirty-Sixth Illinois Infantry had been on the expedition to Newtonia to retrieve the flour and salt. Bennett and Haigh, *Thirty-Sixth Regiment Illinois*, 43, 123.
32. Brigadier General Samuel R. Curtis reported a loss of thirteen killed and fifteen to twenty wounded in the action at Little Sugar Creek on February 17, 1862. *Official Records*, ser. 1, 8:61.
33. Major William D. Bowen was wounded in the wrist at the action of Little Sugar Creek on February 17, 1862. Ibid.
34. According to a report by Brigadier General Samuel R. Curtis, Captain Theodore Switzler was "not dangerously" wounded at the Little Sugar Creek skirmish on February 17, 1862. Ibid.
35. Organized in St. Louis in 1861, the Benton Hussars Cavalry Battalion were assigned to the Fifth Missouri Cavalry in February 1862. Frederick H. Dyer, comp., *A Compendium of the War of the Rebellion*, 2 vols. (1908; repr., Dayton, OH: Morningside Bookshop, 1978), 2:1312.
36. A soldier in the Fifth Missouri Cavalry was murdered after returning to Bentonville to get whiskey. His body was found in an outhouse, and the "search party . . . burned much of Bentonville in retribution." Shea and Hess, *Pea Ridge*, 46.

13. The Pea Ridge Campaign

1. William L. Shea and Earl J. Hess, *Pea Ridge: Civil War Campaign in the West* (Chapel Hill: The Univ. of North Carolina Press, 1992), 331.

2. Shea and Hess, *Pea Ridge*, 60, 68.
3. Ibid., 69.
4. Ibid., 74–75.
5. Ibid., 75–76.
6. Ibid., 93.
7. Ibid., 104.
8. Ibid., 110, 115.
9. Ibid., 114–15.
10. Ibid., 125, 141.
11. Ibid., 218.
12. Ibid., 228.
13. Ibid., 225.
14. Ibid., 231.
15. Ibid., 229.
16. Ibid., 236.
17. Ibid., 236–38, 244.
18. Ibid., 243.
19. Ibid., 244.
20. Ibid., 252–53.
21. Colonel Jefferson C. Davis's division consisted of two brigades made up of units from Indiana and Illinois. Shea and Hess, *Pea Ridge*, 332–33.
22. The Twelfth Missouri Infantry also served in Colonel Nicholas Greusel's brigade. Their casualties during the Battle of Pea Ridge were three killed, twenty-nine wounded, and two missing. Ibid., 331.
23. Colonel Greusel reported that thirty-one men were captured from the Thirty-Sixth Illinois Infantry. United States War Department, *War of the Rebellion: A Compilation of the Official Records of the Union and Confederate Armies*, 128 vols. (1880–1901; repr., Harrisburg, PA: National Historical Society, 1971), ser. 1, 8:227.
24. Colonel William Vandever commanded the Second Brigade of Colonel Eugene A. Carr's Fourth Division. Shea and Hess, *Pea Ridge*, 333.
25. On March 4, Major Joseph Conrad was given detachments from several regiments, including Company F from the Thirty-Sixth Illinois Infantry, and ordered to Maysville along the Arkansas-Indian Territory border. There, they were to block Missourians from joining Sterling Price's Missouri State Guard. The next day, Conrad was ordered to return but discovered that they were cut off from the army. After an exhausting 125-mile march in four days, Conrad's detachment returned on March 9. Ibid., 60, 68, 78, 276.
26. Bennett served in Company E.
27. A reference to Leetown, "where nearly every building in the village served as a hospital." Shea and Hess, *Pea Ridge*, 209.

28. Some Union troops fled from the First Bull Run battlefield on July 21, 1861. Patricia L. Faust, *Historical Times Illustrated Encyclopedia of the Civil War* (New York: Harper & Row, 1986), 92.
29. Private Ira O. Fuller, a native of New Hampshire and a twenty-four-year-old single farmer, enlisted on August 1, 1861. A resident of Bristol in Kendall County, he was five feet, five inches tall with brown hair and gray eyes. *Illinois, U. S., Databases of Illinois Veterans Index, 1775–1995* (Provo, UT: Ancestry.com Operations, Inc., 2015).
30. A native of Ireland, John Ray, a single farmer and resident of Little Rock in Kendall County, enlisted on August 10, 1861. The nineteen-year-old was five feet, seven and a half inches tall with dark hair and black eyes. *Illinois, U. S., Databases of Illinois Veterans Index, 1775–1995* (Provo, UT: Ancestry.com Operations, Inc., 2015).
31. The Thirty-Sixth Illinois Infantry suffered four killed, thirty-seven wounded, and thirty-four missing or captured during the Pea Ridge campaign. Shea and Hess, *Pea Ridge*, 331.
32. There were no Mississippi units at the Battle of Pea Ridge.
33. Lieutenant Colonel Edward S. Joslyn and Major Alonzo H. Barry resigned their commissions in the summer of 1862. L. G. Bennett and William M. Haigh, *History of the Thirty-Sixth Regiment Illinois Volunteers, during the War of the Rebellion* (1876; repr., Marengo, IL: Prairie State Press, Inc., 1999), 221.
34. Lieutenant Arnold Hoeppner, a native of Prussia, was Brigadier General Samuel R. Curtis's "only engineer officer" at the time. In the 1870 census, Hoeppner was listed as an assistant city engineer in St. Louis. *Official Records*, ser. 1, 8:192; *Ninth Census of the United States*, 1870, St. Louis, Missouri, 88.
35. Thirteenth Illinois Infantry.
36. Colonel Cyrus Bussey commanded the Third Iowa Cavalry, a regiment that was overrun early in the fighting on March 7, 1862. Shea and Hess, *Pea Ridge*, 334.
37. Colonel Grenville M. Dodge commanded a brigade in Colonel Eugene A. Carr's division and was wounded in the fighting near Elkhorn Tavern. Ibid., 333.
38. Lieutenant Colonel Francis J. Herron commanded the Ninth Iowa Infantry at the Battle of Pea Ridge. The regiment suffered heavy losses there, and Herron was wounded and captured. For his bravery at Pea Ridge, Herron was promoted to brigadier general and in the 1890s was awarded the Medal of Honor. In the spring of 1863, Herron earned a promotion to major general at the young age of twenty-six. Lieutenant Colonel William P. Chandler of the Thirty-Fifth Illinois Infantry was captured in the fighting near Elkhorn Tavern. The men were exchanged for Colonel Louis Hébert and Major William

F. Tunnard, both of the Third Louisiana Infantry. Ezra J. Warner, *Generals in Blue: Lives of the Union Commanders* (1964; repr., Baton Rouge: Louisiana State Univ. Press, 1992), 228–29; Shea and Hess, *Pea Ridge*, 285, 333.

39. Possibly a reference to Missouri State Guard division commander Colonel John B. Clark, Jr., who was not killed at Pea Ridge. Shea and Hess, *Pea Ridge*, 338.
40. Colonel William Y. Slack commanded the Second Missouri Brigade of the State Guard and was killed near Elkhorn Tavern on March 7. Ibid., 175, 337.
41. A reference to the Thirteenth Illinois rather than the Thirteenth Iowa.
42. The Second Independent Battery, Ohio Light Artillery served in Brigadier General Alexander S. Asboth's division and lost one man killed and two wounded during the Battle of Pea Ridge. Shea and Hess, *Pea Ridge*, 332.
43. Captain W. H. Stark was an acting aide on Brigadier General Samuel R. Curtis's staff. *Official Records*, ser. 1, 8:192.
44. Private George B. Raymond served in Company D of the Thirty-Sixth Illinois Infantry and at this time was an "acting Orderly" for Curtis. Earlier in the campaign, he traveled with a guide from Cross Hollow and delivered an order to Brigadier General Franz Sigel near Bentonville. Along the way, Confederate pickets shot and killed the guide. Later Raymond was promoted to captain in the First Arkansas Infantry. Bennett and Haigh, *Thirty-Sixth Regiment Illinois*, 43, 129, 222.
45. Corporal Andrew L. Scofield was in Company D of the Thirty-Sixth Illinois Infantry, and Private Charles H. Scofield served in Company E. Both companies were from Kendall County and neither man enlisted in Newark. Bennett and Haigh, *Thirty-Sixth Regiment Illinois*, 43–44.

14. Pea Ridge to the Mississippi River

1. Quoted in Robert G. Schultz, *The March to the River: From the Battle of Pea Ridge to Helena, Spring 1862* (Iowa City: Camp Pope Publishing, 2014), 151.
2. William L. Shea and Earl J. Hess, *Pea Ridge: Civil War Campaign in the West* (Chapel Hill: The Univ. of North Carolina Press, 1992), 287.
3. Shea and Hess, *Pea Ridge*, 289.
4. Ibid., 290.
5. Ibid., 291.
6. Ibid., 292.
7. Ibid., 295.
8. Ibid.
9. Ibid., 297–98.
10. Ibid., 300.
11. Ibid., 301.

12. Ibid., 303.
13. Ibid., 301.
14. Ibid., 304.
15. Ibid., 305.
16. *Chicago Tribune*, September 1, 1862.
17. A native of Tennessee, Major General Thomas Carmichael Hindman was a Mexican American War veteran. After becoming an attorney, he served in the Mississippi State legislature then moved to Arkansas where voters elected him twice to the US House of Representatives. He strongly advocated for secession, became colonel of the Second Arkansas Infantry, and was injured at the Battle of Shiloh while fighting as a brigade commander. Hindman arrived in Little Rock, Arkansas, in May 1862 to take command of the Trans-Mississippi District. An administrative genius, Hindman rebuilt the army and its supply system and instituted a host of other reforms. Ezra J. Warner, *Generals in Gray: Lives of the Confederate Commanders* (1959; repr., Baton Rouge: Louisiana State Univ. Press, 1987), 137–38; O. Edward Cunningham, *Shiloh and the Western Campaign of 1862*, ed. Gary D. Joiner and Timothy B. Smith (New York: Savas Beatie, 2009), 243, 402; William L. Shea, *Fields of Blood: The Prairie Grove Campaign* (Chapel Hill: The Univ. of North Carolina Press, 2009), 4–6.
18. Organized in 1861, the Fifth Illinois Cavalry campaigned primarily in Arkansas, Louisiana, and Mississippi during the conflict, losing twenty-eight men killed and mortally wounded and 419 to disease. Frederick H. Dyer, comp., *A Compendium of the War of the Rebellion*, 2 vols. (1908; repr., Dayton, OH: Morningside Bookshop, 1978), 2:1024.
19. Brigadier General Alvin Peterson Hovey, a native of Indiana, and a justice on the Indiana Supreme Court besides other activities, became colonel of the Twenty-Fourth Indiana Infantry at the beginning of the war. He fought at Shiloh and was promoted to brigadier general because of gallantry there. He commanded a division in Curtis's army and later served in the Vicksburg and Atlanta Campaigns. Ezra J. Warner, *Generals in Blue: Lives of the Union Commanders* (1964; repr., Baton Rouge: Louisiana State Univ. Press, 1992), 235–36.
20. Colonel Graham N. Fitch, commander of the Forty-Sixth Indiana Infantry, was involved in a cooperative effort with the navy to resupply Curtis's army in June 1862. Bennett probably referenced an incident during that time. Schultz, *The March to the River*, 233–35.
21. First Sergeant John J. Winchel enlisted on August 31, 1861, in Marion County in Company B of the Eleventh Indiana Infantry. He was killed at Clarendon, Arkansas, on August 13, 1862. *Report of the Adjutant General of*

the State of Indiana (Indianapolis: Samuel M. Douglass, State Printer, 1866), 4:176.

22. Private John A. Rader enlisted on December 8, 1861, in Company H of the Fifth Illinois Cavalry at Appleton, Missouri. He died of wounds received at Clarendon, Arkansas, on August 12, 1862. Illinois GenWeb (website), accessed March 23, 2021, https://civilwar.illinoisgenweb.org/acm/cavoo5-h.html.
23. Private Robert T. Larne enlisted on August 31, 1861, in Cumberland County in Company A of the Fifth Illinois Cavalry. Guerrillas shot him at Little Cypress, Arkansas, on August 15, 1862. Illinois GenWeb (website), accessed March 23, 2021, https://civilwar.illinoisgenweb.org/acm/cavoo5-a.html.
24. Much of this section is based on Brigadier General Alvin. P. Hovey's reports that may be found in *Official Records*, ser. 1, 13:206–7.

15. Recruiting in Dixie, Part One

1. Bennett Compiled Service Record, Thirty-Sixth Illinois Infantry. Copy accessed from the Lyman G. Bennett Papers, R0274, State Historical Society of Missouri, Rolla Center.
2. Mark K. Christ, *Civil War Arkansas, 1863: The Battle for a State* (Norman: Univ. of Oklahoma Press, 2010), 248–49.
3. Bennett Compiled Service Record, Fourth Arkansas Cavalry, Fold3.com (website), accessed March 15, 2021.
4. James J. Johnston, *Mountain Feds: Arkansas Unionists and the Peace Society* (Little Rock: Butler Center Books, 2018), 64.
5. Johnston, *Mountain Feds*, 11.
6. Ibid., 19.
7. Ibid., 10–11.
8. Ibid., 11, 59.
9. Ibid., 11.
10. Ibid., 84, 87–88.
11. Ibid., 13.
12. Ibid., 102.
13. Brooks Blevins, *A History of the Ozarks*, vol. 2, *The Conflicted Ozarks* (Urbana: Univ. of Illinois Press, 2019), 2:63.
14. Johnston, *Mountain Feds*, 91.
15. Ibid., 93.
16. Ibid., 12.
17. Ibid., 155.
18. Ibid., 14–15.
19. Brigadier General John Wynn Davidson, a native of Virginia, commanded the District of St. Louis at this time and commanded the Army of Southeast

Missouri. Ezra J. Warner, *Generals in Blue: Lives of the Union Commanders* (1964; repr., Baton Rouge: Louisiana State Univ. Press, 1992), 112.
20. Bennett's description of events is similar to the one in A. W. Bishop, *Loyalty on the Frontier or Sketches of Union Men of the South-West with Incidents and Adventures in Rebellion*, ed. Kim Allen Scott Bishop (Fayetteville: Univ. of Arkansas Press, 2003), 21–23.
21. Dr. James M. Johnson left Arkansas in April 1862 with Isaac Murphy. Johnson became colonel of the First Arkansas Infantry. Johnston, *Mountain Feds*, 54.
22. William Meade Fishback, born in Culpeper County, Virginia, in 1831, graduated from the Univ. of Virginia. Following graduation he taught school and read law, but in 1857 he moved to Springfield, Illinois, where he opened a law practice and became acquainted with Abraham Lincoln. A year later he opened a law practice with Judge Solomon F. Clark in Sebastian County, Arkansas. Fishback was a delegate at the secession convention and voted against secession. He then changed his vote to approval, unlike Isaac Murphy, who remained the only delegate to stick to his no vote. Fishback traveled to Missouri and took the oath of allegiance upon the outbreak of war and made his way to St. Louis, where he edited the *St. Louis Democrat*. He moved to Little Rock when that city was captured in 1863 and began publishing the *Unconditional Unionist* newspaper. In that year he became a colonel with the expectation that he would recruit for the Third Arkansas Infantry. Instead, he helped recruit hundreds of men for the Fourth Arkansas Cavalry. In 1894, Fishback, now a Democrat, was elected governor of Arkansas. Harry W. Readnour, "William Meade Fishback (1831–1903)," *Encyclopedia of Arkansas* (website), accessed March 24, 2021, https://encyclopediaofarkansas.net/entries/william-meade-fishback-103/.
23. Established in 1842, Fort Scott became a significant supply complex and "staging area for Union troops operating into Missouri, Arkansas, and Indian Territory" from Kansas during the Civil War. Leo E. Oliva, *Fort Scott: Courage and Conflict on the Border* (Topeka: Kansas State Historical Society, 1984), 1, 67.
24. Fought on December 7, 1862, the Battle of Prairie Grove "was a resounding strategic victory" for the Union army. William L. Shea, *Fields of Blood: The Prairie Grove Campaign* (Chapel Hill: The Univ. of North Carolina Press, 2009), 265.
25. Possibly a reference to George W. Worthington, the father of two men (James P. Worthington and Wayne O. Worthington) who served in the First Arkansas Infantry. James J. Johnston, ed., "Recruiting in Dixie, Part III," *White River Valley Historical Quarterly* 39 (Spring 2000): 8, endnote iv.

26. Organized in Fayetteville in the spring of 1863, the First Arkansas Infantry campaigned in Arkansas and the Indian Territory. Frederick H. Dyer, comp., *A Compendium of the War of the Rebellion*, 2 vols. (1908; repr., Dayton, OH: Morningside Bookshop, 1978), 2:999.
27. Brigadier General John Sappington Marmaduke led a mounted raid into southern Missouri in late 1862. They unsuccessfully attacked Springfield on January 8, 1863. Louis S. Gerteis, *The Civil War in Missouri: A Military History* (Columbia: Univ. of Missouri Press, 2012), 149–52.
28. Fought on April 18, 1863, Brigadier General William L. Cabell and his Confederate troops attacked the town but retreated after several hours of combat. Russell L. Mahan, *Fayetteville, Arkansas in the Civil War, 1860–1865* (Bountiful, UT: Historical Byways, 2003), 77–78, 83.
29. Richard H. Wimpey enlisted on June 15, 1862, and mustered in as captain of Company D of the First Arkansas Cavalry. He was promoted to major of the regiment on May 1, 1865. Historical Data Systems, Inc., American Civil War Research Database, accessed on March 24, 2021, http://www.civilwardata.com/active/hdsquery.dll?SoldierHistory?U&2391609.
30. The composition of the force at Fayetteville as stated by Bennett is accurate. Mahan, *Fayetteville, Arkansas in the Civil War*, 72.
31. William Lewis Cabell, a native of Virginia, graduated from the United States Military Academy in 1850. Initially serving in Virginia, he transferred to the trans-Mississippi in 1862. Authorities promoted him to brigadier general in January 1863. Cabell commanded a mixed force of about nine hundred men from Arkansas, Missouri, and Texas at the Battle of Fayetteville. Ezra J. Warner, *Generals in Gray: Lives of the Confederate Commanders* (1959; repr., Baton Rouge: Louisiana State Univ. Press, 1987), 41; Mahan, *Fayetteville, Arkansas in the Civil War*, 72.
32. Elhanen J. Searle enlisted at age twenty-four on November 1, 1861. An attorney and a resident of Springfield, Illinois, he served initially as a first lieutenant in Company M of the Tenth Illinois Cavalry. Promoted to captain on July 7, 1862, he became lieutenant colonel of the First Arkansas Infantry on July 7, 1863. Historical Data Systems, Inc., American Civil War Research Database, accessed on March 24, 2021, http://www.civilwardata.com/active/hdsquery.dll?SoldierHistory?U&1260698.
33. Elijah D. Ham enlisted on March 10, 1863, and was commissioned as major of the First Arkansas Infantry. He resigned his commission on March 22, 1864. Historical Data Systems, Inc., American Civil War Research Database, accessed on March 24, 2021, http://www.civilwardata.com/active/hdsquery.dll?SoldierHistory?U&2385954.
34. Colonel James C. Monroe commanded the First Arkansas Cavalry (Confederate). Mahan, *Fayetteville, Arkansas in the Civil War*, 72.

35. Joseph S. Robb enlisted on August 24, 1862, at Springfield, Missouri, and became a First Lieutenant in Company L of the First Arkansas Cavalry. Later promoted to captain, he died of wounds on January 20, 1864. Historical Data Systems, Inc., American Civil War Research Database, accessed on March 24, 2021, http://www.civilwardata.com/active/hdsquery.dll?SoldierHistory?U&2389530.
36. Colonel Marcus LaRue Harrison reported casualties of five killed and approximately seventeen wounded and sixteen captured. United States War Department, *War of the Rebellion: A Compilation of the Official Records of the Union and Confederate Armies*, 128 vols. (1880–1901; repr., Harrisburg, PA: National Historical Society, 1971), ser. 1, 22, pt. 1:306, 308.
37. Cabell reported that his loss did "not exceed 20 killed, 30 wounded, and 20 missing." *Official Records*, ser. 1, 22, pt. 1:311.
38. On April 23, 1863, Harrison reported to Curtis that he had no artillery, a shortage of horses, and no more supplies. Curtis ordered the evacuation of Fayetteville the next day. *Official Records*, ser. 1, 22, pt. 2:246; Mahan, *Fayetteville, Arkansas in the Civil War*, 86.
39. Bennett had previously visited the Wilson's Creek battlefield on February 23, 1862.
40. Thousands of soldiers passed by the Wilson's Creek battlefield while traveling on Telegraph Road. Many soldiers wrote their name on a piece of limestone and added it to the rock cairn that supposedly marked the location of Brigadier General Nathaniel Lyon's death. Kip A. Lindberg and Thomas P. Sweeney, *"A Scene of Horrors": A Medical and Surgical History of the Battle of Wilson's Creek, August 10, 1861* (Cassville, MO: Litho Printers and Bindery, 2013), 94.
41. The First Arkansas Light Artillery was organized in Fayetteville and Springfield over an eight-month period in 1863. They served in Arkansas and the Indian Territory during the conflict. Dyer, *Compendium*, 2:999.
42. Denton D. Stark enlisted at age twenty-one on September 12, 1861, as a first lieutenant in Company H of the Thirty-Seventh Illinois Infantry. He transferred to the First Arkansas Cavalry and then became captain of the First Arkansas Light Artillery on April 1, 1863. He resigned his commission on November 17, 1864. Historical Data Systems, Inc., American Civil War Research Database, accessed on March 24, 2021, http://www.civilwardata.com/active/hdsquery.dll?SoldierHistory?U&2390372.
43. A reference to either the Eighth Missouri Cavalry or the Eighth Regiment State Militia Cavalry. Dyer, *Compendium*, 2:1308–9.
44. Possibly Shade Waldron, a native of Tennessee, listed in 1880 as a fifty-one-year-old farmer, married, and with four children in the home. *Tenth Census of the United States*, 1880, Madison County, Ark., 26.

45. Possibly a reference to Colonel DeWitt Clinton Hunter, a former colonel of the Eleventh Missouri Infantry. After resigning his commission, he recruited a cavalry regiment. Arkansas-Louisiana Seventh-Day Adventist History, accessed on February 19, 2022, https://arklasdahistory.org/dewitt-clinton-hunter/.
46. Walter Brashear was a thirty-year-old farmer in Pope County in 1870. A native of Arkansas, he was married and had four daughters. *Ninth Census of the United States*, 1870, Pope County, Ark., 4. According to James J. Johnston, an expert on Arkansas Unionists, Bennett misidentified Brashear. Walter's father, Mortimer M. Brashear, encouraged his son to join the Confederate army and then desert. Mortimer recruited ninety-six men for Company E of the Second Arkansas Cavalry. Johnston, *Mountain Feds,* 175, 275n275.

16. Recruiting in Dixie, Part Two

1. United States War Department, *War of the Rebellion: A Compilation of the Official Records of the Union and Confederate Armies*, 128 vols. (1880–1901; repr., Harrisburg, PA: National Historical Society, 1971), ser. 1, 22, pt. 2:761.
2. Bennett Compiled Service Record, Fourth Arkansas Cavalry, Fold3.com (website), accessed March 15, 2021.
3. Frederick H. Dyer, comp., *A Compendium of the War of the Rebellion*, 2 vols. (1908; repr., Dayton, OH: Morningside Bookshop, 1978), 2:999; *Official Records*, ser. 1, 34, pt. 1:858–59.
4. *Official Records*, ser. 1, 34, pt. 3:481, 689; David Sesser, "Fourth Arkansas Cavalry," *Encyclopedia of Arkansas* (website), accessed March 24, 2021, https://encyclopediaofarkansas.net/entries/fourth-arkansas-cavalry-12096/.
5. Pension records on Lyman G. Bennett, RG T-288, Invalid Record Application 1200162, National Archives.
6. *Official Records*, ser. 1, 34, pt. 2:631; 34, pt. 4:323.
7. Lieutenant Gideon M. Waugh resided in Olathe, Kansas, when he was commissioned as a second lieutenant in Company G of the Second Kansas Cavalry on January 7, 1862. In January 1864 he became lieutenant colonel of the Second Arkansas Infantry but resigned his commission in June 1865. Historical Data Systems, Inc., American Civil War Research Database, accessed on March 25, 2021, http://www.civilwardata.com/active/hdsquery.dll?SoldierHistory?U&1773558.
8. The Second Kansas Cavalry campaigned extensively in Arkansas, the Indian Territory, and Missouri, losing sixty-four men killed or mortally wounded. Dyer, *Compendium,* 2:1181–82.
9. Possibly a reference to John C. Priddy, who enlisted at age thirty-seven in Company I of the First Arkansas Infantry. Historical Data Systems, Inc.,

American Civil War Research Database, accessed on March 29, 2021, http://www.civilwardata.com/active/hdsquery.dll?SoldierHistory?U&2389198.
10. A reference to Walter's father, Mortimer M. Brashear, who recruited actively. James J. Johnston, *Mountain Feds: Arkansas Unionists and the Peace Society* (Little Rock: Butler Center Books, 2018), 175.
11. Probably a reference to Charles Bird, a former county judge and state legislator from Stone County, Missouri. He "raised a company to protect the area from Jayhawkers and bushwhackers. He was later joined by Isaac Bledsoe, who gave his name to the guerrilla band." James J. Johnston, ed., "Recruiting in Dixie, Part III," *White River Valley Historical Quarterly* 39 (Spring 2000): 9, endnote xviii.
12. Colonel Joseph Orville Shelby, born in 1830 in Kentucky, migrated to Missouri, where he became quite wealthy. A fine cavalry leader, Shelby experienced combat many times in Arkansas and Missouri. The Confederate Congress promoted him to brigadier general in December 1863. Ezra J. Warner, *Generals in Gray: Lives of the Confederate Commanders* (1959; repr., Baton Rouge: Louisiana State Univ. Press, 1987), 273–74.
13. Private Henry C. Kelly enlisted on November 11, 1861, in Company I of the Second Kansas Cavalry. He was commissioned as first lieutenant in Company E of the Second Arkansas Cavalry in December 1863 and died on March 25, 1864, of wounds received at Bellefonte, Arkansas. Historical Data Systems, Inc., American Civil War Research Database, accessed on March 25, 2021, http://www.civilwardata.com/active/hdsquery.dll?SoldierHistory?U&1774339.
14. James R. Vanderpool was thirty years old when he enlisted on February 27, 1863, in Fayetteville. He commanded Company C of the First Arkansas Infantry. Historical Data Systems, Inc., American Civil War Research Database, accessed on March 25, 2021, http://www.civilwardata.com/active/hdsquery.dll?SoldierHistory?U&2391014.
15. James Harrison Love farmed in Searcy County, Arkansas, before the war. He organized an independent cavalry company that spent much of the war campaigning in that county. Daniel E. Sutherland, "Guerrillas: The Real War in Arkansas," in *Civil War Arkansas: Beyond Battles and Leaders,* ed. Anne J. Bailey and Daniel E. Sutherland (Fayetteville: Univ. of Arkansas Press, 2000), 136–37.
16. Possibly a reference to Private Jonas B. Spivey, who was forty-five years old when he enlisted in Fayetteville on February 18, 1863. He served in Company I of the First Arkansas Infantry. Historical Data Systems, Inc., American Civil War Research Database, accessed on March 25, 2021, http://www.civilwardata.com/active/hdsquery.dll?SoldierHistory?U&2390319.
17. Possibly a reference to Private David King. The twenty-two-year-old enlisted

in Company C of the First Arkansas Infantry in Fayetteville on February 12, 1863. King deserted the regiment on October 20, 1863, in Fort Smith, Arkansas. Historical Data Systems, Inc., American Civil War Research Database, accessed on March 25, 2021, http://www.civilwardata.com/active/hdsquery.dll?SoldierHistory?U&2387142.

18. Captain William J. Heffington from Yell County, Arkansas, commanded Company I of the First Arkansas Infantry. Killed on August 15, 1863, in western Arkansas, A. W. Bishop described him as "'Wild Bill,' a cool, daring, intelligent woodsman, who, unwillingly in the rebel service, had remained there long enough to become disgusted with it . . . had rallied the bold spirits of the neighborhood, appeared with a band of followers, loyal all to the Stripes and Stars." A. W. Bishop, *Loyalty on the Frontier or Sketches of Union Men of the South-West with Incidents and Adventures in Rebellion*, ed. Kim Allen Scott Bishop, (Fayetteville: Univ. of Arkansas Press, 2003), 164.
19. The Second Arkansas Infantry served exclusively in Arkansas and mustered out in August 1865. Dyer, *Compendium*, 2:999–1000.
20. A reference to P. T. Barnum, known primarily for the Barnum & Bailey Circus.
21. Silas Casey and William Joseph Hardee graduated from the United States Military Academy, and both authored tactical textbooks. Casey became a Union general, and Hardee became a Confederate general. Ezra J. Warner, *Generals in Blue: Lives of the Union Commanders* (1964; repr., Baton Rouge: Louisiana State Univ. Press, 1992), 74–75; Warner, *Generals in Gray*, 124–25.
22. Jack Chrisman was a "guerrilla chieftain" in Arkansas. Daniel E. Sutherland, *A Savage Conflict: The Decisive Role of Guerrillas in the American Civil War* (Chapel Hill: The Univ. of North Carolina Press, 2009), 67.
23. Samuel Farmer lived in Johnson County, Mulberry Township according to the 1850 census. He was a farmer. *Seventh Census of the United States*, 1850, Johnson County, Ark., 247.
24. Elizabeth Farmer was listed as thirty-seven years old in the 1850 census. In that enumeration, she had nine children: eight boys (including one set of twins) and a seven-year-old daughter named Isabella. *Seventh Census of the United States*, 1850, Johnson County, Ark., 247.
25. Samuel Farmer enlisted in Fayetteville on October 17, 1861, as a first lieutenant in the Sixteenth Arkansas Infantry and received a promotion to major in December 1861. Farmer resigned his commission in May 1862 after his regiment fought at the Battle of Pea Ridge. His resignation letter, accompanied by a surgeon's statement, testified that he was disabled due to illness. Farmer Compiled Service Record, Fourth Arkansas Cavalry, Fold3.com (website), accessed March 25, 2021.

26. The Mississippi River strongholds of Vicksburg and Port Hudson were surrendered respectively on July 4 and July 9, 1863. Patricia L. Faust, *Historical Times Illustrated Encyclopedia of the Civil War* (New York: Harper & Row, 1986), 597, 784.

27. Major General William S. Rosecrans and his Army of the Cumberland advanced toward Chattanooga, Tennessee, in June and July 1863 during the Tullahoma campaign. Faust, *Historical Times Illustrated Encyclopedia*, 764–65.

28. Major General Sterling Price commanded a division during the Battle of Helena. Edwin Cole Bearss, *The Campaign for Vicksburg* (Dayton, OH: Morningside House, Inc., 1991), 3:1243.

29. Major General Thomas C. Hindman, former commander of the Trans-Mississippi Department, had been relieved by Holmes and was not present at the Battle of Helena. Warner, *Generals in Gray*, 138.

30. Lieutenant General Theophilus H. Holmes led an approximately 7,600-man force in an attack on Helena, Arkansas on July 4, 1863, that resulted in about 1,638 Confederate casualties. Many of these soldiers were killed or wounded before the guns of Fort Curtis and from fire by the gunboat, *Tyler*. Bearss, *The Campaign for Vicksburg*, 3:1231–32, 1242–43.

31. "Gaily The Troubadour" was written by English composer Thomas Haynes Bayly in the 1820s. Contemplator.com, accessed on March 29, 2021, https://www.contemplator.com/england/gaily.html.

32. The "personification of darkness in Greek mythology." Merriam-Webster online, accessed on March 29, 2021, https://www.merriam-webster.com/dictionary/Erebus.

33. John F. Fesperman enlisted on March 10, 1863, and became first lieutenant of Company F of the First Arkansas Infantry. He was killed at Jasper, Arkansas, on August 1, 1863. Historical Data Systems, Inc., American Civil War Research Database, accessed on March 29, 2021, http://www.civilwardata.com/active/hdsquery.dll?SoldierHistory?U&2385295.

34. John Cecil, a Tennessee native, served initially as a private in Company D of Captain John M. Harrell's battalion but later was the "guerrilla captain of Co. E, Harrell's Battalion." James J. Johnston, ed., "Recruiting in Dixie, Part III," *White River Valley Historical Quarterly* 39 (Spring 2000): 10, endnote xxxvii.

35. Probably a reference to Lord Bryon's poem, "Mazeppa," written in 1819. Mazeppa was an officer in King Charles XII's Swedish army. Captured, Mazeppa was tied to a wild horse and rescued after a dangerous ride. Oxford Reference, accessed March 29, 2021, https://www.oxfordreference.com/view/10.1093/oi/authority.20110803100142801?rskey=6wkiSX&result=6.

36. Possibly a reference to James Adare, a forty-three-year-old farmer and native of Kentucky who resided in Newton County, Arkansas. He and his wife, Margaret, had several children living with them in 1860. *Eighth Census of the United States*, 1860, Newton County, Ark., 12.
37. Possibly this is a reference to Colonel M. LaRue Harrison.
38. There is no documentation in the *Official Records* of this skirmish at Jasper.
39. Pleasant Houston Spears served briefly in the Confederate army as a second lieutenant then obtained authorization to recruit for Company D of the Second Arkansas Cavalry. Bennett reported Spears's previous service in the Confederate army to Major General John M. Schofield, who then refused to authorize a commission for Spears. James J. Johnston, ed., "Recruiting in Dixie, Part III," White River Valley Historical Quarterly 39 (Spring 2000): 11, endnote xli.
40. Organized in 1862, the Second Arkansas Cavalry campaigned in Arkansas, Kansas, and Missouri during the conflict. Dyer, *Compendium*, 2:998.
41. According to a report submitted by First Lieutenant Edgar A. Barker of the Second Kansas Cavalry, the attack on Fayetteville started about five minutes after Vanderpool's men arrived, and they were easily routed. Bennett's commentary was considerably more detailed than the one report contained in the *Official Records*. *Official Records*, ser. 1, 22, pt. 1:595.
42. John T. Worthington enlisted on August 7, 1862, at age thirty-nine and was commissioned captain of Company H of the First Arkansas Cavalry. He became involved in a scandal in February 1863 when he was exposed as a bigamist. Promoted to major in February 1865, he was "killed at the head of his men while leading a charge against a column of bushwhackers" at Kingston, Arkansas, on March 11, 1865. Historical Data Systems, Inc., American Civil War Research Database, accessed on March 29, 2021, http://www.civilwardata.com/active/hdsquery.dll?SoldierHistory?U&2391728 Compiled Service Record, First Arkansas Cavalry, Fold3.com (website), accessed March 29, 2021.
43. Major General James G. Blunt, a physician and a native of Maine, saw much campaigning in Arkansas and the Indian Territory in 1862 and 1863. The First Arkansas Infantry joined his army at Fort Gibson in August 1863 and aided in the capture of Fort Smith. Warner, *Generals in Blue*, 37–38; Dyer, *Compendium*, 2:999.
44. No other report of this skirmish has been located.
45. Joseph G. Peevy commanded Company B of Hunter's Missouri Infantry. James J. Johnston, ed., "Recruiting in Dixie, Part III," *White River Valley Historical Quarterly* 39 (Spring 2000): 11, endnote xlvi; James E. McGhee, *Guide to Missouri Confederate Units, 1861–1865* (Fayetteville: Univ. of Arkansas Press, 2008), 225.

17. Mapping in Kansas and Missouri

1. Patricia L. Faust, *Historical Times Illustrated Encyclopedia of the Civil War* (New York: Harper & Row, 1986), 602–3.
2. United States War Department, *War of the Rebellion: A Compilation of the Official Records of the Union and Confederate Armies*, 128 vols. (1880–1901; repr., Harrisburg, PA: National Historical Society, 1971), ser. 1, 41, pt. 1:464–523.
3. Calvin D. Cowles, comp., *The Official Military Atlas of the Civil War* (1891–1895; repr., New York: Arno Press, 1978), plate LXVI.
4. Kyle S. Sinisi, *The Last Hurrah: Sterling Price's Missouri Expedition of 1864* (Lanham, NY: Rowman & Littlefield, 2015), 384n4.
5. Drew Gilpin Faust, *This Republic of Suffering: Death and the American Civil War* (New York: Vintage Books, 2009), 9, 63.
6. The Corps of Topographical Engineers became a part of the Corps of Engineers in 1863. Faust, *Historical Times Illustrated Encyclopedia*, 244.
7. First Lieutenant George T. Robinson was commissioned on July 7, 1864, in Company E of the Eleventh Kansas Cavalry and mustered out on August 7, 1865. Historical Data Systems, Inc., American Civil War Research Database, accessed on March 30, 2021, http://www.civilwardata.com/active/hdsquery.dll?SoldierHistory?U&1870955.
8. The battles of Big Blue and Westport were fought on October 22 and 23 respectively in the Kansas City, Missouri, area and resulted in Union victories against Major General Sterling Price's army. Faust, *Historical Times Illustrated Encyclopedia*, 602.
9. Private Jacob Miller, a resident of Menomonie, Wisconsin, enlisted in the Third Wisconsin Cavalry on February 27, 1864, and mustered out on November 25, 1865. Historical Data Systems, Inc., American Civil War Research Database, accessed on March 30, 2021, http://www.civilwardata.com/active/hdsquery.dll?SoldierHistory?U&2037233.
10. Charles Ransford Jennison became commander of the Seventh Kansas Cavalry on October 29, 1861. Jennison resigned his commission in May 1862 when James G. Blunt was promoted to brigadier general over him. Jennison rejoined the military in the fall of 1863 and commissioned colonel of the Fifteenth Kansas Cavalry. During Price's Missouri Expedition, complaints about alleged plundering by Jennison's regiment led to his dishonorable discharge in June 1865. Roger D. Hunt, *Colonels in Blue: Missouri and the Western States and Territories: A Civil War Biographical Dictionary* (Jefferson, NC: McFarland and Co. Inc., 2019): 177.
11. Organized in 1862, the Eleventh Kansas Cavalry campaigned in Kansas, Missouri, Nebraska, and eventually into the Dakota and Wyoming territories,

losing sixty-one men killed or mortally wounded during their service. Frederick H. Dyer, comp., *A Compendium of the War of the Rebellion*, 2 vols. (1908; repr., Dayton, OH: Morningside Bookshop, 1978), 2:1184.

12. John Murphy was commissioned as first lieutenant in Company B of the Fifteenth Kansas Cavalry on September 17, 1863, and later transferred to Company C. He mustered out on October 19, 1865. Historical Data Systems, Inc., American Civil War Research Database, accessed on March 30, 2021, http://www.civilwardata.com/active/hdsquery.dll?SoldierHistory?U&1887330.

13. Major General John S. Marmaduke commanded a cavalry division during Price's Missouri Expedition. *Official Records*, ser. 1, 41, pt. 1:641.

14. Located south of the Mound City Road along the north part of the Mine Creek battlefield, the W. H. Ragain house served as a collection point for Confederate prisoners after the battle. General Marmaduke was captured during the battle and taken to Major General Samuel R. Curtis at the Ragain house. Lumir F. Buresh, *October 25th and the Battle of Mine Creek* (Kansas City, MO: Lowell Press, 1977), 133.

15. Charles White Blair, a native of Ohio, enlisted in 1861 at age thirty-two. He served in the Second Kansas Infantry and the Second Kansas Cavalry before receiving a commission as colonel of the Fourteenth Kansas Cavalry on November 20, 1863. He mustered out on August 11, 1865. Historical Data Systems, Inc., American Civil War Research Database, accessed on March 30, 2021, http://www.civilwardata.com/active/hdsquery.dll?SoldierHistory?U&1762622.

16. L. B. Judson was a fifty-four-year-old farmer who resided in Kendall County, Illinois, in 1860. *Eighth Census of the United States*, 1860, Kendall County, Ill., 284.

17. A reference to the Battle of the Marmaton River, fought on October 25, 1864. Command confusion and other problems in the Federal army allowed the Confederates to retreat across the Marmaton. Sinisi, *The Last Hurrah*, 300–304.

18. Encumbered by wagons for much of the expedition, Price ordered the destruction of a large part of his supply train soon after the Battle of the Marmaton River. Ibid., 304–5.

19. Lyman's sister, Caroline (Carrie) Bennett, was born in New York and was about twenty-three years old in 1865. *Eighth Census of the United States*, 1860, Kendall County, Ill., 239.

20. Established in 1848, Fort Kearney was an important post on the Overland Trail in Nebraska. Robert W. Frazer, *Military Forts and Presidios and Posts Commonly Called Forts West of the Mississippi River to 1898* (Norman: Univ. of Oklahoma Press, 1977), 87.

21. Newtonia, Missouri, was the site of two battles: one on September 30, 1862, and the other on October 28, 1864. Larry Wood, *The Two Civil War Battles of Newtonia* (Charleston, SC: The History Press, 2010), 59, 113.
22. Located on the Grand River, Fort Gibson was founded in the Indian Territory in 1824 and depended on supplies from Fort Scott during the Civil War. Frazer, *Military Forts*, 120.
23. Located near the Arkansas River, Fort Smith was founded in 1817. Ibid., 16.
24. A native of New York, Clara Bulkley was a fifty-three-year-old widow who resided in the household of Martin Bulkley in Leavenworth County, Kansas, in 1865. Kansas, US, State Census Collection, 1855–1925 (Provo, UT: Ancestry.com Operations, Inc., 2009), accessed March 30, 2021.
25. Colonel William Fletcher Cloud, a native of Ohio, served as a brigade commander in the Army of the Frontier in Arkansas and Missouri for much of 1862 and 1863. In the fall of 1864, he was appointed to Major General Curtis's staff. Later, he commanded a sub-district in the Department of Missouri and was finally mustered out in October 1865. Hunt, *Colonels in Blue*, 166–67.
26. Born in 1831, Isaac Smith Kalloch, a native of Maine, was a pastor's son. Kalloch served at the Tremont Baptist Church in Boston, Massachusetts, but moved to New York after allegations of sexual improprieties. He founded the First Baptist Church in Leavenworth, Kansas. Later, he moved to San Francisco, California, where he became involved in politics and was shot by a newspaper editor. He survived the attack but died in 1887 in Washington Territory. "Rev. Isaac Smith Kalloch," accessed March 30, 2021, http://kalloch.org/rev_isaac.htm.

18. Fort Leavenworth, Kansas, to Fort Kearney, Nebraska Territory

1. Merrill J. Mattes, *The Great Platte River Road: The Covered Wagon Mainline via Fort Kearney to Fort Laramie* (1969; repr., Lincoln: Univ. of Nebraska Press, 1987), xx–xxv.
2. Mattes, *Great Platte River Road*, xx–xxv.
3. Probably a reference to the Sixteenth Kansas Cavalry that was organized at Fort Leavenworth in 1863 and 1864. The regiment campaigned against Native Americans in Kansas in early 1865. Frederick H. Dyer, comp., *A Compendium of the War of the Rebellion*, 2 vols. (1908; repr., Dayton, OH: Morningside Bookshop, 1978), 2:1185.
4. Federal troops marched into Charleston, South Carolina, on February 18, 1865, and Wilmington, North Carolina, fell four days later. Patricia L. Faust, *Historical Times Illustrated Encyclopedia of the Civil War* (New York: Harper & Row, 1986), 131, 831.

5. Brigadier General Robert Byington Mitchell migrated to Kansas in 1855 and commanded the Second Kansas Infantry at the Battle of Wilson's Creek, where he fell wounded. Late in the war, Mitchell successively commanded the District of Nebraska, the District of North Kansas, and the District of Kansas. Ezra J. Warner, *Generals in Blue: Lives of the Union Commanders* (1964; repr., Baton Rouge: Louisiana State Univ. Press, 1992), 329.

6. Colonel Robert R. Livingston of the First Nebraska Cavalry commanded the Eastern Sub-District of Nebraska when Bennett met him. United States War Department, *War of the Rebellion: A Compilation of the Official Records of the Union and Confederate Armies*, 128 vols. (1880–1901; repr., Harrisburg, PA: National Historical Society, 1971), ser. 1, 48, pt. 1:88.

7. A resident of Nebraska City, Nebraska, Lee P. Gillette enlisted on June 15, 1861, and was commissioned as a first lieutenant in the First Nebraska Infantry. Gillette was wounded at the Battle of Shiloh on April 7, 1862. His regiment converted to a mounted one in November 1863, and Gillett accepted a promotion as captain of Company A of the First Nebraska Cavalry. Historical Data Systems, Inc., American Civil War Research Database, accessed on March 31, 2021, http://www.civilwardata.com/active/hdsquery.dll?SoldierHistory?U&2137900.

8. Stephen W. Moore enlisted on July 3, 1861, and became a private in Company H of the First Nebraska Infantry. Promoted to second lieutenant in October 1862, he transferred into Company H of the First Nebraska Cavalry. Historical Data Systems, Inc., American Civil War Research Database, accessed on March 31, 2021, http://www.civilwardata.com/active/hdsquery.dll?SoldierHistory?U&2147734.

19. Fort Kearney to Denver

1. Gregory F. Michno, *Encyclopedia of Indian Wars: Western Battles and Skirmishes, 1850–1890* (Missoula, MT: Mountain Press Publishing Co., 2003), 159.
2. Elliott West, *The Contested Plains: Indians, Goldseekers, and the Rush to Colorado* (Lawrence: Univ. Press of Kansas, 1998), 307.
3. Robert Huhn Jones, *Guarding the Overland Trails: The Eleventh Ohio Cavalry in the Civil War*, Frontier Military Series, 24 (Spokane, WA: Arthur H. Clark Co., 2005), 205.
4. Jones, *Guarding the Overland Trails*, 205.
5. Ibid., 193–94, 196.
6. West, *Contested Plains*, 307.
7. Jones, *Guarding the Overland Trails*, 204.
8. Ibid., 204, 206–7.
9. Ibid., 206–7.

10. Organized in 1861 and 1862, the Eleventh Ohio Cavalry guarded overland trails and mail routes in the West and engaged in actions against Native Americans. The first battalion was mustered out in April 1865. Frederick H. Dyer, comp., *A Compendium of the War of the Rebellion*, 2 vols. (1908; repr., Dayton, OH: Morningside Bookshop, 1978), 2:1479.
11. Captain Thomas J. Majors commanded a detachment of three companies of the First Nebraska Cavalry at Plum Creek Station. United States War Department, *War of the Rebellion: A Compilation of the Official Records of the Union and Confederate Armies*, 128 vols. (1880–1901; repr., Harrisburg, PA: National Historical Society, 1971), ser. 1, 48, pt. 1:1040.
12. Probably a reference to Mullahla's (or Mullally's) Station, which had been manned by two companies of the First Nebraska Cavalry in late October 1864. *Official Records*, ser. 1, 41, pt. 4:377; Eugene F. Ware, *The Indian War of 1864* (1911; repr., New York: St. Martin's Press, 1960), 80.
13. Captain Charles F. Porter commanded Company A of the First Battalion Nebraska Cavalry rather than the Seventh Iowa Cavalry. *Official Records*, ser. 1, 48, pt. 1:1040.
14. An Irishman, Major George M. O'Brien of the Seventh Iowa Cavalry commanded the post at Cottonwood Springs that consisted of recruits for his regiment and Company C of the First Battalion Nebraska Cavalry. O'Brien was breveted lieutenant colonel and brigadier general of US Volunteers in March 1865. *Official Records*, ser. 1, 48, pt. 1:1040; Ware, *Indian War*, 443.
15. A reference to First Lieutenant George T. Robinson, Major General Samuel R. Curtis's chief engineer. Kyle S. Sinisi, *The Last Hurrah: Sterling Price's Missouri Expedition of 1864* (Lanham, NY: Rowman & Littlefield, 2015), 384n4.
16. Explorers Warren and Ferdinand V. Hayden and William F. Raynolds passed through the Black Hills in separate expeditions in the late 1850s and reported on the existence of gold there. However, the public paid little attention at the time to their mention of gold. The 1868 Treaty of Fort Laramie placed the Black Hills inside of the Sioux reservations, and six years later gold was rediscovered there. This discovery of gold proved to be one of the causes of the Great Sioux War in 1876. Paul L. Hedren, *Powder River: Disastrous Opening of the Great Sioux War* (Norman: Univ. of Oklahoma Press, 2016), 17, 35.
17. Many travelers commented on Jack Morrow's large ranch house. Ware, *Indian War*, 70–71, 442.
18. Captain John Wilcox of Company B of the Seventh Iowa Cavalry commanded at O'Fallon's Bluff as of February 28, 1865. *Official Records*, ser. 1, 48, pt. 1:1040.
19. Captain Edward B. Murphy, a native of Canada, commanded Company A of the Seventh Iowa Cavalry at Alkali Station. Murphy had a reputation for

aggressiveness and was considered one of the better officers in the Seventh Iowa. *Official Records*, ser. 1, 48, pt. 1:1040; Ware, *Indian War*, 471.

20. Lieutenant Talbot of the Seventh Iowa Cavalry apparently had excellent eyesight. According to Eugene F. Ware, who served in the Seventh (and missed encountering Bennett by mere weeks), Talbot galloped his horse by a line of telegraph poles near Fort Kearney and used two Colt pistols to put a bullet in eleven of twelve poles. Later, Ware recorded that Talbot spotted "a lone buffalo, and killed it." Ware, *Indian War*, 32, 348.

21. A reference to Beauvais Station mentioned in Bennett's next entry.

22. Captain Harrison W. Cremer and his C Company of the Seventh Iowa Cavalry were stationed at Beauvais Station. *Official Records*, ser. 1, 48, pt. 1:1040.

23. Captain Nicholas J. O'Brien commanded Fort Rankin near Julesburg. He and his men had fought against the Cheyenne and Arapaho when they had attacked Julesburg on January 7 and February 2, 1865. Jean Afton, David Fridtjof Halaas, and Andrew Masich, *Cheyenne Dog Soldiers: A Ledgerbook History of Coups and Combat* (Boulder and Denver: Univ. Press of Colorado and Colorado Historical Society, 1997), 74, 76.

24. Julesburg, Colorado Territory, and nearby Fort Rankin were attacked by Cheyenne and Arapaho on February 2, 1865. The stage station and several other buildings were destroyed in the attack. Afton, *Cheyenne Dog Soldiers*, 296.

25. The Wisconsin Ranch, fifty-six miles west of Julesburg, was attacked and destroyed on January 15, 1865. The location of this ranch, though, does not fit the mileage mentioned by Bennett. Instead, Bennett probably referenced Antelope Station, west of Julesburg, that was burned by Cheyenne and Arapaho on January 28, 1865. Afton, *Cheyenne Dog Soldiers*, xxx, 294–95.

26. Eighteen miles west of Julesburg, Buffalo Springs Ranch had been attacked on January 29, 1865. Ibid., 295.

27. This was possibly a reference to the First Colorado Mounted Militia. *Official Records*, ser. 1, 48, pt. 1:1042.

28. Located thirty-three miles west of Julesburg, Lillian Springs Ranch had been destroyed on January 27, 1865. Afton, *Cheyenne Dog Soldiers*, 295.

29. On January 28, 1865, Native Americans drove off 650 head of cattle belonging to a J. A. Moore, and a small group of soldiers from the First Colorado Cavalry pursued them. Ibid., 295.

30. Attacked on January 26, 1865, Bennett correctly noted that the Indians were driven away at Washington or Moore Ranch. Ibid., 295.

31. Valley Station and Fort Rankin managed to hold out during the series of attacks by Cheyenne and Arapaho in January 1865. Ibid., 228.

32. A reference to Thomas Chivington, who drowned on June 24, 1866, in the North Platte River about forty miles west of Fort Halleck. "Thomas

Chivington," Find a Grave (website), accessed April 20, 2021, https://www.findagrave.com/memorial/216422341/thomas-chivington.

33. Colonel John M. Chivington was one of the most controversial figures in the history of the Civil War West. An ordained Methodist clergyman, Chivington became commander of the First Colorado Cavalry in April 1862. As commander of the Military District of Colorado, he ordered the attack on the peaceful Cheyenne village at Sand Creek in November 1864. This assault resulted in the massacre of approximately 150 Native Americans, many of whom were noncombatants. Roger D. Hunt, *Colonels in Blue: Missouri and the Western States and Territories: A Civil War Biographical Dictionary* (Jefferson, NC: McFarland and Co. Inc., 2019), 157; Michno, *Encyclopedia of Indian Wars*, 159.

34. American, or Morrison's, Ranch, sixty-eight miles west of Julesburg, was attacked twice in January 1865. After the second attack on January 15, the Morrison family was found missing. Afton, *Cheyenne Dog Soldiers*, 294.

35. Godfrey's Ranch was attacked on January 14, 1865, with four civilians driving off the Native Americans. Ibid., 294.

36. Seventy-five men for the Second Colorado Cavalry had been recruited in Denver, and they engaged in several actions against Indians in January and February 1865. These men were under the command of Lieutenant Albert Walter. Christopher M. Rein, *The Second Colorado Cavalry: A Civil War Regiment on the Great Plains*, Campaigns and Commanders Series, vol. 69 (Norman: Univ. of Oklahoma Press, 2020), 174–76.

37. Born in New York in July 1837, John H. Coryell and his family lived in Kane County, Illinois, in 1860, farming there. He still lived in Colorado in 1900, when as a widower he resided with his daughter and son-in-law in Greeley. *Eighth Census of the United States*, 1860, Kane County, Ill., 228; *Twelfth Census of the United States*, 1900, Colorado, Weld County, 114B.

38. Colonel Thomas Moonlight, a Scottish immigrant, commanded the District of Colorado. Earlier in the war, he had served as Major General James G. Blunt's chief of staff and then commanded the Eleventh Kansas Cavalry. Moonlight was breveted brigadier general of US Volunteers when he mustered out in July 1865. *Official Records*, ser. 1, 48, pt. 1:1041; Sinisi, *Last Hurrah*, 161; Hunt, *Colonels in Blue*, 162.

39. Unidentified. Coryell's wife was identified as Pheobe in the 1860 census. *Eighth Census of the United States*, 1860, Kane County, Ill., 228.

40. Cherry Creek overflowed its banks on May 19, 1864, causing significant property damage and taking the lives of eight people. Duane A. Smith, *The Birth of Colorado: A Civil War Perspective* (Norman: Univ. of Oklahoma Press, 1989), 227–28.

20. A Tour of Colorado Gold Mines

1. Elliott West, *The Contested Plains: Indians, Goldseekers, and the Rush to Colorado* (Lawrence: Univ. Press of Kansas, 1998), 115.
2. West, *Contested Plains*, 105, 116.
3. Ibid., 238.
4. Duane A. Smith, *The Trail of Gold and Silver: Mining in Colorado, 1859–2009* (Boulder: Univ. Press of Colorado, 2009), 20.
5. Sandra Dallas, *Colorado Ghost Towns and Mining Camps* (Norman: Univ. of Oklahoma Press, 1985), 44.
6. Dallas, *Colorado Ghost Towns*, 28, 30.

21. To Fort Laramie and Back to Denver

1. David E. Wagner, *Powder River Odyssey: Nelson Cole's Western Campaign of 1865: The Journals of Lyman G. Bennett and Other Eyewitness Accounts*, Frontier Military Series, 27 (Norman, OK: Arthur H. Clark Co., 2009), 24–25.
2. Wagner, *Powder River Odyssey*, 27.
3. Ibid., 242.
4. Quoted in ibid., 236.
5. Bennett visited Brant Coryell in Wisconsin in 1857 and wrote in his diary that Brant told many "amusing stories," and "I never saw a man that could give the minutest particulars of a story, & with so much coolness & *sang froid* as Brant Coryell." Lyman G. Bennett, diary, 10 January 1857, Lyman G. Bennett Papers, R0274, The State Historical Society of Missouri Research Center-Rolla.
6. Captain William H. Evans and his Company F of the Eleventh Ohio Cavalry were stationed at Camp Collins along with Company B, commanded by Captain Wesley Love. United States War Department, *War of the Rebellion: A Compilation of the Official Records of the Union and Confederate Armies*, 128 vols. (1880–1901; repr., Harrisburg, PA: National Historical Society, 1971), ser. 1, 48, pt. 1:1040.
7. First Lieutenant Jeremiah H. Triggs lived in Lancaster, Iowa, when he enlisted in the Seventh Iowa Cavalry at age twenty-two. Commissioned as a second lieutenant, he was promoted to first lieutenant on August 24, 1864. At some point in 1865 he was promoted to captain. Historical Data Systems, Inc., American Civil War Research Database, accessed on April 27, 2021, http://www.civilwardata.com/active/hdsquery.dll?SoldierHistory?U&1976480.
8. Camp Walbach was established in September 1858, but the army abandoned it seven months later in April 1859. It was located "on Lodgepole Creek, east of Cheyenne Pass, about twenty miles east of the present town of Laramie." Robert W. Frazer, *Forts of the West: Military Forts and Presidios and Posts*

Commonly Called Forts West of the Mississippi to 1898 (Norman: Univ. of Oklahoma Press, 1977), 186.

9. Major Thomas L. Mackey of the Eleventh Ohio Cavalry commanded at Fort Laramie. *Official Records*, ser. 1, 48, pt. 1:1041.
10. Bennett first heard about gold discoveries in the Black Hills on March 8, 1865.
11. Organized in the spring of 1862, the soldiers of the Eleventh Kansas Cavalry served along the Missouri-Kansas Border until early 1865 when they were dispatched to various locations in the Dakota, Nebraska, and Wyoming territories. Frederick H. Dyer, comp., *A Compendium of the War of the Rebellion*, 2 vols. (1908; repr., Dayton, OH: Morningside Bookshop, 1978), 2:1184.
12. Preston B. Plumb, a resident of Emporia, Kansas, and a newspaper editor, enlisted on May 17, 1864, and was commissioned as major of the Eleventh Kansas Cavalry. Later promoted to lieutenant colonel, he mustered out on September 13, 1865. Historical Data Systems, Inc., American Civil War Research Database, accessed on April 20, 2021, http://www.civilwardata.com/active/hdsquery.dll?SoldierHistory?U&1869200.
13. Captain Jacob S. Shuman enlisted on August 2, 1863, and commanded Company H of the Eleventh Ohio Cavalry. Shuman mustered out on July 14, 1866. Eugene F. Ware, *The Indian War of 1864* (1911; repr., New York: St. Martin's Press, 1960), 460.
14. Camp Mitchell, also known as Fort Mitchell, was established in August 1864 not far from Scotts Bluff. Frazer, *Forts of the West*, 88.
15. Richmond, Virginia, fell to Union forces on April 3, 1865. Patricia L. Faust, *Historical Times Illustrated Encyclopedia of the Civil War* (New York: Harper & Row, 1986), 632.
16. A large force of natives attacked Mud Springs Station from February 4 through February 6, 1865. The battle finally ended when soldiers released a horse herd that drew the warriors away from the station. Lieutenant Colonel William O. Collins of the Eleventh Ohio Cavalry dispatched reinforcements to Mud Springs from Camp Mitchell and then went there himself on February 5, arriving during one of the lulls in the fighting. Jean Afton, David Fridtjof Halaas, and Andrew Masich, *Cheyenne Dog Soldiers: A Ledgerbook History of Coups and Combat* (Boulder and Denver: Univ. Press of Colorado and Colorado Historical Society, 1997), 296; Robert Huhn Jones, *Guarding the Overland Trails: The Eleventh Ohio Cavalry in the Civil War*, Frontier Military Series, 24 (Spokane, WA: Arthur H. Clark Co., 2005), 197–98.
17. Originally from Ireland, Patrick E. Connor's family immigrated to New York City when he was a child. A combat veteran of the Mexican American War, Connor participated in the 1849 California Gold Rush. He was appointed colonel of the Third California Infantry when the Civil War

began and became commander of the District of Utah. In February 1865, Major General Grenville M. Dodge's Department of Missouri merged with Connor's department, and Connor was then ordered to Denver. At the end of March 1865, Connor took command of the new District of the Plains at Fort Kearney. Ezra J. Warner, *Generals in Blue: Lives of the Union Commanders* (1964; repr., Baton Rouge: Louisiana State Univ. Press, 1992), 87; Jones, *Guarding the Overland Trails*, 215, 220.

18. Possibly a reference to Captain William T. Clark of the First Nebraska Cavalry. He enlisted on July 25, 1861, at age thirty and was promoted to first lieutenant on June 18, 1863. Clark had only been a captain since February 3, 1865. Historical Data Systems, Inc., American Civil War Research Database, accessed on April 27, 2021, http://www.civilwardata.com/active/hdsquery.dll?SoldierHistory?U&2147704.

22. After the War

1. *Pictorial and Genealogical Record of Greene County, Missouri: Together with Biographies of Prominent Men* (Chicago: Goodspeed Bros., 1893), 205.
2. *Ninth Census of the United States*, 1870, Illinois, Kendall County, 4.
3. "Edgar Agaz Bennett," Find a Grave (website), accessed December 3, 2020, https://www.findagrave.com/memorial/33070798/edgar-agaz-bennett; Pension records on Lyman G. Bennett, RG T-288, Invalid Record Application 1200162, National Archives.
4. "William Bennett," Find a Grave (website), accessed December 3, 2020, https://www.findagrave.com/memorial/153980944/william-bennett.
5. "Carrie M. Bennett," Find a Grave (website), accessed December 3, 2020, https://www.findagrave.com/memorial/6633847/carrie-m-bennett.
6. Newton Bateman and Paul Selby, eds., *Encyclopedia of Illinois and History of Kendall County* (Chicago: Munsell Publishing Co., 1914), 1:760, 822; *Pictorial and Genealogical Record*, 205; *Biographical Directory of the Voters and Tax-Payers of Kendall County, Illinois* (Chicago: Geo. Fisher & Co., 1876), 21, 34.
7. Bateman and Selby, *Encyclopedia of Illinois*, 1:489.
8. Ibid.
9. Ibid.
10. Ibid., 821; *Pictorial and Genealogical Record*, 205.
11. Bateman and Selby, *Encyclopedia of Illinois*, 1:744.
12. L. G. Bennett, *Map of Winona County, Minnesota*. Chicago: lith. by C. Shober & Co., 1867. https://www.loc.gov/item/97685001/.
13. *Ninth Census of the United States*, 1870, Illinois, Kendall County, 4.
14. *Pictorial and Genealogical Record*, 200.
15. *Ninth Census of the United States*, 1870, Illinois, Kendall County, 3.

16. L. G. Bennett and William M. Haigh, *History of the Thirty-Sixth Regiment Illinois Volunteers, during the War of the Rebellion* (1876; repr., Marengo, IL: Prairie State Press, Inc., 1999), iv.
17. *Pictorial and Genealogical Record*, 205.
18. Ibid.
19. Ibid.
20. Ibid.
21. Ibid.
22. Illinois, US, County Marriage Records, 1800–1940 (Lehi, UT: Ancestry.com Operations, Inc., 2016), accessed May 4, 2021.
23. *Pictorial and Genealogical Record*, 200.
24. *Springfield Leader*, January 5, 1888.
25. "Carrie M. Bennett," Find a Grave (website), accessed December 3, 2020, https://www.findagrave.com/memorial/6633847/carrie-m-bennett.
26. *Springfield Leader*, January 5, 1888.
27. United States Department of the Interior, National Register of Historic Places Registration Form, Edward M. and Della C. Wilhoit House, Springfield, MO. https://dnr.mo.gov/shpo/nps-nr/04001384.pdf.
28. National Archives and Records Administration, RG 49, United States Department of the Interior, Field Notes of Surveys for California, Minnesota, and Oklahoma and Indian Territories, 1881–1898.
29. *Pictorial and Genealogical Record*, 205.
30. Pension records on Lyman G. Bennett, RG T-288, Invalid Record Application 1200162, National Archives.
31. *The Journal-Gazette*, December 8, 1899.
32. Pension records, Lyman G. Bennett.
33. *Cassville Republican*, March 10, 1904.
34. *Springfield Daily Republican*, March 3, 1904.
35. Ibid.
36. Ibid.

Bibliography

Manuscripts
The Huntington Library, Art Museum, and Botanical Gardens.
    Bennett, Lyman G. "Personal Reminiscences: Border Sketches & Recruiting in Dixie."
Kansas Historical Society, Topeka.
    Bennett, Lyman G. "Route of the Army of the Southwest from March 1st to July 10, 1862."
Library of Congress. Washington, DC.
    Abraham Lincoln Papers.
    Bennett, L. G. *Map of Winona County, Minnesota.* Chicago: lith. by C. Chober & Co., 1867. https://www.loc.gov/item/97685001/.
The State Historical Society of Missouri Research Center–Rolla.
    Lyman G. Bennett Papers, R0274.

Newspapers
Illinois
Chicago Tribune (1862)

Missouri
Cassville Republican (1904)
The Journal-Gazette (1899)
Springfield Daily Republican (1904)
Springfield Leader (1888)

Public Documents
Hewitt, Janet B., Noah Andre Trudeau, and Bryce A. Suderow, eds. *Supplement to the Official Records of the Union and Confederate Armies.* 100 vols. Wilmington, NC: Broadfoot Publishing Co., 1994.
National Archives and Records Administration. M313. Index to War of 1812 Pension Application Files. Accessed on Fold3.com.
National Archives and Records Administration. Pension records on Lyman G. Bennett. RG T-288. Invalid Record Application 1200162.
National Archives and Records Administration. RG 49. United States Department of the Interior. Field Notes of Surveys for California, Minnesota, and Oklahoma and Indian Territories, 1881–1898.

National Archives and Records Administration. RG 94. Compiled Service Records of Volunteer Union Soldiers Who Served in Organizations from the State of Arkansas, First Arkansas Cavalry. Accessed on Fold3.com.

National Archives and Records Administration. RG 94. Compiled Service Records of Volunteer Union Soldiers Who Served in Organizations from the State of Arkansas, Fourth Arkansas Cavalry. Accessed on Fold3.com.

National Archives and Records Administration. RG 94. Compiled Service Records of Volunteer Union Soldiers Who Served in Organizations from the State of Illinois, Thirty-Sixth Illinois Infantry. Copy accessed from the Lyman G. Bennett Papers, R0274, The State Historical Society of Missouri Research Center-Rolla.

National Archives and Records Administration. RG 94. Compiled Service Records of Volunteer Union Soldiers Who Served in Organizations from the State of Missouri, Phelps' Independent Missouri Regiment. Accessed on Fold3.com.

National Archives and Records Administration. RG 94. Compiled Service Records of Volunteer Union Soldiers Who Served with the United States Colored Troops, Fifty-Fourth Regiment United States Colored Troops. Accessed on Fold3.com.

National Geospatial-Intelligence Agency. *Wilson's Creek Staff Ride: Fieldbook*. 3rd ed. St. Louis: NGA, 2012.

Report of the Adjutant General of the State of Indiana. Indianapolis: Samuel M. Douglass, State Printer, 1866.

United States Census Bureau. *Eighth Census of the United States*, 1860. Provo, UT: Ancestry.com Operations, 2005.

United States Census Bureau. *Ninth Census of the United States*, 1870. Provo, UT: Ancestry.com Operations, 2005.

United States Census Bureau. *Seventh Census of the United States*, 1850. Provo, UT: Ancestry.com Operations, 2005.

United States Census Bureau. *Tenth Census of the United States*, 1880. Provo, UT: Ancestry.com Operations, 2005.

United States Census Bureau. *Twelfth Census of the United States*, 1900. Provo, UT: Ancestry.com Operations, 2005.

United States Department of the Interior. National Register of Historic Places Registration Form. Edward M. and Della C. Wilhoit House. Springfield, MO. https://dnr.mo.gov/shpo/nps-nr/04001384.pdf.

United States War Department. *War of the Rebellion: A Compilation of the Official Records of the Union and Confederate Armies*. 128 vols. 1880–1901. Reprint, Harrisburg, PA: National Historical Society, 1971.

Books

Afton, Jean, David Fridtjof Halaas, and Andrew Masich. *Cheyenne Dog Soldiers: A Ledgerbook History of Coups and Combat.* Boulder and Denver: Univ. Press of Colorado and Colorado Historical Society, 1997.

Allardice, Bruce S. *More Generals in Gray.* Baton Rouge: Louisiana State Univ. Press, 1995.

Bateman, Newton, and Paul Selby, eds. *Encyclopedia of Illinois and History of Kendall County.* 2 vols. Chicago: Munsell Publishing Co., 1914.

Bearss, Edwin Cole. *The Campaign for Vicksburg.* 3 vols. Dayton, OH: Morningside House, Inc., 1991.

Bennett, L. G., and William M. Haigh. *History of the Thirty-Sixth Regiment Illinois Volunteers, during the War of the Rebellion.* 1876. Reprint, Marengo, IL: Prairie State Press, Inc., 1999.

Biographical Directory of the Voters and Tax-Payers of Kendall County, Illinois. Chicago: Geo. Fisher & Co., 1876.

Bishop, A. W. *Loyalty on the Frontier or Sketches of Union Men of the South-West with Incidents and Adventures in Rebellion.* Edited by Kim Allen Scott Bishop. Fayetteville: Univ. of Arkansas Press, 2003.

Blevins, Brooks. *A History of the Ozarks.* Vol. 2, *The Conflicted Ozarks.* Urbana: Univ. of Illinois Press, 2019.

Blocker, Jack S., Jr. *American Temperance Movements: Cycles of Reform.* Boston: Twayne Publishers, 1989.

Boatner, Mark M., III. *The Civil War Dictionary.* New York: David McKay Co., Inc., 1959.

Bollet, Alfred Jay. *Civil War Medicine: Challenges and Triumphs.* Tucson, AZ: Galen Press, Ltd., 2002.

Buresh, Lumir F. *October 25th and the Battle of Mine Creek.* Kansas City, MO: Lowell Press, 1977.

Christ, Mark K. *Civil War Arkansas, 1863: The Battle for a State.* Norman: Univ. of Oklahoma Press, 2010.

Cowles, Calvin D., comp. *The Official Military Atlas of the Civil War.* 1891–1895. Reprint, New York: Arno Press, 1978.

Cullum, George W. *Biographical Register of the Officers and Graduates of the U. S. Military Academy at West Point, N. Y.* 2nd ed. New York: D. Van Nostrand, 1868.

Cullum, George W. *Biographical Register of the Officers and Graduates of the U. S. Military Academy at West Point, N. Y.* 3rd ed. Boston: Houghton, Mifflin, and Co., 1891.

Cunningham, O. Edward. *Shiloh and the Western Campaign of 1862.* Edited by Gary D. Joiner and Timothy B. Smith. New York: Savas Beatie, 2009.

Dallas, Sandra. *Colorado Ghost Towns and Mining Camps*. Norman: Univ. of Oklahoma Press, 1985.
Daniel, Larry J. *Battle of Stones River: The Forgotten Conflict between the Confederate Army of Tennessee and the Union Army of the Cumberland*. Baton Rouge: Louisiana State Univ. Press, 2012.
Devine, Shauna. *Learning from the Wounded: The Civil War and the Rise of American Medical Science*. Chapel Hill: The Univ. of North Carolina Press, 2014.
Dyer, Frederick H., comp. *A Compendium of the War of the Rebellion*. 2 vols. 1908. Reprint, Dayton, OH: Morningside Bookshop, 1978.
Fahey, David M. *Temperance & Racism: John Bull, Johnny Reb, and the Good Templars*. Lexington: The Univ. Press of Kentucky, 1996.
Faust, Drew Gilpin. *This Republic of Suffering: Death and the American Civil War*. New York: Vintage Books, 2009.
Faust, Patricia L. *Historical Times Illustrated Encyclopedia of the Civil War*. New York: Harper & Row, 1986.
Fellman, Michael. *Inside War: The Guerrilla Conflict in Missouri during the American Civil War*. New York: Oxford Univ. Press, 1989.
Frazer, Robert W. *Military Forts and Presidios and Posts Commonly Called Forts West of the Mississippi River to 1898*. Norman: Univ. of Oklahoma Press, 1977.
Gerteis, Louis S. *The Civil War in Missouri: A Military History*. Columbia: Univ. of Missouri Press, 2012.
Gerteis, Louis S. *Civil War St. Louis*. Modern War Studies. Lawrence: Univ. Press of Kansas, 2001.
Hedren, Paul L. *Powder River: Disastrous Opening of the Great Sioux War*. Norman: Univ. of Oklahoma Press, 2016.
Hess, Earl J., Richard W. Hatcher, III, William Garrett Piston, and William L. Shea. *Wilson's Creek, Pea Ridge & Prairie Grove: A Battlefield Guide with a Section on Wire Road*. Lincoln: Univ. of Nebraska Press, 2006.
Hicks, E. W. *History of Kendall County Illinois, from the Earliest Discoveries to the Present Time*. Aurora, IL: Knickerbocker & Hodder, 1877.
History of Benton, Washington, Carroll, Madison, Crawford, Franklin, and Sebastian Counties, Arkansas. Chicago: Goodspeed Publishing Co., 1889.
Holcombe, R. I., ed. *History of Greene County, Missouri, Written and Compiled from the Most Authentic Official and Private Sources*. St. Louis: Western Historical Co., 1883.
Holcombe, Return I. *History of the First Regiment Minnesota Volunteer Infantry, 1861–1864*. 1916. Reprint, Ann Arbor, MI: St. Croix Valley Civil War Round Table, 2006.
Hunt, Roger D. *Colonels in Blue: Missouri and the Western States and Territories: A Civil War Biographical Dictionary*. Jefferson, NC: McFarland and Co. Inc., 2019.

Johansson, M. Jane, ed. *Albert C. Ellithorpe, the First Indian Home Guards and the Civil War on the Trans-Mississippi Frontier*. Baton Rouge: Louisiana State Univ. Press, 2016.

Johnston, James J. *Mountain Feds: Arkansas Unionists and the Peace Society*. Little Rock: Butler Center Books, 2018.

Jones, Robert Huhn. *Guarding the Overland Trails: The Eleventh Ohio Cavalry in the Civil War*. Frontier Military Series, 24. Spokane, WA: Arthur H. Clark Co., 2005.

Knight, James R. *The Battle of Pea Ridge: The Civil War Fight for the Ozarks*. Charleston, SC: The History Press, 2012.

Lindberg, Kip A., and Thomas P. Sweeney. *"A Scene of Horrors": A Medical and Surgical History of the Battle of Wilson's Creek, August 10, 1861*. Cassville, MO: Litho Printers and Bindery, 2013.

Linklater, Andro. *Measuring America: How the United States Was Shaped by the Greatest Land Sale in History*. New York: Plume Books.

Mahan, Russell L. *Fayetteville, Arkansas in the Civil War, 1860–1865*. Bountiful, UT: Historical Byways, 2003.

Marvel, William. *Lincoln's Mercenaries: Economic Motivation among Union Soldiers during the Civil War*. Baton Rouge: Louisiana State Univ. Press, 2018.

Mattes, Merrill J. *The Great Platte River Road: The Covered Wagon Mainline via Fort Kearney to Fort Laramie*. 1969. Reprint., Lincoln: Univ. of Nebraska Press, 1987.

McElfresh, Earl B. *Maps and Mapmakers of the Civil War*. New York: Harry N. Abrams, Inc., 1999.

McGhee, James E. *Guide to Missouri Confederate Units, 1861–1865*. Fayetteville: Univ. of Arkansas Press, 2008.

McPherson, James M. *For Cause & Comrades: Why Men Fought in the Civil War*. New York: Oxford Univ. Press, 1997.

Michno, Gregory F. *Encyclopedia of Indian Wars: Western Battles and Skirmishes, 1850–1890*. Missoula, MT: Mountain Press Publishing Co., 2003.

Nichols, Bruce. *Guerrilla Warfare in Civil War Missouri*. Jefferson, NC: McFarland and Co., Inc., 2012.

Oliva, Leo E. *Fort Scott: Courage and Conflict on the Border*. Topeka: Kansas State Historical Society, 1984.

Patterson, Dan, and Clinton Terry. *Surveying in Early America: The Point of Beginning*. Cincinnati: Univ. of Cincinnati Press, 2020.

Pictorial and Genealogical Record of Greene County, Missouri: Together with Biographies of Prominent Men. Chicago: Goodspeed Bros., 1893.

Piston, William Garrett, and Richard W. Hatcher, III. *Wilson's Creek: The Second Battle of the Civil War and the Men Who Fought It*. Chapel Hill: The Univ. of North Carolina Press, 2000.

Piston, William Garrett, and Thomas P. Sweeney. *Portraits of Conflict: A Photographic History of Missouri in the Civil War.* Fayetteville: Univ. of Arkansas Press, 2009.

Poe, Edgar Allan. *Poetry and Tales.* Library of America. New York: Literary Classics of the United States, 1984.

Rein, Christopher M. *The Second Colorado Cavalry: A Civil War Regiment on the Great Plains.* Vol. 69, Campaigns and Commanders Series. Norman: Univ. of Oklahoma Press, 2020.

Rutkow, Ira M. *Bleeding Blue and Gray: Civil War Surgery and the Evolution of American Medicine.* New York: Random House, 2005.

Schultz, Robert G. *The March to the River: From the Battle of Pea Ridge to Helena, Spring 1862.* Iowa City: Camp Pope Publishing, 2014.

Shannon, Mary Eulalie. *Poems.* Cincinnati: Moore, Wilstach & Keys, 1854.

Shea, William L., and Earl J. Hess. *Pea Ridge: Civil War Campaign in the West.* Chapel Hill: The Univ. of North Carolina Press, 1992.

Shea, William L. *Fields of Blood: The Prairie Grove Campaign.* Chapel Hill: The Univ. of North Carolina Press, 2009.

Shelburne, Orville B. *From Presidio to the Pecos River: Surveying the United States-Mexico Boundary along the Rio Grande, 1852 and 1853.* Norman: Univ. of Oklahoma Press, 2020.

Sinisi, Kyle S. *The Last Hurrah: Sterling Price's Missouri Expedition of 1864.* Lanham, NY: Rowman & Littlefield, 2015.

Smith, Duane A. *The Trail of Gold and Silver: Mining in Colorado, 1859–2009.* Boulder: Univ. Press of Colorado, 2009.

Sutherland, Daniel E. *A Savage Conflict: The Decisive Role of Guerrillas in the American Civil War.* Chapel Hill: The Univ. of North Carolina Press, 2009.

The Union Army: A History of Military Affairs in the Loyal States 1861–1865. 9 vols. 1908. Reprint, Wilmington, NC: Broadfoot Publishing Co., 1998.

Wagner, David E. *Powder River Odyssey: Nelson Cole's Western Campaign of 1865: The Journals of Lyman G. Bennett and Other Eyewitness Accounts.* Frontier Military Series, 27. Norman, OK: Arthur H. Clark Co., 2009.

Ware, Eugene F. *The Indian War of 1864.* 1911. Reprint, New York: St. Martin's Press, 1960.

Warner, Ezra J. *Generals in Blue: Lives of the Union Commanders.* 1964. Reprint, Baton Rouge: Louisiana State Univ. Press, 1992.

Warner, Ezra J. *Generals in Gray: Lives of the Confederate Commanders.* 1959. Reprint, Baton Rouge: Louisiana State Univ. Press, 1987.

Welsh, Jack D. *Medical Histories of Union Generals.* Kent, OH; Kent State Univ. Press, 1996.

West, Elliott. *The Contested Plains: Indians, Goldseekers, and the Rush to Colorado.* Lawrence: Univ. Press of Kansas, 1998.

Wood, Larry. *The Siege of Lexington Missouri: The Battle of the Hemp Bales*. Charleston, SC: The History Press, 2014.

Wood, Larry. *The Two Civil War Battles of Newtonia*. Charleston, SC: The History Press, 2010.

Articles

Bland, Richard A. "Marcus LaRue Harrison (1830–1890)." In *Encyclopedia of Arkansas*. Central Arkansas Library System, 2018. https://encyclopediaofarkansas.net/entries/marcus-larue-harrison-1665/.

Bradbury, John F., Jr. "'Buckwheat Cake Philanthropy': Refugees and the Union Army in the Ozarks." *Arkansas Historical Quarterly* 57, no. 3 (Autumn 1998): 233–54.

Bradbury, John F., Jr. "Fort Wyman and the Defenses of Rolla." *Pioneer Times* 7, no. 3 (July 1983): 247–52.

Bradbury, John F., Jr. "Fort Wyman, Fort Dette and the Defense of the Railhead." *PCHS [Phelps County Historical Society] Newsletter*, new series no. 44 (October 2011): 2–19.

Bradbury, John F., Jr. "'Good Water & Wood but the Country is a Miserable Botch': Flatland Soldiers Confront the Ozarks." *Missouri Historical Review* 90 (January 1996): 166–86.

Cain, Marvin R. and John F. Bradbury, Jr. "Union Troops and the Civil War in Southwestern Missouri and Northwestern Arkansas." *Missouri Historical Review* 88 (October 1993): 29–47.

Davenport, Helen C. "Recruiting in Dixie." *Christian County [Missouri] Historian* IV (1991): 14–16, 28–30, 43–45, 57–59, 61–73.

"DeWitt Clinton Hunter." *Arkansas-Louisiana Seventh-Day Adventist History*. https://arklasdahistory.org/dewitt-clinton-hunter/. Accessed on Feb. 19, 2022.

Hanson, David J. "Sons of Temperance: A Temperance Brotherhood & Insurance Company." Alcohol Problems and Solutions. https://www.alcoholproblemsandsolutions.org/sons-of-temperance-a-temperance-brotherhood/. Accessed on May 17, 2021.

Hughes, Michael A. "Wartime Gristmill Destruction in Northwest Arkansas and Military Farm Colonies." In *Civil War Arkansas: Beyond Battles and Leaders*, edited by Anne J. Bailey and Daniel E. Sutherland, 31–45. Fayetteville: Univ. of Arkansas Press, 2000.

Johnston, James J., ed. "Recruiting in Dixie, Part I." *White River Valley Historical Quarterly* 39 (Fall 1999): 13–18.

Johnston, James J., ed. "Recruiting in Dixie, Part II." *White River Valley Historical Quarterly* 39 (Winter 2000): 3–14.

Johnston, James J., ed. "Recruiting in Dixie, Part III." *White River Valley Historical Quarterly* 39 (Spring 2000): 3–11.

Piston, William Garrett. "Struggle for the Trans-Mississippi." *North & South* 11, no. 5 (October 2009): 14–21, 67.

Readnour, Harry W. "William Meade Fishback (1831–1903)." In *Encyclopedia of Arkansas*. Central Arkansas Library System, 2022. https://encyclopediaofarkansas.net/entries/william-meade-fishback-103/.

"A School Teacher: A Letter Sent by Lyman Bennett." *White River Valley Historical Quarterly* 4, no. 1 (Fall 1970): 14–16.

Sesser, David. "Fourth Arkansas Cavalry." In *Encyclopedia of Arkansas*. Central Arkansas Library System, 2020. https://encyclopediaofarkansas.net/entries/fourth-arkansas-cavalry-12096/.

Starr, Stephen Z. "The Grand Old Regiment." *The Wisconsin Magazine of History* 48, no. 1 (Autumn 1964): 21–31.

Sutherland, Daniel E. "Guerrillas: The Real War in Arkansas." In *Civil War Arkansas: Beyond Battles and Leaders*, edited by Anne J. Bailey and Daniel E. Sutherland, 133–53. Fayetteville: Univ. of Arkansas Press, 2000.

Wilson, Mark R. "Chicago Tribune." In *Encyclopedia of Chicago*. Chicago Historical Society, 2005. http://www.encyclopedia.chicagohistory.org/pages/275.html.

Electronic Databases

Find a Grave. https://www.findagrave.com.

Historical Data Systems, Inc. American Civil War Research Database. https://www.civilwardata.com.

Illinois, US, Databases of Illinois Veterans Index, 1775–1995. Provo, UT: Ancestry.com Operations, Inc., 2015.

Illinois, US Compiled Marriages, 1851–1900. Provo, UT: Ancestry.com Operations, Inc., 2005.

Illinois GenWeb. https://illinoisgenweb.org.

Kansas, US, State Census Collection, 1855–1925. Provo, UT: Ancestry.com Operations, Inc., 2009.

Library of Congress. *Chronicling America*. https://chroniclingamerica.loc.gov.

National Park Service. *Soldiers and Sailors Database*. https://www.nps.gov/civilwar/soldiers-and-sailors-database.htm.

Springfield-Greene County Library District. *Community & Conflict: The Impact of the Civil War in the Ozarks*. https://ozarkscivilwar.org.

Index

Page numbers in **boldface** refer to illustrations.

Adare, James, 338n36
Adare, Margaret, 338n36
African Americans, treatment of, 169, 240–41
Alkali Station, Nebraska Territory, 269, 344n19
American Ranch, Colorado Territory, 272–73, 275, 345n34
Appomattox Courthouse, Virginia, 284, 291
Arapaho, 263–64, 284, 344nn23–25, 345n31
Arkansas, 41, 72, 79, 142, 149, 171–72, 182, 184, 190, 203–4, 213, 218–19, 235, 251, 308n22, 309n21, 314n34, 315n42, 318n17, 322n4, 326n25, 329nn17–18
Arkansas: counties, 40–41; military governor: 181, 184, 186, 311n39; refugees, 187, 192, 195–97, 229–30; Secession Convention, 185; secessionists, 183; soldiers, 178, 196–97, 200
Arkansas Peace Society, 182–83
Arkansas River, 179, 184, 191, 204, 214–15, 341n20
Arkansas troops (CSA): 1st Cavalry, 332n34; 1st Mounted Rifles, 155; 2nd Mounted Rifles, 154; 2nd Infantry, 329n17; 16th Infantry, 154, 336n25; 32nd Infantry, 183
Arkansas troops (USA): 1st Cavalry, 193–94, 207, 309n21, 314n38, 332n29, 333n35, 333n42, 338n42; 1st Infantry, 190, 194, 229, 328n44, 331n21, 331n25, 332n26, 332nn32–33, 335n14, 335n16, 336nn17–18, 337n33, 338n43; 1st Light Artillery, 333nn41–42; 2nd Cavalry, 215, 228, 334n46, 335n13, 338n40; 2nd Infantry, 336n19; 3rd Infantry, 331n22; 4th Cavalry, 182, 203, 207, 232–33, 235, 331n22, 336n25
Arlington, Illinois, 24
Army of Tennessee, 337n29
Army of the Frontier, 322n4
Army of the Southwest, 127–28, 133, 141–42, 153, 171, 173, 177
Army of the Tennessee, 173, 320n6
Army of the West (Confederate), 172
Asboth, Brig. Gen. Alexander Sandor (USA), 67, 156, 162, 314n39; division, 67, 71, 86, 317n14, 328n42
Ashland, Minnesota Territory, 3–4
Atchison, Kansas, 253–55
Aurora, Illinois, 8–9, 11–12, 18–19, 24, 86, 312n23
Axley, Rebecca, 86, 309n21, 318n16

Baldwin, Capt. Elias B. (USA), 9, 302n20, 305n20
Baldwin, Capt. Melvin B. (USA), 43, 305n20, 310n28
Barker, Lt. Edgar A. (USA), 338n41
Barnes, Dr. Joseph K. (USA), 95, 319n26

Barnum, P. T., 105, 215, 336n20
Barry, Maj. Alonzo H. (USA), 56, 104, 164, 312n20, 327n33
Batesville, Arkansas, 171–73
Beauvais Station, Nebraska Territory, 270, 344n21, 344n22
Bennett, Caroline (Carrie) (sister), 24, 248, 305n19, 340n19
Bennett, Carrie (daughter), 295–96
Bennett, Charles (brother), 295, **297**, 321n4
Bennett, Charles M. (father), 1, 11, 14, 19, 61, 89, 124, 126, 128, 261, 295–96, 318n19
Bennett, Edgar (Eddie) Agaz (son), 295–96, 298
Bennett, Franklin (brother), 1
Bennett, Guy (also known as George), 45, 69, 101, 124–25, 311n40
Bennett, James, 1
Bennett, Louisa (sister), 24, 305n19
Bennett, Louisa Canfield (mother), 1, 14–15, 17, 19, 51, 79, 124–26, 296, 316n52
Bennett, Lyman G.: appointed corporal, 11; arrest and confinement, 116–18; assignment to Maj. Gen. Curtis's staff, 142, 150, 153, 167; attitudes toward Native Americans, 1, 253, 255–56, 263, 287, 289; battle of Pea Ridge participation, 153, 157–63; birth and youth, 1–2; childhood home, **107**, 111; circuit clerk, 2, 295, 303n23; civilian employee of the Engineer Corps, 239; coauthor, *History of The Thirty-Sixth Regiment Illinois Volunteers,* 296; correspondence with friends and relatives, 13, 26, 31, 33, 62, 88–89, 105, 109, 149, 164, 169, 181, 284, 291; correspondence with wife, 43, 45, 48, 58, 61, 65, 68, 74, 85, 88, 110, 112, 116, 130, 137, 143, 163, 167–68, 170, 255, 261; courting, 4–5; death, reflections on, 20; death of, 298; departure from family, 19–20, 24, 126–28, 252–53; description of Arkansans, 193; description of Missouri, 27, 60, 70, 78–79, 130, 136, 138, 144–45, 149, 173–77; description of Missourians, 30, 35–36, 47, 60, 63–64, 72, 76, 79, 133–34, 169, 189, 193, 197, 200–201, 206; diary, 83, 86–87, 170–71, 235, 238; enlistment, 7–9; expedition to Newtonia, Missouri, 142, 146–48; farming, 1, 3, 10, 296; finances, 2, 4, 7, 15, 61, 109, 115, 121, 150, 154, 165, 168–69, 239, 244, 250, 293, 295–96; Fort Curtis, engineering work at, 219; Fort Wyman, engineering work at, 39, 41–43, 46–47, 49, 51, 61, 66; furlough, 10, 121, 123–25; guard duty, 11–12, 17, 43; health, 62, 65, 84, 87–90, 93, 95, 99, 100–103, 105, 109, 112, 118, 120, 221, 223–24, 296, 298; homesickness, 32–33, 36, 65–66, 80, 84, 109; hospital stay, 87–88, 90–92, 94, 97, 99, 109–10, 115, 120–21; hunting, 132; injuries, postwar, 298; inspections of military posts, 253, 260–61, 265–70, 272, 285, 287–89; invalid pension, 296–97; Kendall County Agricultural Society, secretary, 2; map memoir, 55, 173–77; mapmaking technique,

54–55; mapping of battlefields, 153, 165, 167–68, 235, 239, 242, 244–46, 248; mapping of route of the Army of the Southwest, 171, 177; marriage, 5, 20; medical treatments, 89–90, 97, 118; moves to Illinois, 1; moves to Springfield, Missouri, 296; newspaper articles, writing of, 65, 167; ordered to work in St. Louis by Halleck, 56, 80; organizational memberships, 2, 110, 296; Pea Ridge battlefield visit, 231; photographs of, 11, 129–**130**, **207**, **297**; poetry, writing of, 83, 90–91, 94–95, 102–3, 105; Powder River Expedition, 283–84, 295; promotions, 182, 203; railroad travel, 27, 123, 129, 188; reconnaissance in Arkansas, 177–78; recruiting commission, 182, 232–33, 235; recruiting Unionists, 182, 187, 197–98, 200, 203–32, 235; religion, reflections on, 109, 112–13, 119–20; resigns commission, 203, 232, 235; retrieval of Wilson Judson's body, 236, 244; spelunking, 71–74, 315n45; teacher, 2, 4; temperance activities, 2, 15, 30; topographical engineer assignment in St. Louis, 85–86, 109, 116, 118, 120–21, 181, 183–84, 186–87; topographical engineering, description of, 54–55, 171–72, 247; Wilson's Creek battlefield visits, 142–45, 197, 333n39; worship service attendance, 11, 15, 20, 37, 46, 66, 106, 109, 112–14, 119, 124, 251; wounded, 203, 298

—surveying: in Kansas, 236, 244, 246–48; for Kendall County, 5, 295; postwar, 296; prewar, 2–4; in Rolla, Missouri, 54, 56, 58–60, 62–64, 66, 68–74, 76–80, 87, 109, 121; in vicinity of Pea Ridge battlefield, 164–65, 167, 169, 248

Bennett, Martin (brother), **297**
Bennett, Melissa (wife), 5, 10–15, 17, 19–20, 42, 47, 63, 66, 75, 81, 85, 123–26, 128, 169, 171, 252–53, 275, 295–**297**; death, 298; dental problems, 18; Kansas, 235, 250; letter writing, 34, 43, 47, 62, 68, 76, 91–92, 101, 110, 120, 163, 165, 275; loneliness, 109–10
Bennett, Minnie (daughter), 5, 11–14, 17, 19–20, 47, 76, 124, 126, 128, 298
Bennett, William (son), 295
Benton Barracks, Missouri, 83, 85, 101, 116, 118, 317n12, 318n24
Bentonville, Arkansas, 142, 150, 153, 155–56, 164–65, 325n36, 328n44; rearguard action at, 153–157, 163
Bibbins, E., 126, 321n5
Biddle, Catharine C., 84, 318n23
Biddle, Charles, 318n23
Biddle, Elizabeth Newbold, 84, 318n23
Biddle, Hannah Stokes, 84, 318n23
Biddle, Miss, 93–94, 105–6, 113, 115, 121
Biddle, Nicholas, 83, 90
Biddle, Thomas, 84
Biddle Sisters, 83–84, 90, 109, 111; visit to, 106, 129
Bierce, Ambrose (USA), 53
Big Blue, battle of, 339n8
Big Piney River, Missouri, 134
Big Rock, Illinois, 125
Big Sugar Creek, 241

Bigby, Medora, 318n16
Bird, Charles, 206, 208–9, 335n11
Birge, Col. John W. (USA), 319n30
bison, 263, 267, 269–70
Black Hawk, Colorado Territory, 277, 279–81, 292
Black Hills, Dakota Territory, 268, 288, 343n16
Blair, Col. Charles White Blair (USA), 244, 246, 340n15
Blair, Francis P., Jr., 21
Bloomington, Illinois, 123
Blunt, Maj. Gen. James G. (USA), 229, 239, 338n43, 339n10, 345n38
Bobs Knob, Missouri, 164, **166**
Boonville, Missouri, 21, 23
Boston Mountains, 187, 194, 204, 209, 213, 218
Bowen, Maj. William D. (USA), 34, 42, 75, 150, 307n19, 309n23, 315n47
Brackett, Capt. Albert Gallatin (USA), 304n31
Brashear, Mortimer M., 201, 205, 209, 211–12, 214–15, 220, 226, 334n46, 335n10
Brashear, Walter, 201, 205, 334n46
Bridger, Jim, 295
Briggs, Pvt. Frank (USA), 86–87, 108, 118, 129, 317n15
Bristol, Illinois, 9–10, 12–13, 129, 302n21, 306n8, 307n17, 321n10, 327n29
Brown, Buck (CSA), 228, 230
Buffalo River, Arkansas, 213, 226–27
Buffalo Springs Ranch, Colorado Territory, 271, 292, 344n26
Bulkley, Clara, 235, 250, 341n24
Bullock, Mr., 288
burial of soldiers, 33, 42–43, 51, 77, 81, 100, 142, 223, 244, 310n28

Burlington, Colorado Territory, 285
bushwhackers, 147, 190, 193, 198–99, 200, 204–5, 208, 214, 220, 229
Bussey, Col. Cyrus (USA), 165, 327n36
Butterfield, John, 142
Butterfield Overland Mail, 142, 264

Cabell, Brig. Gen. William L. (CSA), 194, 242, 332n28, 333n37
Cairo, Illinois, 85, 110
California, 259, 341n26
California troops: 3rd Infantry, 348n17
Calvary Presbyterian Church, Springfield, Missouri, 296
Camp, Capt. Samuel C. (USA), 18–19, 304n33, 304n37
Camp Collins, Colorado Territory, 285–86, 346n6
Camp Hammond, Illinois, 8, 10, 12, 17, 20, 23–24, 117
Camp Jackson, Missouri, 21
Camp Mitchell, Nebraska Territory, 289–90, 347n14
Camp Rolla, Missouri, 42
Camp Walbach, Dakota Territory, 286, 347n8
Canfield, Oliver, 1
Cape Girardeau, Missouri, 172
Carnisal, Col. (USA), 10
Carrollton, Arkansas, 210
Carthage, Missouri, 67
Casey, Maj. Gen. Silas (USA), 216, 336n21
Cassville, Missouri, 142, 145, 147–49, 163, 168–69, 192–93, 196, 201, 220, 229, 231, 325n29; description of, 173; hospitals at, 168
Cecil, Capt. John (CSA), 220–22, 225, 337n34

Central City, Colorado Territory, 277, 279–80
Chandler, Lt. Col. William P. (USA), 167, 327n38
Chappel, Lt. Edward S. (USA), 42–43, 309n24
Charleston, South Carolina, capture of, 260, 342n4
Chattanooga, Tennessee, 218, 337n27
Cherokee County, Kansas, 5
Cheyenne, 263–64, 284, 344nn23–25, 345n31, 345n33
Chicago, Burlington, and Quincy Railroad, 8, 10, 23, 302n18, 309n21
Chicago, Illinois, 10, 17, 32, 123, 128–29, 170, 177
Chicago Tribune, xx, 116, 142, 149, 167, 177, 322n11
Chickamauga, battle of, 295, 306n8, 323n5
chiggers, 181, 200
Chimney Rock, Nebraska Territory, 283, 287, 290–91
Chivington, Col. John M. (USA), 263, 272, 345n33
Chivington, Thomas, 272, 345n32
Chrisman, Jack (CSA), 216, 336n22
Clarendon, Arkansas, 173, 178, 329n21, 330n22
Clark, Capt. William T. (USA), 292, 348n18
Clark, Col. John B., Jr. (Missouri State Guard), 167, 328n39
Clark, Lt. William H. (USA), 9, 55, 303n24, 316n55
Clark, Solomon F., 331n22
Clarksville, Arkansas, 211
Cloud, Col. William Fletcher (USA), 251, 341n25
Cloud, Samuel P., 325n27

Cole, Col. Nelson D. (USA), 284
Cole, J. J., 125, 321n2
Coleman, Capt. John S. (USA), 50, 81, 311n48
Collins, Lt. Col. William O. (USA), 290, 347n16
Collins, Solomon, 41, 309n20
Colorado Gold Rush, 277
Colorado Territory, 263, 277–78
Colorado troops: 1st Cavalry, 272, 344n29; 1st Mounted Militia, 273, 344n27; 2nd Cavalry, 274, 345n36
Connecticut, 1, 18
Connor, Maj. Gen. Patrick E. (USA), 283, 291–92, 348n17
Conrad, Maj. Joseph (USA), 326n25
Coryell, Brant, 284, 346n5
Coryell, John H., 264, 275, 277, 284, 293, 345n37
Coryell, Libbie, 275, 346n39
Coryell, Pheobe, 346n39
Cottonwood Springs Post, Nebraska Territory, 267, 269, 343n14
Courthouse Rock, Nebraska Territory, 283, 290–91
Crane's Creek, Missouri, skirmish, 141, 144, 324n21
Crawford County, Missouri, 69
Cremer, Capt. Harrison W. (USA), 344n22
Crimean War, 2
Cromwell, Corp. David G. (USA), 48, 136, 311n45
Cross Hollow, Arkansas, 142, 150, 155–56, 193, 230–32, 328n44
Cross Timber Hollow, Missouri, skirmish at, 141
Cullum, Brig. Gen. George Washington (USA), 85, 317n10

Curtis, Maj. Gen. Samuel Ryan (USA), 104, 112, 144, 159, 164, 171, 176–77, 183–84, 187, 195, 232, 235–36, 239, 242, 250, 264, 320n34; campaign to the Mississippi River, 172–73; engineering staff, 142, 153, 167, 327n34; Pea Ridge campaign, 127–28, 138, 141, 153; Price's Missouri Expedition, 239

Daily Missouri Democrat, 172, 260
Dakota Territory, 284, 340n11, 347n11
Dardanelle, Arkansas, 203
Davidson, Brig. Gen. John Wynn (USA), 183, 330n19
Davis, Brig. Gen. Jefferson C. (USA), 325n30; division, 149, 156, 231, 325n30
Deerfield, Missouri, 244
Democratic Party, 331n22
Denver City, Colorado Territory, 251, 263–64, 272–75, 277–78, 281, 283–84, 291–93, 345n36, 348n17
Department of Kansas, 235, 264
Department of Missouri, 183–84, 186, 264, 315n49, 341n25
Department of the West, 21, 302n19, 307n22, 317n11
Dickey, Corp. Orrin (USA), 131–33, 322n12
Dillon, Missouri, 64, 66, 71
District of Colorado, 345n33, 345n38
District of Kansas, 342n5
District of Nebraska, 342n5
District of North Kansas, 342n5
District of Southwest Missouri, 127, 309n21
District of St. Louis, 330n19
District of the Plains, 283, 348n17
District of Utah, 348n17

District of West Florida, 314n39
Dix, Dorothea, 83
Dodge, Col./Maj. Gen. Grenville M. (USA), 53–54, 61, 65, 69, 71, 76, 80, 84, 86, 142, 150, 162, 167, 264, 283, 312n21, 313n26, 314n35, 315n47, 327n37, 348n17; commander of Rolla, 39, 59; introduces Bennett to Samuel Curtis, 142; orders survey of Rolla, 55–56, 77
Dodge, Minnesota Territory, 5
Dodson, Capt. (USA), 87, 317n15, 318n17
Douglas County, Missouri, 41, 308n17
Drake, C. P., 18–19, 304n36
Drew, Sadie, 311n40
Dug Springs, Missouri, skirmish, 144, 324n20
Dyke, Lt. Charles F. (USA), 35, 307n21

Edwards, Pvt. Clark W. (USA), 150, 325n31
Elgin, Illinois, 19, 43, 310n28
Ellis, Col. Calvin A. (USA), 144, 324n22
Emancipation Proclamation, 181, 184–86
erysipelas, 99, 319n27
Evans, Capt. William H. (USA), 285, 346n6

Faribault, Minnesota Territory, 3
Farmer, Elizabeth, 217, 336n24
Farmer, Samuel, 217, 220, 227–29, 231, 336n23, 336n25
Faust, Drew Gilpin, 236
Fayel, William, 171
Fayetteville, Arkansas, 141, 150, 187, 193, 198, 227–29, 332n26, 335n14, 335nn16–17; attack on, 228–29,

338n41; battle of, 192, 322n30, 322n31; evacuation of, 195, 333n38
Fesperman, Lt. John F. (USA), 220, 222, 337n33
Ficklin's Ranch, Nebraska Territory, 290
First Bull Run, battle of, 8–9, 21, 327n28
Fish, Capt. Charles D. (USA), 9, 35, 48–49, 55, 61, 80, 303n22
Fishback, Col. William Meade (USA), 187, 189–90, 192, 196–98, 200, 331n22; description of, 189
Fitch, Col. Graham N. (USA), 178, 329n20
flag: hospital, 158; secession, 60; United States, 7, 9, 33, 60, 96–**98**, 147, 196, 231, 310n28
Follett, Catharine, 100, 319n28
Follett, Pvt. Dwight (USA), 92, 318n24
forage, 80, 141, 164, 167, 172, 179, 231, 252
Forsyth, Missouri, 172, 176–77, 185, 205
Fort Curtis, Arkansas, 219, 337n30
Fort Donelson, Tennessee, 307n16
Fort Gibson, Indian Territory, 229, 236, 249, 319n32, 338n43, 341n22
Fort Halleck, Idaho Territory, 345n32
Fort Kearney, Nebraska Territory, 236, 248, 250–51, 253, 255–56, 259–60, 263, 265–67, 341n20, 344n20, 348n17; defenses at, 260–61; skunk infestation, 261
Fort Laramie, Dakota Territory, 289, 295, 347nn8–9; defenses at, 283, 288
Fort Leavenworth, Kansas, 235–36, 239, 248–50, 253, 255, 292, 341n3, 341n24

Fort Leavenworth Road, 253
Fort Rankin, Colorado Territory, 344nn23–24
Fort Ridgley, Minnesota Territory, 3
Fort Scott, Kansas, 187, 236, 239, 244, 246–49, 341n22
Fort Smith, Arkansas, 29, 167, 187, 198, 229, 232, 236, 247, 249, 335n17, 338n43, 341n23
Fort Snelling, Minnesota Territory, 317n11
Fort Sumter, South Carolina, 7
Fort Wyman, Missouri, 39–40, 130–32, **135**, 309n21; construction, 39; work of prisoners, 39, 45, 48, 51, 61, 69
Fowler, Rev., 11
Fox River Regiment. *See* Illinois troops: 36th Infantry
France, 2
Freeman, Thomas Roe (Missouri State Guard), 67, 75, 77, 79, 81, 313n26, 314nn34–35, 315n47
Fremont, Maj. Gen John C. (USA), 9, 33, 36, 49, 59, 302n19, 307n16, 307n22, 314n39, 317n12
Fremont's Springs, Nebraska Territory, 269
Fuller, Pvt. Ira O. (USA), 159, 327n29

Gadfly, Missouri, 145–48
Gaither Mountain, Arkansas, 210, 212–13
Galesburg, Illinois, 24
Gasconade River, Missouri, 36, 44, 47, 73, 79, 136, 190
Geneva, Illinois, 8–9
Gillette, Capt. Lee P. (USA), 260–61, 342n7
Gillman, Mr., 268

Gillman's Station, Nebraska Territory, 267
Godfrey's Ranch, Colorado Territory, 273, 292, 345n35
Golden City, Colorado Territory, 278, 281
good death ethos, 236
Gorton, L. C., 126, 321n4
Gorton, Susan, 4
Granada, Kansas, 256
Grand Army of the Republic, 296, 298
Gratiot Street Prison, Missouri, 321n7
Great Britain, 2
Greene, Thomas, 69, 314n40
Greene County, Missouri, 295–96, 314n40, 315n43
Gregory, John H., 277
Gregory mines, Colorado Territory, 277, 280–81
Greusel, Col. Nicholas (USA), 8–9, 19, 55, 60–61, 65, 80, 112, 156–57, 159, 161, 164, 168, 302n18, 305n18, 315n48, 326n23; brigade, 153, 326n22; Pea Ridge, battle of, 156–57, 159, 161
Greusel, Jane, 315n48
gristmills, 137, 145, 147

Haigh, William M., 23, 296
Hake House, Colorado Territory, 279
Halleck, Maj. Gen. Henry W. (USA), 29, 56, 76–77, 80, 86–87, 102, 119, 127, 172, 315n49, 317n10
Ham, Maj. Elijah D. (USA), 194, 199, 332n33
Hammond, Mr., 8, 10
Hanson, Pvt. Judson W. (USA), 33, 307n17
Hardee, Lt. Gen. William Joseph (CSA), 216, 336n21

Harney, Gen. William S. (USA), 21
Harrison, Lt. Marcus LaRue (USA), 39, 41, 43, 48–49, 56, 61, 70, 81, 83–84, 86–87, 224, 314n38, 316n51, 318n16, 338n37; engineering work, 39, 42, 58, 66; friendship with Lyman G. Bennett, 64, 68, 72, 78, 80, 91, 110, 121, 145–46, 168; military command, 194–96, 198, 229, 333n36
Harrison, Rebecca Axley, 318n16
Hatch, Pvt. James S. (USA), 137, 323n5
Hayden, Ferdinand V., 343n16
Hayden, Warren, 343n16
Hazelwood Cemetery, Springfield, Missouri, 296, 298
Hébert, Col. Louis (CSA), 154
Heffington, Capt. William J. (USA), 336n18
Helena, Arkansas, 173, 177–79, 203; battle of, 219, 337n28
Herron, Lt. Col. Francis J. (USA), 167, 189, 327n38
Hindman, Maj. Gen. Thomas Carmichael (CSA), 177, 179, 219, 329n17, 337n29
Hobbs, Capt. Albert M. (USA), 9, 33, 295, 303n23
Hocker, Rev., 14
Hoeppner Lt. Arnold (USA), 142, 183, 327n34
Holland, Lt. Col. Colley B. (USA), 145–49, 168, 324n24
Holmes, Lt. Gen. Theophilus H. (CSA), 337n29, 337n30
Horne, Pvt Jacob, 143, 324n17
Hotchkiss, Jed (CSA), 53
Houston, Missouri, 58, 60
Hovey, Brig. Gen. Alvin Peterson (USA), 178, 329n19, 330n24

Hunter, Col. DeWitt Clinton (CSA), 201, 334n45
Hunter, Maj. Gen. David (USA), 22, 307n22, 319n26
Huntsville, Arkansas, 185, 227

Illinois, 1, 7–8, 27, 30, 46, 53, 64, 69–70, 79, 101, 110, 113, 123, 190, 235, 251, 270, 296, 320n2, 325n30, 326n21
Illinois troops: 5th Cavalry, 177–78, 329n18, 330nn22–23; 7th Infantry, 8; 13th Infantry, 14, 27–28, 39, 47, 67, 70, 129, 303n28, 307n12, 319n32, 327n35, 328n41; 15th Cavalry, 318n17; 15th Infantry, 33, 307n16; 25th Infantry, 159; 35th Infantry, 112, 325n38; 36th Infantry (*see* Illinois troops: 36th Infantry); 37th Infantry, 159, 333n42; 44th Infantry, 315n41; 56th Infantry, 306n11; 88th Infantry, 307n14; Birge's Western Sharpshooters, 100, 319nn29–30; Dodson's Kane County Independent Company, 87, 108, 318n17; Sherer's Independent Cavalry Company, 59, 63–64, 302n12, 312n23
—36th Infantry, 71, **96**, 112, 309n21; alcohol consumption, 17, 25–26, 31, 63, 75; band, 111–12; battle of Pea Ridge, 154–55, 157–63, 327n31; camp violence, 16–17, 42–43; Company A (Elgin Company), 32, 42, 48; Company B, 154, 156, 159, 161; Company C (Young America Guards), 9–10, 14, 302n20; Company D (Newark Company), 10, 16, 20, 159, 169, 307n14, 325n31; Company E (Bristol Company), 9–10, 137, 154, 158–59, 161, 295, 302n21, 307n17, 318n18, 323n5, 324n19; Company F, 153, 157, 163; Company G, 154, 314n38; Company H (Woodstock Company), 35–37, 43, 307n21, 333n42; Company I (Oswego Company), 17–19, 24, 31, 37, 42–43; cooking arrangements petition, 49; counterfeiting, 35–36; deaths from disease, 70, 81–82; desertion, 37; dress parade, 16, 19–20, 43, 67, 80; drill, 14–17, 30, 32, 40; foraging, 42; German soldiers in, 16–17; gristmills operated by, 137; leaves Camp Hammond, 24; medical examinations, 16; mustered, 10; organization, 8–9; punishments, 63; rations, 25, 33, 35, 48–49; religious services, 11, 14, 20; sexual behavior, 36; sickness, 14, 36, 43; skirmish at Bentonville, 153–57; uniforms, 18–20; weapons, 20

Independence County, Arkansas, 182
Independence-St. Joe Road, 253
Independent Order of Odd Fellows, 296
Indiana troops: 24th Infantry, 329n19
Insley, Capt. (USA), 236
International Order of Good Templars (IOGT), 2–3, 296
Iowa troops: 3rd Cavalry, 158, 327n36; 4th Infantry, 51, 53, 71; 7th Cavalry, 343nn13–14, 344nn18–19, 344n20, 344n22, 346n7; 9th Infantry, 327n38

Jack Morrow's Ranch, Nebraska Territory, 269, 343n17
Jackson, Andrew, 41
Jackson, Gov. Claiborne F., 21, 314n36
Jacksonport, Arkansas, 172
Jasper, Arkansas, 210–15, 219, 222, 224–25
Jefferson City, Missouri, 33, 35
Jenks, Capt. Albert (USA), 64, 313n29
Jennison, Col. Charles Ransford (USA), 239, 339n10
Johnson, Col. James M. (USA), 187, 190, 193, 196, 198–99, 331n21
Johnson County, Arkansas, 215, 336n23
Joliet, Illinois, 128
Joslyn, Lt. Col. Edward S. (USA), 16, 27, 75, 104, 156–57, 164, 304n32, 327n33
Joslyn, Ted, 59
Judson, L. B., 244, 340n16
Judson, Lt. Wilson (USA), 101, 236, 244, 247, 319n32
Julesburg, Colorado, 264, 268, 270–71, 275, 283, 286, 288, 291–92, 344nn23–26, 344n28, 345n34
Junction Station, Colorado Territory, 274–75, 292

Kalloch, Isaac Smith, 251, 341n26
Kansas, 51, 239, 240–41, 248–49, 251, 256–58, 274, 340n11, 342n5, 347n11
Kansas City, Missouri, 235, 339n8
Kansas Rangers, 310n37, 313n26
Kansas River, 250
Kansas troops: 2nd Cavalry, 340n15; 2nd Infantry, 340n15, 342n5; 7th Cavalry, 339n10; 11th Cavalry, 289, 339n7, 340n11, 345n38; 14th Cavalry, 340n15; 15th Cavalry, 339n10, 340n12; 16th Cavalry, 341n3
Kearney City, Nebraska Territory, 265
Keetsville, Missouri, 141, 146, 149, 162, 165, 167, 231, 324n26,
Keit, Thomas, 231
Kelly, Pvt./Lt. Henry C. (USA), 206, 208–11, 226–28, 335n13
Kelton, Col. John C. (USA), 85, 87–88, 121, 317n11
Kendall County, Illinois, 1, 5, 7, 233, 236, 266, 295, 303n23, 303n26, 306n8, 313n30, 321n2, 327nn29–30, 328n45, 340n16
Kendall County Agricultural Society, Illinois, 2
Kendall County Free Press, 58, 311n42, 312n22
Kennedy, Mr., 3
Kennekuk, Kansas, 255
Ketchum, Pvt. Justus G. (USA), 129, 321n10
King, Pvt. David (USA), 335n17
Knobelsdorff, Col. Charles (USA), 70, 315n41

La Fair, Mrs., 292
Lake, Private Eugene (USA), 86, 168, 317n14
Lancaster, Iowa, 346n7
Lancaster, Kansas, 255
Larne, Pvt. Robert T. (USA), 178, 330n23
Lawrence, Kansas, 311n40
Lawrence County, Arkansas, 182
Leavenworth, Kansas, 251, 283, 295, 341n26
Leavenworth County, Kansas, 341n24

Lebanon, Missouri, 42, 44, 67, 70, 127–28, 131–32, 136–37, 149, 191, 309n22
Lee, Gen. Robert E. (CSA), 283
Lexington, Missouri, 23, 28, 61, 83, 305n21
Lillian Springs Ranch, Colorado Territory, 272, 292, 344n28
Lincoln, Abraham, 7–8, 21, 181, 186, 310n39, 314n33, 320n34, 331n22
Little Piney Creek, Missouri, 133–34, 136, 189, 216
Little Rock, Arkansas, 172–73, 178–79, 182, 203, 215, 329n17, 331n22; arsenal, 317n13
Little Rock, Illinois, 9, 16, 49, 327n30
Little Sugar Creek, Arkansas, 141–42; skirmish at, 141, 325nn32–34
Living Springs Station, Colorado Territory, 274–75, 293
Livingston, Col. Robert R. (USA), 260–61, 342n6
Logan, Senator John A., 295
Louisiana troops: 3rd Infantry, 327n38
Love, Capt. James Harrison (CSA), 210–12, 222, 335n15
Love, Capt. Wesley (USA), 346n6
Lovejoy, Owen, 181, 184, 186
Lyon, Brig. Gen. Nathaniel (USA), 21, 23, 102, 144, 197, 320n33, 333n40
Lyon, Chaplain George G. (USA), 46, 120, 311n41
Lyon, Edgar, 5, 65, 125, 313n30
Lyon, George, 5
Lyon, Mary, 5
Lyon, Melissa Emma (wife). *See* Bennett, Melissa (wife)
Lyon, Mr., 5
Lyon, Thomas, 5

Mackey, Maj. Thomas L. (USA), 347n9
Madison, Wisconsin, 2
Magazine Mountains, 187, 204
Majors, Capt. (USA), 265, 268–70
Majors, Capt. Thomas J. (USA), 287, 343n11
Marmaduke, Maj. Gen. John S. (CSA), 191, 242, 244, 332n27, 340n13
Marmaton River, 235, 245–46, 248–49, 340nn17–18
Martin, Pvt. John (USA), 31, 88, 130, 306n11
Marvel, William, 7
Marysville, Kansas, 253, 257
Masonic Order, 86, 296
Maysville, Arkansas, 153, 326n25
McBride, James Haggin (Missouri State Guard), 41, 308n19
McCulloch, Maj. Gen. Benjamin (CSA), 70–71, 85, 141, 144, 315n42; death of, 154, 159, 167
McDowell, R. (friend), 3, 43, 110, 310n29, 320n1
McDowell Medical College, 321n7
McIntosh, Brig. Gen. James M. (CSA), 153–54, 167
McPherson, Lt. Col. James B. (USA), 120–21, 320n6
medical treatments, blistering, 89–90, 99–100, 318n21
Memphis, Tennessee, 172
Mendota, Illinois, 24
Menomonie, Wisconsin, 339n9
Merrill, Capt. Orville B. (USA), 17, 43, 104, 112, 304n33
Miller, Pvt. Jacob (USA), 235, 239, 241–42, 244–48, 250, 339n9
Mine Creek, Kansas, battle of, 235, 242; battlefield, 235, 242; dead on, 242
Minnesota City, Minnesota Territory, 3

370 | Index

Minnesota Territory, 2–4
Minnesota troops: 1st Infantry, 100, 319n31
Mississippi, 172, 303n22, 312n20, 322n3, 327n32, 329n17
Mississippi River, 2, 21, 23, 29, 81, 172, 177, 181, 307n12, 337n26
Missouri, 20–21, 23, 32, 39–40, 50, 60, 71–72, 76, 84, 111, 118, 124–25, 127, 162, 172, 181, 191, 210, 213, 231, 235, 296, 307n16, 307n22, 308n23, 317n13, 318n17, 320n3, 322n3, 331n22, 332n27, 332n31, 335n12, 338n40
Missouri Democrat, 100, 171, 260
Missouri Home Guard, 67
Missouri River, 23
Missouri State Guard, 21, 23, 29, 127–28, 133, 141, 305n21, 308nn22–23, 314n34, 314n36; uniform, 323n7
Missouri troops (CSA): 11th Infantry, 334n45, 338n45
Missouri troops (USA): 1st Cavalry, 324n22; 1st Battalion Cavalry, 34, 307n19, 309n23, 315n47, 316n53; 3rd Infantry, 26, 306n24; 5th Cavalry, 325nn35–36; 6th Cavalry, 47, 322n3, 324n26, 325n29; 7th Infantry, 8, 34, 307n18; 8th Cavalry, 201, 333n43; 8th State Militia Cavalry, 333n43; 9th Infantry, 317n11; 10th Cavalry, 307n19; 12th Infantry, 60, 153, 156, 159, 313n25, 320n2, 326n22; 24th Infantry, 40; 25th Infantry, 112; 44th Infantry, 112; Benton Hussars Cavalry Battalion, 325n35; Fremont Battalion Cavalry, 44, 148, 309n22, 310n30, 310n34, 324n26; John S.

Phelps' Independent Regiment, 40, 71, 81, 137, 145, 148, 183, 231, 310n39, 324n17, 324n24
Mitchell, Brig. Gen. Robert Byington (USA), 260, 342n5
Moneka, Kansas, 241
Monroe, Col. James C. (USA), 194, 332n34
Montana Territory, 284
Montgomery, Capt. Bacon (USA), 44–45, 81, 146, 310n30
Montgomery, Illinois, 9–10, 14–15
Montgomery County, Arkansas, 203–4, 214
Moonlight, Col. Thomas (USA), 275–76, 345n38
Moore, Lt. David K. (USA), 146–47, 324n25
Moore, J. A., 344n29
Moore, Lt. Stephen W. (USA), 261, 342n8
Mound City, Kansas, 241–42, 244, 340n14
Mount Comfort, refugees, 195
Mount Pleasant, Kansas, 254
Mountain Feds, 182, 206, 209, 215–16, 219, 222
Mud Spring Station, Nebraska Territory, 290, 347n16
Mullally's Station, Nebraska Territory, 266, 343n12
Mulligan, Col. James A. (USA), 23, 25, 305n21
murder, 187, 198–99, 209, 216, 241, 325n30, 325n36
Murphy, Capt. Edward B. (USA), 269, 271, 344n19
Murphy, Isaac, 181, 185–86, 331n21
Murphy, Lt. John (USA), 241, 340n12

Nebraska City, Nebraska Territory, 293, 342n7
Nebraska troops: 1st Cavalry, 342n6, 342n8, 343nn11–12, 348n18; 1st Battalion Cavalry, 343nn13–14; 1st Infantry, 342n8
Nelson, Maj. Gen. William, 325n30
New Madrid, Missouri, 172
New York, 1–2, 5, 296, 304n32, 306n8, 307n17, 309n21, 311n40, 315n49, 317n10, 320n34, 321n4, 340n19, 341n24, 345n37
New York City, New York, 302n18, 314n33, 348n17
Newark, Illinois, 10–11, 37, 303n26, 328n45
Newton County, Arkansas, 145, 213–14, 338n36
Newton County, Missouri, 145
Newtonia, Missouri, 142, 146–47, 246–49, 325nn27–28, 325n31, 341n21
Nixforstay, Hans, 239

O'Brien, Capt. Nicholas J. (USA), 344n23
O'Brien, Maj. George M. (USA), 343n14
O'Fallons Bluff Post, Nebraska Territory, 269, 344n18
Ohio, 89, 92, 100, 102, 113, 287, 320n34, 340n15, 341n25
Ohio troops: 2nd Cavalry, 113, 320n3; 2nd Independent Light Battery, 160, 168, 317n14, 328n42; 2nd Infantry, 320n34; 11th Cavalry, 289, 343n10, 346n6, 347n9, 347n13, 347n16; 22nd Infantry, 318n24
Oketo, Kansas, 257
Oklahoma Territory, 296
Olathe, Kansas, 249, 334n7

Omaha, Nebraska Territory, 265, 284, 293
Orr, Sample, 49, 311n46
Osage Springs, Arkansas, 150
Osawatomie, Kansas, 239–40
Osterhaus, Maj. Gen. Peter J. (USA), 153–54, 320n2; division of, 112, 128, 153
Oswego, Illinois, 1, 4, 7, 14–15, 21, 31, 86, 123–26, 149, 233, 273, 306nn10–11, 321n2
Overland Stage Company, 255
Overland Trail, 253, 283
Owatonna, Minnesota Territory, 3
Ozark, Missouri, 194–95, 205
Ozark County, Missouri, 309n20
Ozark Mountains, 182, 201, 203

Panic of 1857, 2, 4, 277
Paoli, Kansas, 239–40, 249
Paris, Kansas, 241
Pea Ridge, battle of, 142, 153–55, 157–63, 312n21, 314n39, 315n42, 320n2, 320n34, 324n23, 327n32, 328n39; artillery, 158–61; casualties, 161–62; Confederate prisoners, 160–61; hospital, pursuit of Confederate army, 162; Leetown hospital, 154, 158, 326n27; sounds, 159–61
Pearce, Henry (USA), 225
Pearce, Luther (USA), 225
Peevy, Capt. Joseph G. (CSA), 231, 338n45
Perry, Miss, 4
Perryville, battle of, 296, 303n24
Pettis, Spencer D., 84
Phelps, Col. John S. (USA), 45–46, 48, 51, 62, 69, 163, 181, 184, 186, 310n39

Phelps, Mary, 84–85, 317n9
Phelps County, Missouri, 30
Philadelphia, Pennsylvania, 84, 318n23
Pickwick Place Addition, Springfield, Missouri, 296
Pierce, Capt. William P. (USA), 32, 307n14
Plainfield, Illinois, 126
Platte River, 253, 260, 263–65, 270–71, 274, 276–77, 285–86, 289, 291, 293
Plum Creek Post, Nebraska Territory, 265–66, 343n11
Plumb, Lt, Col. Preston B. (USA), 289, 347n12
Pocahontas, Arkansas, 85
Pope, Maj. Gen. John (USA), 283
Pope County, Arkansas, 201, 214, 226, 334n46
Port Hudson, Louisiana, 218, 337n26
Porter, Capt. Charles F. (USA), 267–68, 343n13
Powder River Expedition, 283–84, 295
prairie dogs, 270
Prairie Grove, battle of, 183, 189, 218, 331n24
Price, Capt. (USA), 15, 303n30
Price, Col. Thomas (CSA), 163
Price, Maj. Gen. Sterling (CSA), 308n23; Fall 1861 campaigns, 21, 23, 36, 46–47, 70–71, 127; Helena, battle of, 219, 337n28; Missouri Expedition, 235, 239, 246, 339n8, 339n10, 340n13; Pea Ridge campaign, 128, 131–32, 138–39, 141–44, 146–47, 150, 156–57, 163, 165, 167, 231, 326n25
Priddy, John C. (USA), 204–5, 213, 216–17, 223, 334n9

Quincy, Illinois, 23, 25

Rader, Pvt. John A. (USA), 178, 330n22
Ragain, Mary, 242, 244
Ragain, W. H., 340n14
Randolph County, Arkansas, 182
Rawles, James, 5, 301n64
Ray, Pvt. John (USA), 161, 327n30
Raymond, Pvt. George B. (USA), 169, 190, 328n44
Raynolds, William F., 343n16
Redwood Agency, Minnesota Territory, 4
Republican Party, 9, 21, 183, 296, 307n13
Revolutionary War, 1
Rhode Island, 1, 321n2
Rich, Capt. Josephus G. (USA), 51, 62, 311n49
Richmond, Virginia, capture of, 290, 347n15
Ritchey, Mathew, 147, 325n27
Robb, Lt./Capt. Joseph S. (USA), 194, 229, 333n35
Robinson, Lt. George T. (USA), 236, 239, 246, 248, 261, 268, 291, 339n7, 343n15
Rocky Mountains, 121, 263, 273, 275, 277, 281
Rolla, Missouri, 23, 26–28, 30, 34, 44, 47, 50–51, 54–56, 70–71, 75, 84–86, 88–89, 104, 110–12, 120, 127, 129, 131, 133, 136–38, 146, 149, 165, 181, 188–90, 264, 310n39, 315n45, 324n25; prisoners at, 46–48, 51, 59, 68; refugee center, 40–41, 47, 62, 70, 143, 182–83; supply depot, 29, 70
Rosecrans, Maj. Gen. William S. (USA), 218, 337n27

Roseman, Pvt. James (USA), 67, 131–32, 314n38
Roseman, Pvt. Wilbur F. (USA), 67, 314n38

Salem, Missouri, 50, 58, 62–63, 75–76, 81, 84, 315n47, 316n53
Saline County, Arkansas, 206
San Francisco, California, 142, 341n26
Sand Creek Massacre, Colorado Territory, 263–64, 345n33
Sanderson, Ann, 321n1
Sanderson, Miss, 4
Sanderson, T., 125, 321n1
Schryver, Peter, musician (USA), 89, 318n18
Scofield, Corp. Andrew L. (USA), 169–70, 328n45
Scofield, Maj. Gen. John M. (USA), 338n39
Scofield, Pvt. Charles H. (USA), 169–70, 328n45
Scotts Bluff, Nebraska Territory, 283, 289, 347n14
Searcy, Arkansas, 171
Searcy County, Arkansas, 183, 335n15
Searle, Lt. Col. Elhanen J. (USA), 194, 332n32
Sellers, Lt. Alfred H. (USA), 35, 307n21
Seneca, Kansas, 257
Shawnee Mission, Kansas, 239
Shelby, Col. Joseph Orville (CSA), 206, 335n12
Shuman, Capt. Jacob S. (USA), 289, 347n13
Sigel, Brig. Gen Franz (USA), 21, 23, 39, 59, 66, 102, 127, 144, 150, 154–56, 162, 314n33, 324n23, 328n44; division of, 127, 153–54, 159–61,
Sinisi, Kyle S., 236

Sioux, 284, 287, 343
Slack, Col. William Y. (Missouri State Guard), 167, 328n40
Smith, Capt. Henry A. (USA), 63–64, 313n28
Sons of Temperance, 15, 303n29
Southern District of Kansas, 239
Southwest Branch of the Pacific Railroad, 23, 29, 188
Spears, Pleasant Houston (CSA, USA), 226, 338n39
Spivey, Pvt. Jonas B. (USA), 214, 335n16
Springfield, Illinois, 123, 331n22
Springfield, Missouri, 23, 29, 35, 44, 49, 59, 69, 72, 85, 111, 123, 127–28, 131, 133, 137, 141–43, 148–49, 163, 165, 181, 184, 192–93, 197–98, 201, 203–5, 210, 213, 231, 296, 307n15, 311n46, 314n40, 333n35; battle on Jan. 8, 1863, 332n27; description of, 138, 191; hospitals, 43; prisoners at, 198–200; property damage, 138–39; refugees, 40, 183, 196; skirmish on Feb. 13, 1862, 138, 322n13, 323n6
Springfield Road, 31, 42, 56, 60, 63–65, 72, 158, 162, 165, 190
St. Charles, Illinois, 100
St. Louis, Missouri, 9, 21, 23, 25, 27–29, 35–36, 42, 50, 56, 60, 64, 110, 118–19, 127, 142, 182–83, 198, 235, 306n24, 313n25, 314n33, 316n54, 317nn11–12, 320nn2–3, 325n35, 327n34, 331n22; arsenal, 21, 26, 84, 129–30, 305n22; hospitals, 84; Ladies' Union Aid Society, 84; mud, 109, 114, 129–30; prisoners of war, 41, 74; Planter's House, 110, 116; St. Charles Hotel, 84–86, 110, 118, 129

St. Louis Democrat, 331n22
Stafford, Henry, 273–74
Stark, Capt. Denton D. (USA), 198, 333n42
Stark, Capt. W. H. (USA), 169, 328n43
Steele, Brig. Gen. Frederick, division of (USA), 172
Stockton City, Minnesota Territory, 3
Stonax, Lt. George F. (USA), 47, 311n43
Stone County, Missouri, 335n11
Stones River, battle of, 303n23, 304n33, 314n38, 323n5
Strickler's Station, Arkansas, 141
Strother, D. H. (USA), 53
Sugar Creek, Arkansas, 149–50, 155–57, 164–65, 167, 230–31
Summers, Col. William W. (CSA), 310n35
Sutherland, Lt. William F. (USA), 65, 104, 313n31
Sweeney, Brig. Gen. Thomas (USA), 23
Switzler, Capt. Theodore A. (USA), 44, 81, 84, 150, 310n36, 325n34

Tabbot, Maj. (USA), 272
Talbot, Lt. John (USA), 270, 287, 344n20
Telegraph Road, 127, 141–42, 154, 167, 228, 230, 333n40
Tennessee, 76, 183, 309n20, 315nn42–43, 329n17, 333n44, 337n34
Texas, 315n42, 332n31
Texas County, Missouri, expedition, 67, 313n26, 314n35
Thom, Col. George (USA), 80, 85, 87–89, 105, 110, 115–16, 118, 120–21, 316n54
Tipton, Missouri, 99
Todd, Musician William (USA), 66, 313n32
topographical engineering, 53, 55

Totten, Maj. James (USA), 85, 317n13
Transit Railroad Company, 3–4
trans-Mississippi, 21, 29, 83, 172, 332n31
Trans-Mississippi Department, 337n29
Trans-Mississippi District, 329n17
Treaty of Fort Laramie, 343n16
Triggs, Lt. Jeremiah H. (USA), 285, 287–88, 346n7
Twin Springs, Kansas, 241, 249

U. S. Sanitary Commission, 83
Union Hotel, 239
Unionists, 21, 49, 137, 143, 175, 181, 213, 217, 223; persecution of, 41, 45, 47, 59–61, 199–200, 209, 212, 214, 217, 220, 225–27, 325n27; recruits, 187, 195–201, 204; refugees, 40–41, 47, 59, 70–71, 189
United States troops: 6th Cavalry, 317n11; 54th Colored Infantry, 236, 319n32
Utah Territory, 259

Valley Station, Colorado Territory, 272, 275, 292, 345n31
Van Buren, Arkansas, 172
Van Dorn, Maj. Gen. Earl (CSA), 153, 172, 231
Vanderpool, Capt. James R. (USA), 209, 211, 213–15, 220, 222, 224–26, 228–29, 335n14, 338n41
Vandervoort, Mary Biddle, 84, 318n23
Vandever, Col. William, brigade of (USA), 157, 326n24
VanPelt, Lt. John (USA), 17, 304n35
Vera Cruz, Missouri, 41, 308n17
Vicksburg, Mississippi, 173, 218–19, 317n14, 322n3, 329m19, 337n26
Voss, Sgt. Gustav (USA), 19, 304n38

Wabasha, Minnesota Territory, 3
Wagner, Corp. Hiram (USA), 66, 95, 168, 311n44
Wagner, David E., 283
Waldron, Shade, 201, 333n44
Walker, Lt. William (USA), 8, 76, 81, 131, 304n34; controversy, 8, 17, 24, 37, 42–43, 47, 65, 80, 104, 305n18, 309n25
Walker, Pvt. George W. (USA), 31–32, 34, 45, 62, 70–71, 73, 76, 79, 91, 131, 133, 165, 167, 306n10
War Eagle Mills, Arkansas, 157
Washington, D.C., 19, 186, 200
Washington Ranch, Colorado Territory, 272, 345n30
Waugh, Lt./Lt. Col. Gideon M. (USA), 204, 334n7
Waynesville, Missouri, 136
Webb, Col. (USA), 10, 16
Wells, B. F., 100, 319n29
Wells, Pvt. Edwin (USA), 319n31
Wells, Pvt. George A. (USA), 319n31
Wells, Pvt. Henry G. (USA), 319n31
Western Sanitary Commission, 83
Westport, Missouri, battle of, 235–36, 339n8
Wet Glaze, Missouri (skirmish), 44, 71, 130, 309n22, 310n31
White River, Arkansas, 41, 172, 177–79, 187, 201, 203, 205, 209, 214, 221
Whitham, Pvt. Joseph (USA), 17, 304n35
Whitney, Corp. Daniel (USA), 144, 324n19
Wilcox, Capt. John (USA), 344n18
Willett, Sgt. William J. (USA), 31, 306n8
Williams, 254, 259, 261, 266–68, 275, 284–85, 287, 292

Wilmington, North Carolina, capture of, 260, 342n4
Wilson's Creek, battle of, 23, 34, 67, 143–44, 314n37, 315n42, 317n13, 320n33, 323n7, 342n5; battlefield, burials on, 143, 324n18; monument to Lyon, 197
Wimpey, Richard H., Capt./Maj. (USA), 193, 332n29
Winchel, Sgt. John J. (USA), 178, 329n21
Winona, Minnesota Territory, 2–4, 86, 295
Wisconsin, 4, 346n5
Wisconsin Ranch, Colorado Territory, 271, 344n25
Wisconsin troops: 3rd Cavalry, 339n9
Wood, Capt./Lt. Col. Samuel N. (USA), 46–47, 136, 310n37, 313n26, 314n35, 322n3
Woodstock, Illinois, 19, 307n21
Worthington, Capt. John T. (USA), 229–30, 338n42
Worthington, George W., 190, 331n25
Worthington, James P., 190, 331n25
Worthington, Wayne O., 190, 331n25
Wright, Maj./Col. Clark (USA), 44, 84, 148, 310n34, 324n26, 325n29
Wright County, Missouri, 47
Wyman, Col. John B. (USA), 31, 35–36, 39, 42–45, 71, 104, 303n28

Yellow Medicine Agency, Minnesota Territory, 4
Yellville, Arkansas, 211
Yorkville, Illinois, 295
Young, Surg. Delos W. (USA), 62, 313n27